Lecture Notes in Control and Information Sciences

Edited by M. Thoma

Vol. 43: Stochastic Differential Systems
Proceedings of the 2nd Bad Honnef Conference
of the SFB 72 of the DFG at the University of Bonn
June 28 – July 2, 1982
Edited by M. Kohlmann and N. Christopeit
XII, 377 pages. 1982.

Vol. 44: Analysis and Optimization of Systems
Proceedings of the Fifth International
Conference on Analysis and Optimization of Systems
Versailles. December 14–17, 1982
Edited by A. Bensoussan and J. L. Lions
XV, 987 pages, 1982

Vol. 45: M. Arató
Linear Stochastic Systems
with Constant Coefficients
A Statistical Approach
IX, 309 pages. 1982

Vol. 46: Time-Scale Modeling of Dynamic Networks
with Applications to Power Systems
Edited by J. H. Chow
X, 218 pages. 1982

Vol. 47: P. A. Ioannou, P. V. Kokotovic
Adaptive Systems with Reduced Models
V, 162 pages. 1983

Vol. 48: Yaakov Yavin
Feedback Strategies for Partially
Observable Stochastic Systems
VI, 233 pages, 1983

Vol. 49: Theory and Application of Random Fields
Proceedings of the IFIP-WG 7/1
Working Conference
held under the joint auspices of the
Indian Statistical Institute
Bangalore, India, January 1982
Edited by G. Kallianpur
VI. 290 pages. 1983

Vol. 50: M. Papageorgiou
Applications of Automatic Control Concepts
to Traffic Flow Modeling and Control
IX, 186 pages. 1983

Vol. 51: Z. Nahorski, H.F. Ravn, R.V.V. Vidal
Optimization of Discrete Time Systems
The Upper Boundary Approach
V, 137 pages 1983

Vol. 52: A. L. Dontchev
Perturbations, Approximations and Sensitivity Analysis
of Optimal Control Systems
IV, 158 pages. 1983

Vol. 53: Liu Chen Hui
General Decoupling Theory of Multivariable
Process Control Systems
XI, 474 pages. 1983

Vol. 54: Control Theory for Distributed
Parameter Systems and Applications
Edited by F. Kappel, K. Kunisch,
W. Schappacher
VII, 245 pages. 1983.

Vol. 55: Ganti Prasada Rao
Piecewise Constant Orthogonal Functions
and Their Application to Systems and Control
VII, 254 pages. 1983.

Vol. 56: Dines Chandra Saha, Ganti Prasada Rao
Identification of Continuous
Dynamical Systems
The Poisson Moment Functional
(PMF) Approach
IX, 158 pages. 1983.

Vol. 57: T. Söderström, P. G. Stoica
Instrumental Variable Methods
for System Identification
VII, 243 pages. 1983.

Vol. 58: Mathematical Theory of
Networks and Systems
Proceedings of the MTNS-83 International
Symposium
Beer Sheva, Israel, June 20–24, 1983
Edited by P. A. Fuhrmann
X, 906 pages. 1984

Vol. 59: System Modelling and Optimization
Proceedings of the 11th IFIP Conference
Copenhagen, Denmark, July 25-29, 1983
Edited by P. Thoft-Christensen
IX, 892 pages. 1984

Vol. 60: Modelling and Performance
Evaluation Methodology
Proceedings of the International Seminar
Paris, France, January 24–26, 1983
Edited by F. Bacelli and G. Fayolle
VII, 655 pages. 1984

Vol. 61: Filtering and Control of Random
Processes
Proceedings of the E.N.S.T.-C.N.E.T. Colloquium
Paris, France, February 23–24, 1983
Edited by H. Korezlioglu, G. Mazziotto, and
J. Szpirglas
V, 325 pages. 1984

For information about Vols. 1–42 please contact your bookseller or Springer-Verlag.

Lecture Notes in Control and Information Sciences
Edited by M. Thoma and A. Wyner

IFIP

114

A. Bermúdez (Editor)

Control of Partial Differential Equations

Proceedings of the IFIP WG 7.2 Working Conference
Santiago de Compostela, Spain, July 6–9, 1987

Springer-Verlag Berlin Heidelberg GmbH

Editor of Conference Proceedings of the series:
Computational Techniques in Distributed Systems IFIP-WG 7.2
Irena Lasiecka
Dept. of Applied Mathematics
Thornton Hall
University of Virginia
Charlottesville, VA 22903
USA

Editor
Alfredo Bermúdez
Department of Applied Mathematics
University of Santiago de Compostela
15706 Santiago de Compostela
Spain

ISBN 978-3-540-50495-5 ISBN 978-3-540-46018-3 (eBook)
DOI 10.1007/978-3-540-46018-3

Originally published by International Federation for Information Processing, Geneva, Switzerland in 1989

2161/3020-543210

PREFACE

This volume comprises the Proceedings of the IFIP TC-7/WG-7.2 Conference on Optimal Control of Systems Governed by Partial Differential Equations, held at the University of Santiago de Compostela (Spain), July 6 to 9, 1987.

The Conference was organized by the Department of Applied Mathematics of the University of Santiago de Compostela as an activity of the IFIP WG-7.2. It was devoted to the following topics:

. State constrained optimal control problems.

. Shape optimization.

. Identification of parameters.

. Stabilisation.

. Controlability.

. Numerical methods.

. Industrial applications.

Participants included six main speakers: H.T. Banks, J. Blum, H.O. Fattorini, J.L. Lions and I. Lasiecka and also 33 invited lecturers from nine different countries. I wish to express my acknowledgment to all the authors for their contributions contained in this volume.

I am grateful to the following organizations for their financial support:

. International Federation for Information Processing (IFIP).

. University of Santiago de Compostela. Vicerrectorado de Investigación.

. CAICYT.

. Government of Galicia. Dirección Xeral de Ensino Universitario e Política Científica.

Finally I wish to thank all those who have helped me organize the Conference, especially to the other members of the local organizing committee J. Durany and J.M. Viaño.

A. Bermúdez

CONFERENCE ON OPTIMAL CONTROL OF
SYSTEMS GOVERNED BY PARTIAL
DIFFERENTIAL EQUATIONS

July 6-9, 1987

Santiago de Compostela (Spain)

INTERNATIONAL PROGRAM COMMITTEE

A. Bermudez, Univ. of Santiago de Compostela, Spain

A. Butkovski, Control Institut, Moscow

R. Curtain, Univ. of Groningen, Netherlands

G. Da Prato, Scuola Normale, Pisa, Italy

R. Glowinski, INRIA, Paris, France

K. Hoffman, Univ. of Augsburg, Germany

W. Krabs, Technische Hochschule, Darmstadt, Germany

A.B. Kurzhanski, IIASA, Laxenburg, Austria

I. Lasiecka (Chairman), Univ. of Virginia, USA

J.L. Lions, College de France and CNES, Paris, France

U. Mosco, Univ. of Rome, Rome, Italy

O. Pironneau, INRIA, Paris, France

J.P. Yvon, INRIA, Paris, France

J.P. Zolesio, Univ. of Nice, Nice, France

CONFERENCE ON OPTIMAL CONTROL OF
SYSTEMS GOVERNED BY PARTIAL
DIFFERENTIAL EQUATIONS

July 6-9, 1987

Santiago de Compostela (Spain)

LOCAL ORGANIZING COMMITTEE:

A. Bermúdez, J. Durany, J.M. Viaño

Department of Applied Mathematics. University of Santiago de Compostela. Spain.

PARTICIPANTS:

1. Abergel, F. (Univ. Paris Sud. France)
2. Ahmed, N.U. (Univ. of Otawa. Canada)
3. Banks, H.T. (Brown University. U.S.A.)
4. Bermúdez, A. (Univ. of Santiago. Spain)
5. Blum, J. (I.N.R.I.A. France)
6. Casas, E. (Univ. of Cantabria. Spain)
7. Da Prato, G. (Scuola Normale Superiore. Italy)
8. Delfour, M. (Univ. of Montreal. Canada)
9. Durany, J. (Univ. of Santiago. Spain)
10. Fattorini, H.O. (Univ. of California. U.S.A.)
11. Fernández, L.A. (Univ. of Cantabria. Spain)
12. Fernández Cara, E. (Univ. of Sevilla. Spain)
13. Frankoswska, H. (Univ. of Paris-Dauphine. France)
14. Hager, W. (Pennsylvania State Univ. U.S.A.)
15. Haslinger, J. (Charles University. Checkoslovakia)
16. Henry, J. (I.N.R.I.A. France)
17. Joly, G. (Univ. of Technology of Compiègne. France)
18. Lagnese, J. (Georgetown University. U.S.A.)
19. Lasiecka, I. (Univ. of Virginia. U.S.A.)
20. Lee, E.B. (Univ. of Minnesota. U.S.A.)
21. Leugering, G. (Technische Hoshschule Darmstadt. W. Germany)
22. Lions, J.L. (College of France and C.N.E.S.)
23. Littman, W. (Univ. of Minnesota. U.S.A.)
24. Mignot, F. (Univ. Paris Sud. France)
25. Neittanmaki, P. (Univ. of Jyvaskyla. Finland)

26. Puel, J.P. (Univ. of Pierre and Marie Curie. France)
27. Rosen, I.G. (Univ. of Southern Carolina. U.S.A.)
28. Rousselet, B. (I.N.R.I.A. France)
29. Saguez, C. (SIMULOG. France)
30. Seidman, T. (Univ. of Maryland. U.S.A.)
31. Seoane, M.L. (Univ. of Technology of Compiègne. France)
32. Simon, J. (Univ. of Pierre and Marie Curie. France)
33. Sokolowski, J. (Systems Research Institut. Poland)
34. Triggiani, R. (Univ. of Virginia. U.S.A.)
35. Valle, A. (Univ. of Santiago. Spain)
36. Viaño, J.M. (Univ. of Santiago. Spain)
37. Yvon, J.P. (I.N.R.I.A. France)
38. Zolesio, J.P. (Univ. of Nice. France)

TABLE OF CONTENTS

PART I

PLENARY LECTURES

COMPUTATIONAL TECHNIQUES FOR INVERSE PROBLEMS IN SIZE STRUCTURED STOCHASTIC POPULATION MODELS

H. T. Banks
Center for Control Sciences
Division of Applied Mathematics
Brown University
Providence, R. I. 02912

I. Size Structured Population Models

Our initial interest in inverse problems for size structured populations arose from joint investigations with Lou Botsford involving modeling and control problems for two different aquatic populations: mosquitofish (*Gambusia affinis*) in rice fields [BV], [SKMK] and striped bass (*Morone saxatilis*) [SKMK] in the larval stage before they reach sizes at which recruitment is usually measured. Our first efforts [BBKW], [BM] involved use of McKendrick-Von Foerster type models of the form (here u is population density, x is size in length)

$$\frac{\partial u}{\partial t} + \frac{\partial}{\partial x}(gu) = -\mu u, \quad x_0 < x < x_1, \quad t > 0, \tag{1.1}$$

$$u(0,x) = \Phi(x), \tag{1.2}$$

$$g(t,x_0)u(t,x_0) = \int_{x_0}^{x_1} k(t,\xi)u(t,\xi)d\xi, \tag{1.3}$$

$$g(t,x_1) = 0, \tag{1.4}$$

where μ represents the mortality rate and the parameter g represents individual growth rate which can be treated as deterministic or stochastic in a manner explained in [BBKW]. The boundary condition (1.3) at the "minimum size" x_0 is a general recruitment term since in these models gu is the population flux while the condition (1.4) guarantees that x_1 is the maximum size attainable by any member of the population.

In both the mosquitofish and larval striped bass data, one observes cohort pulses with dispersion as these pulses are propagated along in the time and size plane (e.g., along the characteristics of the above system). As explained in [BBKW], deterministic models of the form (1.1)-(1.4) will exhibit this dispersive feature only under conditions on g ($\partial g/\partial x > 0$) which are not reasonable from a biological viewpoint. However, if one assumes that the individual growth rates have intrinsic variability among individuals (this leads to models with stochastic g), then one obtains the desired dispersion in solutions. One then is led naturally to a class of inverse problems for (1.1)-(1.4) in which one desires to estimate stochastic parameters (mean, variance or even the shape of the density function itself without a priori assumption as to distribution) for the random variable g. The methods we discuss here will be applicable to such problems and will be discussed elsewhere in this particular context.

There are other means by which one can modify models such as (1.1)-(1.4) so that they possess solutions that exhibit dispersion. Mathematically, it is well-known that adding a term of the form $\epsilon(\partial^2 u/\partial x^2)$ to (1.1) will have this effect. However, if this is done in an ad hoc way, it may be difficult to justify from a modeling point of view. For example, if our model is derived from balance law considerations, such a term corresponds to a Fickian term $j = -\epsilon(\partial u/\partial x)$ in the flux of the growth process. That is, the growth flux would be given by $J = gu - \epsilon(\partial u/\partial x)$. But Fickian fluxes are based on gradient driven movement in size or growth. In the models we consider here such an assumption is not biologically plausible; individuals will not shrink just because there are lower population densities at nearby smaller sizes.

One can, however, obtain a biologically plausible model with a second order term if one takes a different modeling approach. This involves assuming that growth is a Markov transition process and results in the classical Fokker-Planck equations. Inverse problems for these models are the focus of this note.

II. The Fokker-Planck Model

Weiss [W] was among the first authors to advocate use of the Fokker-Planck system as an alternative to the McKendrick-Von Foerster equations in size/age structured population models. While Weiss gives a derivation of a model based on "physiological age" that reduces under appropriate assumptions to the Fokker-Planck system, one can also give a derivation [O] based on the paradigm of Brownian motion of particles which is applicable to growth processes. These arguments are based on a Markov transition assumption for the growth process.

Let $\varphi(t,\bar{x};t+\Delta t,x)$ denote the probability density function for the transition from size $\bar{x}$ at time t to size x at time $t + \Delta t$. That is, $\varphi(t,\bar{x};t+\Delta t,x)d\bar{x}$ is the probability the members of the population in the size interval $[\bar{x},\bar{x}+d\bar{x}]$ at time t will move to size x at time $t + \Delta t$. Then if $u(t,x)$ is the population density at time t and size x, we have

$$u(t+\Delta t,x) = \int_{-\infty}^{\infty} u(t,\bar{x})\varphi(t,\bar{x};t+\Delta t,x)d\bar{x} .$$

Using this and some elementary but tedious arguments involving the characteristic function for φ and Taylor's series expansions, one can readily derive the equation

$$\frac{\partial u}{\partial x}(t,x) = -\frac{\partial}{\partial x}(M_1(t,x)u(t,x)) + \frac{1}{2}\frac{\partial^2}{\partial x^2}(M_2(t,x)u(t,x)) - \frac{1}{3!}\frac{\partial^3}{\partial x^3}(M_3(t,x)u(t,x)) + \cdots$$

where for $j = 1,2, \ldots$,

$$M_j(t,x) \equiv \lim_{\Delta t\to 0} \frac{1}{\Delta t}\int_{-\infty}^{\infty} (\xi - x)^j \varphi(t,x;t+\Delta t,\xi)d\xi$$

can be interpreted as moments of the rate of increase in size or the time rate of the moments for the growth process. If one makes the usual Fokker-Planck assumptions $M_j(t,x) \approx 0$ for $j \geqslant 3$, one obtains the Fokker-Planck equation

$$\frac{\partial u}{\partial t} + \frac{\partial}{\partial x}(M_1 u) = \frac{1}{2}\frac{\partial^2}{\partial x^2}(M_2 u). \tag{2.1}$$

As Okubo [O] points out, to use these equations effectively in models, one must know the moments M_1, M_2. In problems of practical interest to population ecologists, one can expect only limited success in determining these moments directly from knowledge of the growth process.

Hence a significant but difficult class of inverse problems involves determining the coefficients in the Fokker-Planck equations from observations of population density changes.

We remark that in general one should expect the moments M_1, M_2 to depend explicitly on time t and size x. If $\varphi(t,x;t+\Delta t,\bar{x}) = \varphi(0,x;\Delta t,\bar{x})$, i.e., the probability of transition in the period $[t,t+\Delta t]$ depends only on Δt, then $M_j(t,x) = M_j(x)$. Hence the coefficients are only size dependent.

On the other hand, if $\varphi(t,x;t+\Delta t,\bar{x}) = \varphi(t,0;t+\Delta t,\bar{x}-x)$, i.e., the probability of transition depends only on the amount of size transition and not the current size (an unlikely occurrence in size structured population models), then one finds

$$\begin{aligned} M_j(t,x) &= \lim_{\Delta t\to 0} \frac{1}{\Delta t}\int_{-\infty}^{\infty} (\xi - x)^j \varphi(t,0;t+\Delta t,\xi-x)d\xi \\ &= \lim_{\Delta t\to 0} \frac{1}{\Delta t}\int_{-\infty}^{\infty} \eta^j \varphi(t,0;t+\Delta t,\eta)d\eta \\ &= M_j(t). \end{aligned}$$

Thus in this case the coefficients are only time dependent.

Of course, if both the above suppositions hold for φ, then the moments M_1, M_2 are constants. This, however, does not appear to be a very important case in population modeling.

For notational convenience in subsequent discussions here, we shall use the notation $g_1(t,x) = M_1(t,x)$, $g_2(t,x) = M_2(t,x)/2$ so that our Fokker-Planck model equations then may be written (where we again include a mortaility term)

$$\frac{\partial u}{\partial t} + \frac{\partial}{\partial x}(g_1 u) = \frac{\partial^2}{\partial x^2}(g_2 u) - \mu u, \quad x_0 < x < x_1. \tag{2.2}$$

In this case the population flux is given by $j = g_1 u - (\partial/\partial x)(g_2 u)$ so that using arguments similar to those behind the McKendrick-VonFoerster model (1.1)-(1.3) we may obtain boundary conditions

$$\left[g_1 u - \frac{\partial}{\partial x}(g_2 u)\right]^{x=x_0} = \int_{x_0}^{x_1} k(t,\zeta)u(t,\zeta)d\zeta \tag{2.3}$$

$$\left[g_1 u - \frac{\partial}{\partial x}(g_2 u)\right]^{x=x_1} = 0, \tag{2.4}$$

to be used with initial conditions

$$u(0,x) = \Phi(x). \tag{2.5}$$

Thus, the inverse problem of interest to us here is given by the least squares formulation (e.g., see [BM], [B]): Minimize

$$J(q) = \sum_{i=1}^{n} \left| u(t_i,\cdot;q) - \hat{u}(t_i,\cdot)\right|^2_{H^0(x_0,x_1)} \tag{2.6}$$

over $q \in Q_{AD}$ where Q_{AD} is an admissible set of parameters $q = (q_1,q_2,q_3,q_4) = (g_1,g_2,k,\mu)$ contained in $Q \equiv L_\infty(\Omega) \times W_\infty^{(1)}(\Omega) \times L_\infty(\Omega) \times L_\infty(\Omega)$, $\Omega \equiv [0,T] \times [x_0,x_1]$. Here $u(\cdot,\cdot;q)$ is the solution of (2.2)-(2.5) for a given q and $\hat{u}$ are given observations of the population densities.

There are many important and unresolved questions related to such inverse problems. Our focus here will be an approximation framework for computational techniques for these problems.

III. Theoretical Framework for Approximation

Approximation techniques and associated questions (e.g., see [B]) of convergence and stability (continuous dependence of parameters on data) can be concisely and elegantly treated using the theoretical framework developed recently in [BI1], [BI2]. In order to do this, one must rewrite the system (2.2)-(2.5) in a weak or variational form using a coercive sesquilinear form. For this we seek solutions u with $u(t) \in V = H^1(x_0,x_1)$ satisfying

$$\langle u_t,\varphi\rangle - \langle g_1 u - D(g_2 u),\varphi\rangle + \langle \mu u,\varphi\rangle - \varphi(x_0)\int_{x_0}^{x_1} k(t,\xi)u(t,\xi)d\xi = 0$$

for all $\varphi \in V$. Here $D = \partial/\partial x$ and $\langle\ ,\ \rangle$ is the usual inner product in $H = H^0(x_0,x_1)$. We define a parameter dependent sesquilinear form $\sigma(t,\cdot,\cdot;q)$: $V \times V \to R^1$ by

$$\sigma(t,\psi,\varphi;q) = -\langle g_1\psi - D(g_2\psi),D\varphi\rangle - \varphi(x_0)\dot{R}(t,\psi) + \langle \mu\psi,\varphi\rangle \tag{3.1}$$

for $q = (g_1,g_2,k,\mu) \in Q$ with $R(t,\psi) \equiv \int_{x_0}^{x_1} k(t,\xi)\psi(\xi)d\xi$. Then our system in variational form can be written

$$\langle u_t,\varphi\rangle + \sigma(t,u,\varphi;q) = 0 \quad \text{for} \quad \varphi \in V \tag{3.2}$$

$$u(0) = \Phi\,. \tag{3.3}$$

It is readily seen that, under appropriate assumptions on Q_{AD}, this sesquilinear form satisfies a uniform (in $q \in Q_{AD}$) coercive inequality. In what follows we shall use $|\cdot|$, $|\cdot|_V$, $|\cdot|_\infty$ to denote the norms in $H = H^0(x_0,x_1)$, $V = H^1(x_0,x_1)$, $L_\infty(x_0,x_1)$, respectively.

<u>Lemma 3.1.</u> *Let* Q_{AD} *be bounded in* Q *and* $Q_{AD} \subset \{q \in Q \mid q_2(t,x) \geq \nu_2 > 0\}$. *Then there exist constants* ω, $c_1 > 0$, $c_2 > 0$ *such that*

(i) $\sigma(t,\varphi,\varphi;q) + \omega|\varphi|^2 \geq c_1|\varphi|_V^2$ *for* $\varphi \in V$, $q \in Q_{AD}$,

(ii) $|\sigma(t,\psi,\varphi;q)| \leq c_2|\psi|_V|\varphi|_V$ *for* $\psi,\varphi \in V$, $q \in Q_{AD}$.

<u>Proof:</u> We consider each term in (3.1) separately. For appropriately chosen constants (depending on the bounds for Q_{AD}) we find:

(a) $\langle q_2 D\varphi,D\varphi\rangle \geq \nu_2|D\varphi|^2$,

(b) $|\langle(-q_1 + Dq_2)\varphi,D\varphi\rangle| \leq \nu_1|\varphi|\,|D\varphi| \leq \nu_1(\frac{1}{4\epsilon}|\varphi|^2 + \epsilon|D\varphi|^2)$ so that
$\langle(-q_1 + Dq_2)\varphi,D\varphi\rangle \geq -(\nu_1/4\epsilon)|\varphi|^2 - \nu_1\epsilon|D\varphi|^2$,

(c) $\varphi(x_0)R(t,\varphi) \leq \nu_3|\varphi|_\infty|\varphi| \leq \nu_3(\frac{1}{4\epsilon}|\varphi|^2 + \epsilon|\varphi|_V^2)$ so that
$-\varphi(x_0)R(t,\varphi) \geq -(\nu_3/4\epsilon)|\varphi|^2 - \nu_3\epsilon|\varphi|_V^2$,

(d) $\langle q_4\varphi,\varphi\rangle \geq -\nu_4|\varphi|^2$.

Hence, for $q \in Q_{AD}$ and $\varphi \in V$ we have

$$\sigma(t,\varphi,\varphi;q) \geq -\omega|\varphi|^2 + c_1|\varphi|_V^2$$

with $\omega \equiv -(\nu_1/4\epsilon) - (\nu_3/4\epsilon) - \nu_4 - \nu_2 + \nu_1\epsilon$ and $c_1 \equiv \nu_2 - \nu_1\epsilon - \nu_3\epsilon$ where $c_1 > 0$ if $\epsilon > 0$ is chosen sufficiently small.

Similar considerations yield the estimate claimed in (ii).

Having established the rather standard inequalities in (i) and (ii), we can now appeal to the usual theory (e.g., see Chapter III of [L]) for existence and uniqueness of solutions to (3.2), (3.3) or equivalently, (2.2)-(2.5). We turn next to Galerkin type approximation techniques similar to those explained in [B], [BK], [BM] among numerous other references.

We let $Z^N \subset V$, $N = 1,2, \dots$, denote a family of finite dimensional spaces and let $P^N: H \to Z^N$ denote the orthogonal projections (in $\langle\ ,\ \rangle$) of H onto Z^N. We assume that this family possesses the approximation properties:

$$\text{For } \varphi \in V,\ P^N\varphi \to \varphi \text{ in } V; \text{ for } \varphi \in H,\ P^N\varphi \to \varphi \text{ in } H \text{ as } N \to \infty. \tag{3.4}$$

As noted in the above references, the usual B-splines (piecewise linear, cubic) satisfy these requirements.

We then replace the original problem involving (2.6) by a sequence of tractable approximating problems:

Minimize over $q \in Q$

$$J^N(q) = \sum_{i=1}^{n} \left| u^N(t_i,\cdot;q) - \hat{u}(t_i,\cdot) \right|_{H^0}^2 \tag{3.5}$$

subject to $u^N(t,\cdot;q) \in Z^N$ satisfying

$$\langle u_t^N,\varphi\rangle + \sigma(t,u^N,\varphi;q) = 0 \quad \text{for} \quad \varphi \in Z^N \tag{3.6}$$

$$u^N(0) = P^N\Phi. \tag{3.7}$$

If we solve these problems, we obtain a sequence $\{\bar{q}^N\}$ of "best" parameters that (we hope) approximate in some sense the sought after solutions for (2.6). In actual fact, a second family of approximations must be made for the elements in Q_{AD}, which is, in general, itself an infinite dimensional function space (the parameters q are generally functions of time t and size x). The added technicalities associated with these approximations are explained in detail in [BM]. For brevity (and without loss of generality) we shall here omit this aspect of the problems, assuming in our discussions that minimizations over Q can be effectively carried out if the state systems involved are finite dimensional.

Proceeding then, it can be readily argued (see for example the discussions in [B]) that to obtain certain convergence and method stability results, it suffices to argue that $u^N(t;q^N) \to u(t;q)$ for arbitrary sequences $\{q^N\}$ in Q_{AD} satisfying $q^N \to q$. This convergence in turn follows readily using the approximation framework ideas developed in [BI1], [BI2]. We first discuss the time independent case $\sigma(t,\varphi,\psi;q) = \sigma(\varphi,\psi;q)$.

For the time independent case, the theory of [BI1], [BI2] can be directly applied to our problem to guarantee the desired convergence. This theory employs a resolvent convergence form of the Trotter-Kato theorem for approximation of linear semigroups. It requires that the sesquilinear form σ satisfy the conditions (i) and (ii) of Lemma 3.1 above as well as a continuity condition in $q \in Q_{AD}$ (which is readily established for the Fokker-Planck sesquilinear form (3.1)). Under the approximation assumption (3.4) on Z^N one then obtains the sought after theoretical results for the Fokker-Planck inverse problems.

For the time dependent case, there are several analogues of the Trotter-Kato theorem for evolution operator systems (for example, see [CP]). Unfortunately, none of these approximation theorems appear readily applicable to the problems under consideration here (for example, conditions such as (C1) of [CP] appear difficult if not impossible to verify in our parameter estimation problems). Nonetheless, it is possible to establish convergence and stability results in the time dependent coefficient case by giving arguments that are essentially equivalent to the variational inequality approach used in [B], [BM]. (This amounts essentially to proving a version of a Trotter-Kato like theorem directly.) Indeed, using this approach, one can develop a general theoretical framework analogous to that of [BI1], [BI2] for time dependent problems. The Fokker-Planck system can then be shown to be readily treated as an example satisfying the requirements of this theory -- see [BRR] for details.

IV. Numerical Examples

To verify the efficacy of the ideas discussed above, we have begun numerical testing of computational packages based on these ideas. More complete numerical studies are currently under way using several versions of moving finite element approximation schemes as well as standard spline schemes. We report briefly on our initial encouraging findings using cubic spline approximation schemes for the states and linear spline approximations for the parameters to be estimated (see [BM] for further explanation).

Several examples in which we knew the true solution u^* of (3.2), (3.3) corresponding to given parameter functions q^* were tested. That is, we prepared "data" from solutions of the equations for given parameter functions. This "data" were used in the algorithms and starting from an initial (incorrect) guess q^0 for the parameters, the software packages were required to find an estimate (converged value) of the true parameters.

In both examples given here the true solution was taken as

$$u^*(t,x) = \begin{cases} e^{-.1t}\sin^2(2\pi(x-.5t)) & 0 \leq x-.5t \leq .5 \\ 0 & .5 \leq x-.5t \leq 1.0, \end{cases}$$

and $g_2^* \equiv 10^{-4}$, $k^* = \mu^* = 0$ were fixed.

Example 1. We took $g_1^*(t,x) = H^*(t)G^*(x)$ with $H^*(t) = .15 - .1\tanh(20t-10)$, $G^*(t) = e^{-3x}$. We attempted to estimate H(t) from an initial guess of $H^0(t) \equiv .15$. In Figure 1 we depict the converged estimate (the dashed line) as compared to the true function H^* (the solid line). The results shown are for 11 cubic elements for the states and 4 linear elements for the parameter function H(t).

Example 2. We chose $g_1^*(t,x) = [.15 - .1\tanh(10t - 5)](1 - x)$ and attempted to estimate $g_1(t,x)$ from an initial guess of $g_1^0(t,x) \equiv .075$. The converged estimate corresponding to 11 cubic state elements and 4 linear elements each in the t and x coordinates (bilinear elements) for the parameter approximations are presented in Figure 2.

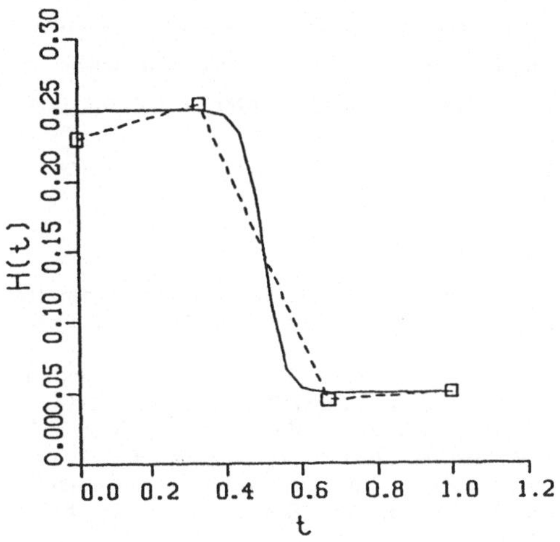

Figure 1

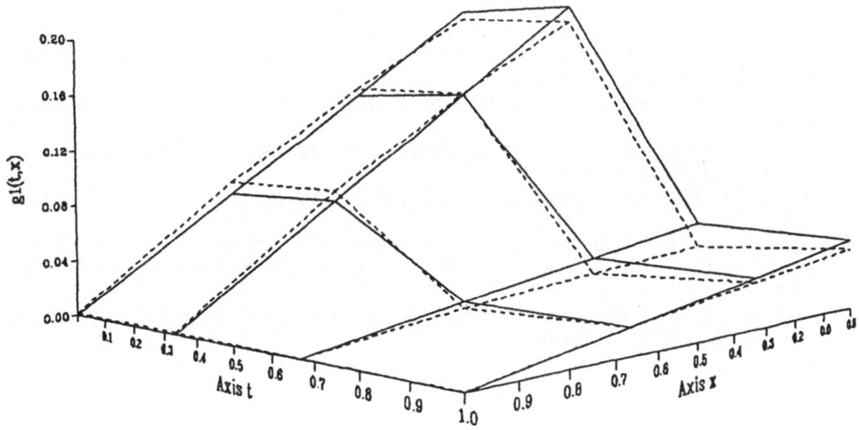

Figure 2

V. Acknowledgements

We are grateful to G. Rosen for discussions concerning the time dependent approximation analogues of the Trotter-Kato theorem mentioned in Section III and to Y. Wang for her assistance in carrying out the numerical experiments reported on in Section IV.

This research was supported in part by the National Science Foundation under NSF Grant MCS-8504316 and in part by the Air Force Office of Scientific Research under Contract F-49620-86-C-011. Part of the research was carried out while the author was a visiting scientist at ICASE, NASA Langley Research Center, Hampton, VA which is operated under NASA Contract NAS1-17070.

VI. References

[B] H.T. Banks, On a variational approach to some parameter estimation problems, in Distributed Parameter Systems, Springer Lec. Notes on Control & Info. Sci. 75 (1985), 1-23.

[BBKW] H.T. Banks, L.W. Botsford, F. Kappel and C. Wang, Modeling and estimation in size structured population models, LCDS/CCS Rep. No. 87-13, Brown Univ., March 1987; Proceedings 2nd Course on Math. Ecology (Trieste, Dec. 8-12, 1986), to appear.

[BI1] H.T. Banks and K. Ito, A theoretical framework for convergence and continuous dependence of estimates in inverse problems for distributed parameter systems, LCDS/CCS Rep. 87-20, March, 1987, Brown University; Applied Math. Lett. Vol. 0, No. 1, June, (1987), 31-35.

[BI2] H.T. Banks and K. Ito, A unified framework for approximation in inverse problems for distributed parameter systems, Control: Theory and Adv. Tech., submitted.

[BK] H.T. Banks and K. Kunisch, The linear regulator problem for parabolic systems, SIAM J. Control and Optimization, 22 (1984), 684-698.

[BM] H.T. Banks and K.A. Murphy, Quantitative modeling of growth and dispersal in population models, LCDS Rep. No. 86-4, Brown University, January (1986); in Math. Topics in Population Biology, Morphogenesis and Neurosciences, Springer LN in Bio. Math., 71 (1987), 98-109.

[BRR] H.T. Banks, S. Reich, and I.G. Rosen, manuscript in preparation.

[BV] L.W. Botsford, B. Vondracek, T.C. Wainwright, A.L. Linden, R.G. Kope, D.E. Reed and J.J. Cech, Jr., Population development of the mosquitofish (*Gambusia affinis*) in rice fields, Env. Biol. Fish., in press.

[CP] M.G. Crandall and A. Pazy, Nonlinear evolution equations in Banach spaces, Israel J. Math. 11 (1972), 57-94.

[L] J.L. Lions, Optimal Control of Systems Governed by Partial Differential Equations, Springer, Heidelberg, 1971.

[O] A. Okubo, Diffusion and Ecological Problems: Mathematical Models, Springer-Verlag, N.Y., 1980.

[SKMK] D.E. Stevens, D.W. Kohlhorst, L.W. Miller, and D.W. Kelley, The decline of striped bass in the Sacramento-San Joaquin estuary, California, Trans. Amer. Fish Soc. 114 (1985), 12-30.

[W] G.H. Weiss, Equations for the age structure of growing populations, Bull. Math. Biophys. 30 (1985), 427-435.

IDENTIFICATION OF FREE BOUNDARIES AND NON-LINEARITIES FOR ELLIPTIC PARTIAL DIFFERENTIAL EQUATIONS ARISING FROM PLASMA PHYSICS

J. BLUM

Laboratoire TIM3, Tour IRMA
Université de Grenoble I
B.P. 68, F-38402 Saint-Martin d'Hères Cedex

This work has been performed in the frame of a collaboration with Association EURATOM-CEA D.R.F.C. (Département de Recherche sur la Fusion Contrôlée), Centre d'Etudes Nucléaires de Cadarache, and with the JET team (Joint European Torus) at Culham (G.B.).

1. THE PHYSICAL PROBLEM

Research in thermonuclear fusion is aimed at realizing an experimental device which permits confinement of the plasma. One such possible device is called a Tokamak, which is based on the principle of magnetic confinement, where the ionised particles are confined within a magnetic field. Figure 1 represents the European Tokamak JET (Joint European Torus). The plasma is confined inside the toroidal vacuum vessel by a magnetic field which is generated by the plasma current I_p and by the poloidal and toroidal field coils. The plasma current I_p is obtained by induction from the currents in the poloidal field coils ; the plasma thus appears as the secondary of a transformer whose poloidal field coils constitute the primary, and the ferromagnetic circuit is the main element of coupling between primary coils and plasma. The poloidal magnetic flux and field are measured by flux loops and magnetic probes, which are located on the vacuum vessel (see fig. 2a). The aim of this paper is to determine the plasma boundary and the plasma current density profile from these magnetic measurements.

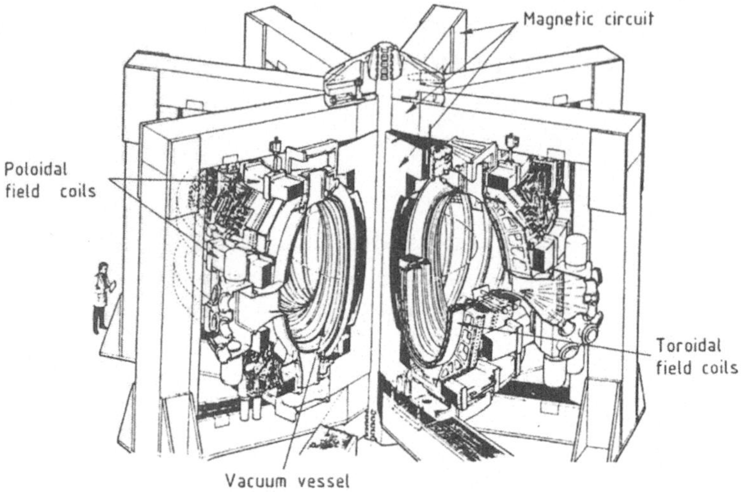

Fig. 1 : The JET Tokamak

The configurations will be assumed to be axisymmetric, i.e. independent of the azimutal angle φ. Then the experimental data will be the poloidal flux Ψ and its derivative $\frac{\partial \Psi}{\partial n}$ (normal to the vacuum vessel), which are measured at discrete points of the cross-section of the vacuum vessel (see fig. 2b). In this way we are faced with solving a Cauchy problem for a two-dimensional elliptic partial differential equation for Ψ. This is an "ill-posed" problem in the terminology of Hadamard, which will be set as a "well-posed" problem by using optimal control theory and regularization techniques.

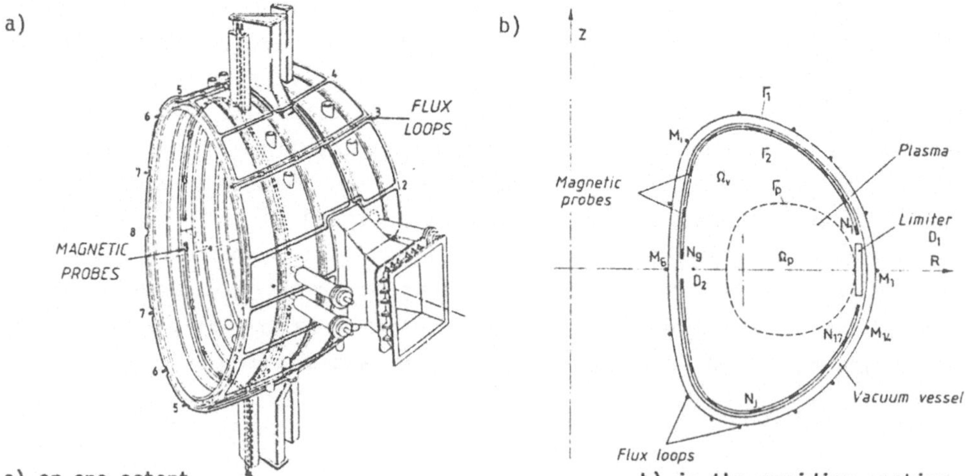

a) on one octant b) in the meridian section

Fig. 2 : Position of the flux loops and magnetic probes in JET

2. THE EQUILIBRIUM EQUATIONS

The equilibrium of the plasma in a Tokamak is governed by the Maxwell equations and by the plasma equilibrium equation :

$$\begin{cases} \vec{\nabla} \cdot \vec{B} = 0 & (1) \\ \vec{\nabla} \times \vec{H} = \vec{j} & (2) \\ \vec{B} = \mu \vec{H} & (3) \\ \vec{\nabla} p = \vec{j} \times \vec{B} & (4) \end{cases}$$

which means that the plasma is in equilibrium when the force $\vec{\nabla} p$ due to the kinetic pressure p is equal to the Lorentz force $\vec{j} \times \vec{B}$.

The configuration is assumed to be axisymmetric, i.e. independent of φ, where (r,z,φ) are cylindrical coordinates (this means in particular that an equivalent axisymmetric magnetic circuit has to be defined). Then, from equation (1) can be defined the poloidal magnetic flux $\Psi(r,z)$ by :

$$\begin{cases} B_r = -\dfrac{1}{r}\dfrac{\partial \Psi}{\partial z} \\ B_z = \dfrac{1}{r}\dfrac{\partial \Psi}{\partial r} \\ \Psi = 0 \text{ on z-axis and at infinity.} \end{cases} \qquad (5)$$

The lines Ψ = constant are the flux lines in the meridian section of the torus and they generate the magnetic surfaces by rotation around the z-axis. By projection of equation (2) on the unit vector $\vec{e}_\varphi$ in the toroidal direction, and by using (3) and (5), the equation for $\Psi(r,z)$ can be written :

$$L\Psi = j_T \qquad (6)$$

with $$L\,. = -\frac{\partial}{\partial r}\left(\frac{1}{\mu r}\frac{\partial .}{\partial r}\right) - \frac{\partial}{\partial z}\left(\frac{1}{\mu r}\frac{\partial .}{\partial z}\right)$$

where j_T is the toroidal component of the current density. If the toroidal component $\vec{B}_T$ of $\vec{B}$ is noted by f/r, then from (4) and from the expressions of $\vec{B}$ and $\vec{j}$, we can show that p and f are constant on the flux lines and that, inside the plasma, one has :

$$j_T\,(r, \Psi) = r\frac{\partial p}{\partial \Psi} + \frac{1}{2\mu_0 r}\frac{\partial f^2}{\partial \Psi} \qquad (7)$$

The equation (6), with j_T given by (7) inside the plasma, is called the Grad-Shafranov equation. The plasma boundary Γ_p is the flux line which is in contact with a limiter D, which prevents the plasma from touching the vacuum vessel (see fig. 2b). Hence Γ_p is defined by :

$$\Gamma_p = \{M \in \Omega_v \text{ such that } \Psi(M) = \sup_D \Psi \} \tag{8}$$

where Ω_v is the vacuum region.

The complete set of non-linear elliptic equations has been solved numerically in /1/ for the whole Tokamak. In the next sections we will only consider these equations inside the vacuum vessel, i.e. in the vacuum region and in the plasma. The first method of identification of the plasma free boundary, which will be presented in section 3, is a fast method which solves the homogeneous linear elliptic equation $L\Psi = 0$ in the vacuum region, with Cauchy boundary conditions on the vacuum vessel. The second method, which will be presented in section 4, identifies from the two Cauchy conditions the profile of the plasma current density $j_T(r,\Psi)$, which is the non-linearity of equation (6).

3. FAST IDENTIFICATION OF THE PLASMA FREE BOUNDARY FROM THE CAUCHY CONDITIONS

Let us assume that the poloidal flux Ψ and the tangential component B_τ of the poloidal field $\vec{B}_p$ are measured not only at discrete points on the vacuum vessel, but on the whole section Γ_v of the vessel (see fig. 3a).

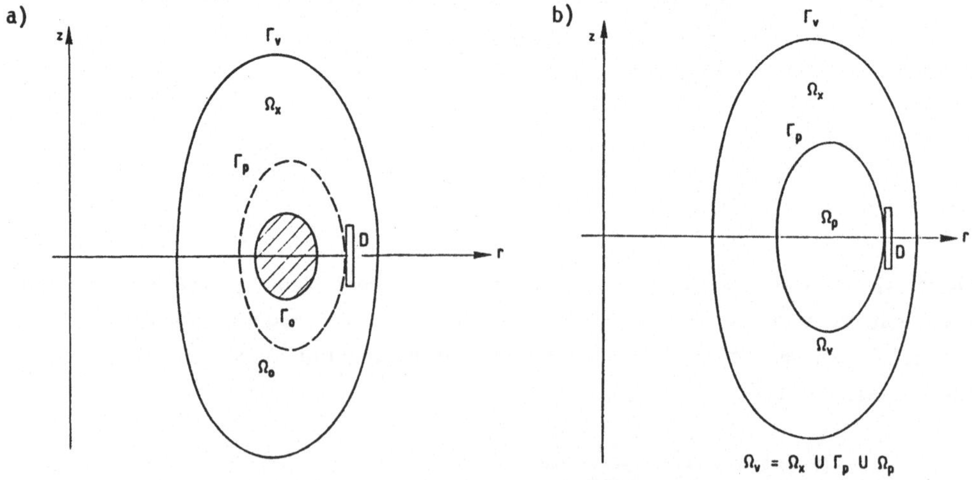

Fig. 3 : Domain of resolution for the identification of the plasma boundary (3a) and of the plasma current density profile (3b)

From (5) it is clear that B_τ is equal to $\frac{1}{r}\frac{\partial\Psi}{\partial n}$, where $\frac{\partial\Psi}{\partial n}$ is the normal derivative of Ψ with respect to Γ_V. There is no current flowing in the vacuum region Ω_X, bounded by Γ_V and by the plasma boundary Γ_p, so that $L\Psi = 0$ in this region. Hence Ψ satisfies :

$$\left\{\begin{array}{ll} \Psi = g & \text{on } \Gamma_V \\ \frac{1}{r}\frac{\partial\Psi}{\partial n} = h & \text{on } \Gamma_V \\ L\Psi = 0 & \text{in } \Omega_X \\ \Psi = \sup\limits_{D} \Psi & \text{on } \Gamma_p \end{array}\right. \tag{9}$$

where Γ_p and hence the domain Ω_X are unknown. The operator L, defined by (6), is a linear elliptic operator as $\mu = \mu_0$ in the vacuum. Problem (9) is "ill-posed" as it consists in a Cauchy problem for an elliptic equation, which is unstable with respect to perturbations of g and h (cf /2/). Let us transform it into a stable problem in a fixed domain. This can be achieved in three stages :

i) we assume that there exists a continuation of Ψ up to a fixed contour Γ_0 (see fig. 3a) in the sense that $L\Psi = 0$ in Ω_0, where Ω_0 is the domain between Γ_V and Γ_0. This continuation exists and is unique in a neighbourhood of Γ_p (cf. /2/). We assume here its existence up to Γ_0,

ii) the Neumann boundary condition is relaxed and transformed into the minimization of the quadratic difference between the calculated and measured values of $\frac{1}{r}\frac{\partial\Psi}{\partial n}$ on Γ_V,

iii) in order to transform it into a stable problem, the cost-function is regularized.

Problem (9) is then modified in the following way : Ψ is related to an arbitrary function v on Γ_0 by :

$$\left\{\begin{array}{ll} \Psi = g & \text{on } \Gamma_V \\ \frac{1}{r}\frac{\partial\Psi}{\partial n} = v & \text{on } \Gamma_0 \\ L\Psi = 0 & \text{in } \Omega_0 \end{array}\right. \tag{10}$$

and problem (P_ε) consists in searching for u_ε such that :

$$J_\varepsilon\,(u_\varepsilon) = \inf_{v\in U} J_\varepsilon\,(v)$$

with $$J_\varepsilon\,(v) = \int_{\Gamma_V} \left(\frac{1}{r}\frac{\partial\Psi}{\partial n} - h\right)^2 d\sigma + \varepsilon \int_{\Gamma_0} v^2\, d\sigma \tag{11}$$

where Ψ is related to v by (10) and where U is the set of the admissible vectors v.

The plasma boundary Γ_p^ε is then the particular equipotential of Ψ_ε(related to u_ε by (10)), which is in contact with the limiter D :

$$\Gamma_p^\varepsilon = \{ M \in \Omega_0 \text{ such that } \Psi_\varepsilon(M) = \sup_D \Psi_\varepsilon \} \tag{12}$$

The determination of Γ_p^ε is a consequence of the resolution of (P_ε), but it does not intervene in its resolution. The physical interpretation of the solution u_ε of (P_ε) is that it is a fictitious surface current density on Γ_0, which represents the plasma current density according to the virtual casing principle /3/. The vector Ψ_ε related to u_ε by (10) satisfies exactly the Dirichlet condition $\Psi = g$, approximately the Neumann condition $\frac{1}{r}\frac{\partial\Psi}{\partial n} \simeq h$, exactly the Maxwell equations in the vacuum ($L\Psi = 0$) and the plasma boundary is smooth because of the ε-regularization. It should be pointed out that the restriction of Ψ_ε to the domain bounded by Γ_p^ε and Γ_0 has no physical meaning, because $L\Psi \neq 0$ inside the plasma, and is just a mathematical continuation.

Problem (9) has thus been transformed into the linear quadratic optimal control problem (P_ε). In the terminology of /4/, v is the control vector, Ψ the state vector, (10) the state equations, (11) the cost-function, the second term of which is the regularizing term. By using the same technique as in /5/, where a similar problem in biomathematics is solved, we can prove the following proposition.

<u>Proposition 1</u> : Let $g \in H^1(\Gamma_V)$, $h \in L^2(\Gamma_V)$, $U = L^2(\Gamma_0)$

Problem (P_ε) has a unique solution u_ε which is stable with respect to g and h :

$$||u_\varepsilon^1 - u_\varepsilon^2||_U \leq \frac{1}{\sqrt{\varepsilon}} \left[||g_1 - g_2||_{H^1(\Gamma_V)} + C\,||h_1 - h_2||_{L^2(\Gamma_V)} \right]$$

where u_ε^1 and u_ε^2 are the solutions of (P_ε) corresponding to two different sets of functions (g_1, h_1) and (g_2, h_2) respectively.
If there exists $u_0 \in U$ such that $\frac{1}{r}\frac{\partial\Psi(u_0)}{\partial n} = h$ then $u_\varepsilon \to u_0$ in U when $\varepsilon \to 0$.

The complete proof of this proposition can be found in /6/. A direct consequence of it is that Ψ_ε is stable with respect to g and h and that, if the continuation of Ψ up to Γ_0 exists, then Γ_p^ε converges towards the real Γ_p when ε converges to 0.

Following /4/ the optimality system for problem (P_ε) is given by :

<u>Proposition 2</u> : If the adjoint state p_ε is defined by :

$$\left\{ \begin{array}{lll} p_\varepsilon = \frac{1}{r}\frac{\partial \Psi_\varepsilon}{\partial n} - h & \text{on } \Gamma_V & \\ L p_\varepsilon = 0 & \text{on } \Omega_0 & \\ \frac{\partial p_\varepsilon}{\partial n} = 0 & \text{on } \Gamma_0 & \end{array} \right. \qquad (13)$$

with Ψ_ε related to u_ε by (10) then the necessary optimality condition for problem (P_ε) can be written :

$$p_\varepsilon = \varepsilon u_\varepsilon \quad \text{on} \quad \Gamma_0 \qquad (14)$$

The optimality system (10) - (13) - (14) can then be solved by a finite element method. If $\mathcal{Y}_\varepsilon$, $\mathcal{P}_\varepsilon$, $\mathcal{U}_\varepsilon$ represent the vectors $\{ \Psi_\varepsilon(M_i) \}$, $\{ p_\varepsilon (M_i) \}$, $\{ u_\varepsilon (M_j) \}$, where M_i are the nodes of the triangulation of Ω_0 and M_j those of Γ_0, then the discrete optimality system can be written :

$$\left\{ \begin{array}{l} A \mathcal{Y}_\varepsilon = B \mathcal{G} + C \mathcal{U}_\varepsilon \\ A \mathcal{P}_\varepsilon = E \mathcal{Y}_\varepsilon - B \mathcal{H} \\ F \mathcal{P}_\varepsilon = \varepsilon \mathcal{U}_\varepsilon \end{array} \right. \qquad (15)$$

where $\mathcal{G}$ and $\mathcal{H}$ are the vectors $\{g(M_k)\}$ and $\{h(M_k)\}$, M_k being the nodes of Γ_V. The matrix A contains the stiffness matrix of the operator L, the matrices B, C and E correspond to the boundary conditions of (10) and (13), F is the identity-matrix on Γ_0 and the zero-matrix in Ω_0. By simple elimination between the three linear systems of (15), we obtain $\mathcal{Y}_\varepsilon$ linearly in terms of $\mathcal{G}$ and $\mathcal{H}$ by :

$$\mathcal{Y}_\varepsilon = J_\varepsilon \mathcal{G} + K_\varepsilon \mathcal{H} \qquad (16)$$

with $J_\varepsilon = (\varepsilon A - CF A^{-1} E)^{-1} \varepsilon B$
$K_\varepsilon = - (\varepsilon A - CF A^{-1} E)^{-1} CF A^{-1} B$

The fact of having only discrete measurements of g and h on the vacuum vessel (see fig. 2) can be handled in the following way. An interpolation of g is made between two discrete measurements g_i on Γ_V in order to have a standard Dirichlet condition on Γ_V and the first term of the cost-function J_ε is modified by taking a discrete sum of quadratic differences over each magnetic probe measuring h_j. Therefore $\mathcal{Y}_\varepsilon$ can be written as in (16), $\mathcal{G}$ and $\mathcal{H}$ representing the vectors of the discrete measurements g_i and h_j and J_ε, K_ε being slightly modified.

Figures 4a and 4b represent the comparison between the "real" plasma boundary (in broken line) and the "identified" one (in continuous line) for a circular plasma (fig. 4a) and for an elliptical one (fig. 4b) with the discrete magnetic probes of JET represented on fig. 2. The "real" plasma boundary is in fact the free boundary obtained by solving the set of equilibrium equations of section 2 in the whole Tokamak, by using the equilibrium code SCED /7/. The values of Ψ and $\frac{1}{r}\frac{\partial\Psi}{\partial n}$ at the positions of the flux loops and magnetic probes are then taken from this simulation and given as input for the identification method. The slight difference between the "real" boundary and the "identified" one observed on fig. 4b is mainly due to the discrete character of the measurements. The choice of ε in the cost-function J_ε depends on the magnitude of the perturbations on g and h. The "identified" boundary Γ_p is, to a large extent, independent of the choice of Γ_0, provided that Γ_0 is inside Γ_p (of course) and is not too small.

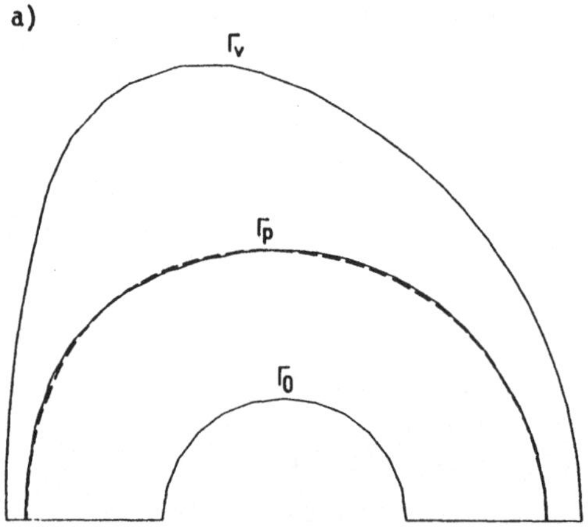

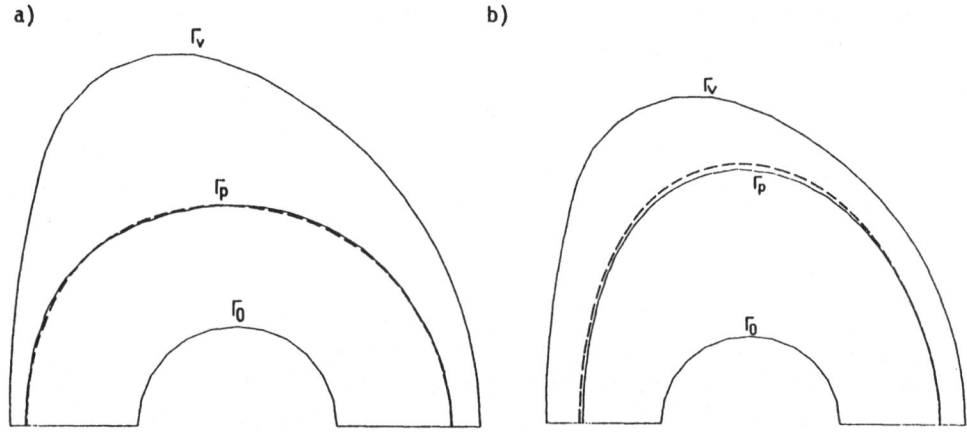

Fig. 4 : Identification of the plasma boundary for a circular case (4a) and for an elliptical one (4b).

This method of identification, which is very fast, as it consists just in multiplying the vectors $\mathcal{G}$ and $\mathcal{H}$ by two matrices, will be used for the real-time control of the plasma shape in Tore Supra, the Tokamak with superconducting toroidal field coils which is being built at Cadarache. One wishes the plasma to pass, at each instant, by a certain number of prescribed points, one of them belonging to the limiter. Therefore one desires that the flux Ψ is equal at all these points and the feedback law has as input the differences between Ψ at each of these points and Ψ at the limiter point. The values of Ψ at these prescribed points are then calculated from the expression (16) in real time by a microprocessor. More details about this isoflux method are given in /8/.

4. IDENTIFICATION OF THE PLASMA CURRENT DENSITY PROFILE FROM THE MAGNETIC MEASUREMENTS

In this section we shall solve the equation for Ψ not only in the vacuum but also in the plasma. We shall exploit the fact that Ψ and $\frac{\partial \Psi}{\partial n}$ are measured on the vacuum vessel in order to identify the plasma current density profile. Figure 3b represents the domain Ω_V, whose boundary is the section Γ_V of the vacuum vessel ; it consists in the vacuum region Ω_X and the plasma Ω_p, separated by the free plasma boundary Γ_p, defined by its contact with the limiter D. From (6), (7) and (8), the equation for Ψ in Ω_V can be written :

$$L\Psi = \left[r\, A(\Psi) + \frac{1}{r} B(\Psi) \right] 1_{\Omega_p} \tag{17}$$

with $\Omega_p = \{ M \in \Omega_V \text{ such that } \Psi(M) > \sup_D \Psi \}$

$$A(\Psi) = \frac{\partial p}{\partial \Psi}, \quad B(\Psi) = \frac{1}{2\mu_0} \frac{\partial f^2}{\partial \Psi}$$

where 1_{Ω_p} is the characteristic function of the plasma Ω_p. Moreover we have the two experimental boundary conditions :

$$\left\{ \begin{array}{l} \Psi = g \\ \\ \frac{1}{r} \frac{\partial \Psi}{\partial n} = h \end{array} \right. \quad \text{on } \Gamma_V \tag{18}$$

It is clear that is not possible to identify from g and h the two functions $A(\Psi)$ and $B(\Psi)$ separately. Therefore we assume that they have the same type of dependence with respect to Ψ, i.e. :

$$\left\{ \begin{array}{l} A(\Psi) = \lambda \frac{\beta}{R_0} C(\Psi) \\ \\ B(\Psi) = \lambda R_0 (1 - \beta) C(\Psi) \end{array} \right. \tag{19}$$

where λ is a normalization parameter, that can be calculated from the value of the total plasma current I_p, where β is a physical parameter to be identified, R_0 a constant (the major radius of the vacuum vessel) and $C(\Psi)$ the profile of current density that has to be determined. This problem can be formulated again as an optimal control problem. The state vector (Ψ, λ) is related to the control vector (β, C) by the state equations :

$$\left\{ \begin{array}{ll} \Psi = g & \text{on } \Gamma_V \\ \\ L\Psi = \lambda\, k_\beta(r)\; C(\Psi)\, 1_{\Omega_p} & \text{in } \Omega_V \\ \\ I_p = \lambda \int_{\Omega_p} k_\beta(r)\, C(\Psi)\, dx & \end{array} \right. \tag{20}$$

with $k_\beta(r) = \frac{r\beta}{R_o} + \frac{R_o(1-\beta)}{r}$, $\Omega_p = \{ M \in \Omega_V \text{ such that } \Psi(M) > \sup_D \Psi \}$

The problem (P') is to determine (β_o, C_o) such that :

$$J'_{\varepsilon'}(\beta_o, C_o) = \inf_{(\beta, C)} J'_{\varepsilon'}(\beta, C) \tag{21}$$

with $J'_{\varepsilon'}(\beta, C) = \int_{\Gamma_V} \left(\frac{1}{r}\frac{\partial\Psi}{\partial n} - h \right)^2 d\sigma + \varepsilon' \int_{\Omega_p} \left(\frac{\partial^2 C}{\partial \Psi^2} \right)^2 dx$

The second term of $J'_{\varepsilon'}$ is a regularization term which obliges the function $C(\Psi)$ to be smooth. Problem (P') has been solved numerically by a sequential quadratic method which consists in linearizing the state equations (20) with respect to $(\Psi, \lambda, \beta, C)$ and in minimizing a quadratic cost-function approximating $J'_{\varepsilon'}$. The algorithm consists then in a sequence of linear quadratic optimal control problems, which are solved by a conjugate gradient method. The resolution of the linearized state equations in Ω_V is performed by a linear finite element method. The resolution of one case by such an algorithm takes approximately 3 s on CRAY-XMP.

If we take, as in section 3, the magnetic measurements from the simulation of one complete equilibrium case, then if A and B correspond to the formulae (19), the parameter β and the function $C(\Psi)$ (supposed to be smooth) can be identified correctly. Let us define the average $\langle \frac{j_T}{r} \rangle$ of the plasma current density j_T divided by r over each flux line $\mathcal{C}$ (where Ψ = constant) :

$$\langle \frac{j_T}{r} \rangle = \int_{\mathcal{C}} \frac{j_T dl}{|\nabla\Psi|} \times \left[\int_{\mathcal{C}} \frac{r\, dl}{|\nabla\Psi|} \right]^{-1} \tag{22}$$

Even if $A(\Psi)$ and $B(\Psi)$ are not given by formulae (19) in the simulation of the full equilibrium code, β and $C(\Psi)$ can be obtained by solving problem (P') and the profile of $\langle \frac{j_T}{r} \rangle$ in terms of Ψ seems to be correctly identified. One example is given in figure 5 where the "real" profile of $\langle \frac{j_T}{r} \rangle$ as a function of Ψ_N (normalized value of Ψ) and the identified profile are represented.

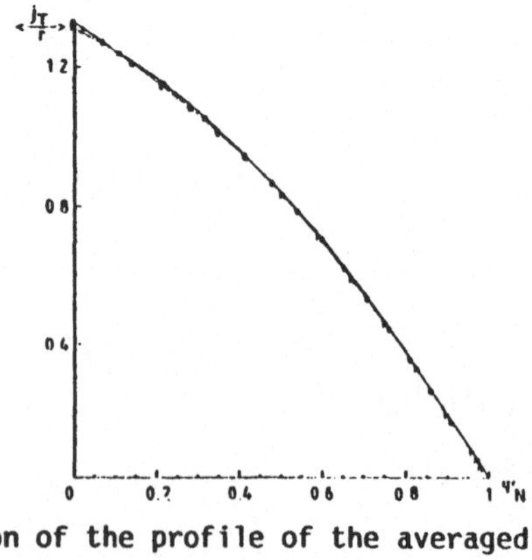

Fig. 5 : Identification of the profile of the averaged current density

The mathematical problem of identifiability of $< \frac{j_T}{r} >$ from g and h is still open. As the plasma boundary Γ_p is well identified and as Ψ is defined up to a constant, this identifiability problem can be formulated as follows :

$$\left\{ \begin{array}{ll} \Psi = 0 & \text{on } \Gamma_p \\ \frac{\partial \Psi}{\partial n} = h_p & \text{on } \Gamma_p \\ L\Psi = j_T\ (r, \Psi) = r\ A(\Psi) + \frac{1}{r}\ B(\Psi) & \text{in } \Omega_p \end{array} \right. \quad (23)$$

where Γ_p is assumed to be known. Is it possible to identify $< \frac{j_T}{r} >$ from the data h_p on Γ_p ? We can give some remarks in particular cases :

i) the <u>circular cylindrical case</u> : the torus is transformed into a cylinder ; the operator L becomes then $(-\Delta)$ and Γ_p is assumed to be a circle. The system (23) becomes :

$$\left\{ \begin{array}{ll} \Psi = 0 & \text{on } \Gamma_p \\ \frac{\partial \Psi}{\partial n} = h_p & \text{on } \Gamma_p \\ -\Delta\Psi = j\ (\Psi) & \end{array} \right. \quad (24)$$

As j is assumed to be positive, from /9/ it is clear that Ψ is a radial function and that h_p must be constant on Γ_p. From h_p, the single information which can be obtained on j is the value of the total plasma current :

$$I_p = \int_{\Omega_p} j(\Psi)\ dx = -\ 2\pi\ a\ h_p \quad (25)$$

where a is the radius of Γ_p.

ii) the <u>large aspect-ratio circular toroidal case</u> : Γ_p is a circle, whose minor radius is a and major radius R. The aspect-ratio R/a (which was infinite in case i) is now assumed to be large, we set $\varepsilon = \frac{a}{R}$. From the analytical equilibrium theory /10/, which is a first order expansion of the equilibrium equations with respect to ε, one obtains :

$$h_p = h_o\ (1 + \varepsilon\ \Lambda\ \cos\theta) \quad (26)$$

with $\Lambda = \beta + \frac{li}{2} - 1$, $h_o = -\ \frac{Ip}{2\pi a}$,

θ being the polar angle with respect to the center of Γ_p and l_i the internal inductance of the plasma per unit length. From the data of h_p, can then be identified two coefficients : I_p and Λ.

iii) the toroidal elliptical case : the circle is now distorted into an ellipse, whose elongation is noted e. It has been proved in /11/ that the parameters β and l_i can now be identified separately with a robustness which increases with e. Three parameters are here identified : I_p, β and l_i.

iv) degenerate toroidal case : it has been pointed out in /12/, /13/ that there exists a configuration, which is oblated, where there is an infinite class of functions $j_T(r,\psi)$ corresponding to the same flux lines and the same function h_p on Γ_p. This is, in some sense, the equivalent case for the operator L of the circular cylindrical case i) for $(-\Delta)$.

The identifiability problem clearly depends on the shape of Γ_p ; it remains open in the general case.

REFERENCES

/ 1/ J. BLUM : Computer Physics Reports 6 (1987) 275-298.

/ 2/ R. COURANT - D. HILBERT : Methods of Mathematical Physics. Interscience Publishers (1962).

/ 3/ V.D. SHAFRANOV - L.E. ZAKHAROV : Nuclear Fusion 12 (1972) 599-601.

/ 4/ J.L. LIONS : Contrôle optimal de systèmes gouvernés par des équations aux dérivées partielles. Dunod (1968).

/ 5/ P. COLLI FRANZONE : in Computing methods in applied sciences and engineering. Ed. R. Glowinski - J.L. Lions. North Holland (1980) 615-633.

/ 6/ J. BLUM : Numerical simulation and optimal control in plasma physics. Dunod (1988).

/ 7/ J. BLUM - J. LE FOLL - B. THOORIS : Computer Physics Communications 24 (1981) 235-254.

/ 8/ J. BLUM - J. LE FOLL - C. LELOUP : Proceedings of the 12th Symposium on Fusion Technology. Varese (1984).

/ 9/ B. GIDAS - W.NI-L. NIRENBERG : Commun. Math. Physics 68, 209-243 (1979).

/10/ V.S. MUKHOVATOV - V.D. SHAFRANOV : Nuclear Fusion 11 (1971) 605.

/11/ L.L. LAO et al. : Nuclear Fusion, Vol. 25, N°10 (1985) 1421.

/12/ B.J. BRAAMS : IPP 5/2 Report (1985). Max-Planck-Institut für Plasmaphysik. Garching bei München (F.R.G.).

/13/ C.M. BISHOP - J.B. TAYLOR : 12th European Conf. on Controlled Fusion and Plasma Physics. Budapest (1985).

CONVERGENCE OF SUBOPTIMAL ELEMENTS IN INFINITE DIMENSIONAL NONLINEAR PROGRAMMING PROBLEMS

H. O. Fattorini
Department of Mathematics
University of California, Los Angeles, California 90024, USA

This work was supported in part by the NSF under grant DMS87-01877

Abstract. We consider a formulation of optimal control problems in infinite dimensional spaces as nonlinear programming problems for functions defined in complete metric spaces which has been developed in detail by the author and H. Frankowska in [10]. This formulation makes possible to unify in a natural way control theory with nonlinear programming theory; in particular we obtain Pontryagin's maximum principle in the form of a Kuhn-Tucker condition. In the same fashion we obtain convergence and robustness results for approximate minima.

§1. *Introduction.* By way of motivation, we consider an ordinary, finite dimensional nonlinear programming problem

(1.1) minimize $f_0(x)$ (minimum = m),

(1.2) subject to $f(x) = (f_1(x),\dots, f_m(x)) = 0.$

There are at least two different approaches to computation (or, rather, to numerical approximation) of solutions.

(a) Use an iterative method (for instance, a quasi-Newton method). Typically, if a minimum $\bar{x}$ satisfies second order sufficient conditions, the method converges if it is started near enough the minimum (which may not be unique). In general, we obtain a rate of convergence for the sequence $\{x_n\}$ of approximations, for instance (see [16]) superlinear convergence

$$\|x_{n+1} - \bar{x}\|/\|x_n - \bar{x}\| \to 0 \quad (n \to \infty).$$

(b) Given a sequence $\{\varepsilon_n\}$, $\varepsilon_n > 0$, $\varepsilon_n \to 0$, construct (by any means) *suboptimal elements*, that is, approximate solutions x_n of the nonlinear programming problem:

$$f_0(x_n) \le m + \varepsilon_n, \tag{1.3}$$

$$\|f(x_n)\| \le \varepsilon_n, \tag{1.4}$$

and show that

$$x_n \to \bar{x} = \text{minimum}, \tag{1.5}$$

where m is the minimum in (1.2).

Several observations on approach (b) are obvious. First, the suboptimal elements x_n can be constructed by means of *penalty methods*, for instance as solutions of

$$\text{minimize } (f_0(x) + \varepsilon^{-2}\|f(x)\|^2) \quad \text{(without constraints)}$$

Convergence of the whole sequence $\{x_n\}$ to a minimum $\bar{x}$ can only be expected if $\bar{x}$ is unique. Otherwise, we can only prove that certain subsequences will converge to (some) minimum $\bar{x}$. However, the treatment is trivial as long as the *feasible set* of the problem is bounded, since we can apply the Bolzano–Weierstrass theorem. No additional conditions on the minimum (such as second order sufficient conditions) are required. In case these conditions are satisfied and the minimum is unique, we can obtain rates of convergence, typically

$$\|x_n - \bar{x}\| = 0(\sqrt{\varepsilon_n}) \tag{1.6}$$

An attractive feature of approach (b) is this: since x_n is not constructed by any particular method, convergence results can be interpreted as *sensitivity* or *robustness* results: small variations of the parameters of the problem (that is, of the functions f_0 and f) will not change much the minimum $\bar{x}$.

§2. *Systems.* Control problems (in finite or infinite dimensional spaces) lead to an infinite dimensional version of the nonlinear programming problem (1.1)–(1.2), where the *control space* (m–dimensional Euclidean space $\mathbf{R}^m$ for (1.1)–(1.2)) is a metric space without any natural linear structure. To fix ideas, we consider a control system described by a quasilinear equation in Hilbert space,

$$y'(t) = Ay(t) + f(t, y(t), u(t)) \quad (0 \le t \le T), \tag{2.1}$$

$$y(0) = y^0, \tag{2.2}$$

(A the infinitesimal generator of a strongly continuous semigroup), although the same approach can be used with other types of systems (see §3). We assume that the controls $u(t)$ are strongly measurable functions with values in a second Hilbert space F, satisfying a constraint of the type

$$u(t) \in U, \tag{2.3}$$

where U, the *control set*, is a bounded subset of F; the *control space* of all such control functions is called $W(0, T; U)$. Under adequate smoothness conditions on the nonlinearity $f(t, y, u)$ (f has a Fréchet derivative $\partial_y f(t, y, u)$ with respect to y such that $f(t, y, u)$ is continuous and $\partial_y(f, y, u)$ is strongly continuous for $0 \leq t \leq T$, $y \in E$, $u \in U$) the initial value problem (2.1)- (2.2) can be locally solved. If *a priori* bounds on the solutions can be obtained, the solutions can be uniquely extended to $0 \leq t \leq T$ and the map

$$u(t) \longrightarrow y(t, u) \tag{2.4}$$

from controls $u \in W(0, T; U)$ to solutions or trajectories $y(t, u)$ of (2.1)-(2.2) corresponding to u satisfies the following three properties:

(i) *Causality:* $y(t, u)$ $(t \leq \bar{t})$ only depends on the values of $u(t)$ for $t \leq \bar{t}$.

(ii) *Continuity:* For $\bar{t}$ fixed the map

$$u(t) \longrightarrow y(\bar{t}, u) \tag{2.5}$$

from $W(0, \bar{t}; U)$ into E is continuous, where $W(0, \bar{t}; U)$ is endowed with the *Ekeland distance*

$$d(u, v) = \text{meas}\,\{t;\ u(t) \neq v(t)\},$$

(iii) *Differentiability* with respect to *spike variations* $u_{s,h,v}(t)$ ($u_{s,h,v}(t) = v$ $(s - h \leq t \leq s)$, $u_{s,h,v}(t) = u(t)$ elsewhere): For every $u \in W(0, \bar{t}; U)$ the limit

$$\lim_{h \to 0+} h^{-1}(y(\bar{t}, u_{s,h,v}) - y(\bar{t}, u)) \tag{2.6}$$

exists in a set $e = e(u)$ of full measure in $0 \leq t \leq \bar{t}$.

In the present case, the limit is

$$\xi(\bar{t}, s, u, v) = S(\bar{t}, s; u)\{f(s, y(s, u), v) - f(s, y(s, u), u(s))\} \tag{2.7}$$

where $S(t, s; u)$ is the solution operator of the linearized equation

$$S'(t, s; u) = (A + \partial_y f(t, y(t, u), u(t)) S(t, s; u),$$

$$S(s, s; u) = I .$$

These properties of the map (2.4) lead to the abstract notion of *system*. A system is, by definition, a map X

$$u(t) \longrightarrow Xu(u) = y(t, u) \tag{2.8}$$

from $W(0, T; U)$ into the space $C(0, T; E)$ of continuous E-valued functions from $[0, T]$ into E that enjoys properties (i), (ii) and (iii). The limit $\xi(t, s, u)$ is called the *derivative* of the system X, u is called the *input* or *control* and $y(t, u)$ is the *output* or *trajectory*. Input-output maps that satisfy (i), (ii) and (iii) are generated not only by initial value problems like (2.1)-(2.2) but also (among other things) by functional differential equations and by boundary control systems of the form

$$y_t(t, x) = Ay(t, x) \qquad ((t, x) \in [0, T] \times \Omega),$$

$$By(t, x) = u(t, x) \qquad ((t, x) \in [0, T] \times \Gamma),$$

$$y(0, x) = y^0(x) \qquad (x \in \Omega),$$

where Ω is a domain in $\mathbf{R}^m$ with boundary Γ, A a differential operator in Ω and B a boundary operator on Γ. For details, see [4].

§3. *Optimal control problems.* These can be formulated for an arbitrary system, whether generated by a distributed parameter system like (2.1)-(2.2), by a different equation or in any other way. The other necessary ingredient is a *cost functional* $y_0(t, u)$, which is simply a system $X_0 : W(0, T; U) \longrightarrow C(0, T; U)$ whose trajectories are real valued. A type of cost functional often used is

$$y_0(t, u) = \int_0^t f_0(\sigma, y(\sigma, u), u(\sigma))\, d\sigma, \tag{3.1}$$

where f_0 is a real valued function. In the *time optimal* problem $f_0 \equiv 1$, so that $y_0(t, u)$ simply measures the time elapsed between the beginning of the process and the arrival to the target.

The *optimal control problem* corresponding to a system X and to a cost functional X_0 is

(3.2) minimize $y_0(t, u)$ (minimum = m),

(3.3) subject to $y(t, u) \in Y$ = target set $\subset E$,

$t > 0$ unrestricted. In this level of generality, if $\bar{u}$ is a solution of the optimal control problem (3.2)–(3.3) then the trajectory $y(t, \bar{u})$ may arrive optimally to the target set Y more than once; we shall denote any of these times by $\bar{t}$ and call it an *optimal arrival time*. Of course, in the time optimal problem, we can arrive optimally to the target only once.

A control $\tilde{u} \in W(0, \tilde{t}, U)$ is called $(\tilde{t}, \varepsilon)$– *suboptimal* if and only if

(3.4) $y_0(\tilde{t}, \tilde{u}) \leq m + \varepsilon$,

(3.5) $\mathrm{dist}(y(\tilde{t}, \tilde{u}), Y) \leq \varepsilon$.

§4. *Infinite dimensional linear programming problems.* We can cast the optimal control problem (3.2)–(3.3) in a form that matches the nonlinear programming problem (1.1)–(1.2) as follows. Denote by $\bar{t}$ any optimal arrival time, and let $f_0(u) = y_0(\bar{t}, u)$, $f(u) = y(\bar{t}, u)$. Then (3.2)–(3.3) becomes

(4.1) minimize $f_0(u)$ (minimum = m),

(4.2) subject to $f(u) \in Y$ = target set $\subset E$.

This leads naturally to the formulation of the following *general nonlinear programming problem,* where V, E are arbitrary sets, Y is a subset of E and f f_0, f are functions, $f_0 : V \to \mathbf{R}$, $f : V \to E$:

(4.3) minimize $f_0(u)$ (minimum = m),

(4.4) subject to $f(u) \in Y$

Obviously, (4.3)–(4.4) generalizes at the same time the finite dimensional nonlinear programming problem (1.1)–(1.2) (where $V = \mathbf{R}^m$, $E = \mathbf{R}$, $Y = \{0\}$) and the optimal control problem (3.2)–(3.3), where $V = W(0, \bar{t}; U)$. In the latter, E and V are, respectively, the prescribed Hilbert space and control set and $f_0(u) = y_0(\bar{t}, u)$, $f(u) = y(\bar{t}, u)$.

A possible definition of suboptimal element for (4.3)–(4.4) is the following: $\tilde{u} \in V$ is ε– *suboptimal* if and only if

(4.5) $f_0(\tilde{u}) \leq m + \varepsilon$,

(4.6) $\quad \mathrm{dist}(f(\tilde{u}), Y) \le \varepsilon.$

Although this definition fits well the situation in §1, it is too restrictive if we want to include also (3.3)–(3.4), where $\tilde{t}$ may not be equal to the optimal time $\bar{t}$ (so that we are using functions $y_0(\tilde{t}, u)$, $y(\tilde{t}, u)$ different from $f_0(u)$, $f(u)$). Accordingly, we can formulate the following problem, companion of (4.3)–(4.4):

Let $\{V_n\}$ be a sequence of arbitrary sets, E a metric space, Y a subset of E, $\{f_{0n}\}$, $\{f_n\}$ two sequences of functions with $f_{0n} : V_n \to \mathbf{R}$, $f_n : V_n \to E$:

Characterize the *suboptimal sequences* $\{u^n\}$, $u^n \in V_n$ such that

(4.7) $\quad f_{0n}(u_n) \le m + \varepsilon_n,$

(4.8) $\quad \mathrm{dist}(f_n(u_n), Y) \le \varepsilon_n$

with $\varepsilon_n \to 0$.

When we apply this setup to the optimal control problem (3.2)–(3.3) we set $V_n = W(0, t_n; U)$; E and Y are, respectively, the prescribed Hilbert space and target sets and $f_{0n}(u) = y_0(t_n, u)$, $f_n(u) = y(t_n, u)$.

Suboptimal sequences can usually be computed by penalty methods, thus the question of whether a sequence $\{u^n\}$ of (t_n, ε_n)–suboptimal controls with $\varepsilon_n \to 0$, $\{t_n\}$ bounded is convergent is of interest. We note, however, that the question cannot be meaningfully answered in the degree of generality of the abstract suboptimal control problem, since each element u^n of the sequence $\{u^n\}$ belongs to a different space V_n. Even in such a case as (4.5)–(4.6) where all the V_n are all subspaces of a same space V and inherit its topology, convergence, when it can be proved, will be in a metric different from that of V (see §6).

§5. *The maximum principle.* The convergence results below depend directly or indirectly on the possibility of proving a *maximum principle* (a first order necessary condition for optimal solutions). This was done in [4] for control problems of the form (3.2)–(3.3) and generalized in [10] to infinite dimensional nonlinear programming problems of the form (4.3)–(4.4). The assumptions on the various entities that make up this last problem are: V is a complete metric space, E is a Hilbert space, the target set Y is closed and f, f_0 are continuous. The first step is the proof of an *approximate maximum principle.* To state it, we recall some notions of set analysis. If Y is an arbitrary set in a Hilbert space E, the *negative polar* (cone) *of* Y is defined by

$$Y^- = \{z \in E;\ \langle z, y\rangle \le 0,\ y \in Y\}.$$

Given a point $\bar{y} \in Y$ the *contingent cone to* Y *at* $\bar{y}$ is the set of all $w \in E$ such that there exists a sequence $\{h_k\}$ of positive numbers with $h_k \to 0$ and a sequence $\{y_k\} \subset Y$ with

$$(y_k - \bar{y})/h_k \to w \quad \text{as } k \to \infty.$$

The *Clarke tangent cone* $C_Y(\bar{y})$ to Y at $\bar{y} \in Y$ is, by definition, the set of all elements $w \in Y$ such that for every sequence $\{h_k\}$ of positive numbers with $h_k \to 0$ and for every sequence $\{\bar{y}_k\} \subset Y$ with $\bar{y}_k \to \bar{y}$ there exists a sequence $\{y_k\}$ in Y such that

$$(y_k - \bar{y}_k)/h_k \to w \quad \text{as } h \to \infty.$$

It can be shown that $C_Y(\bar{y})$ is convex and closed ([1, p. 407]). The cone $N_Y(\bar{y}) = C_Y(\bar{y})^-$ is called the *Clarke normal cone to* Y *at* y.

Let V be a metric space, g an arbitrary function on V. Given a point $u \in V$ where $g(u)$ exists, a vector $\xi \in E$ is called a *(first order) variation of* g *at* u if and only if there exists $\delta > 0$ such that, for all h, $0 < h \leq \delta$, there exists an element $v = v(h) \in V$ with

$$d(v(h), u) \leq h\,,$$

$$\lim_{h \to 0+} h^{-1}(g(v(h)) - g(u)) = \xi$$

The set of such vectors will be denoted by $\partial g(u)$ and called the *variation set of* g *at* u. For instance, if g is a function defined in $\mathbf{R}^m$, any directional derivative at u belongs to $\partial f(u)$. If X is a system as defined in §2 and we set $f(u) = y(\bar{t}, u)$ for $\bar{t}$ fixed, we verify easily that for any control $\bar{u} \in W(-k, \bar{t}; U)$ and any $s \in e(u)$ ($e(u)$ the set in postulate (c)) and $v \in U$, the derivative

$$\xi(\bar{t}, s, u, v)$$

belongs to $\partial f(u)$; to show this it suffices to take $v(h) = u_{s,h,v}$ where $u_{s,h,v}$ is the spike perturbation used in the definition of ξ.

Finally, we recall Kuratowski's lim inf Z_n, where $\{Z_n\}$ is a sequence of subsets of a metric space V:

$$\liminf_{n \to \infty} Z_n = \{z;\ \lim_{n \to \infty} \operatorname{dist}(z, Z_n) = 0\}\,.$$

The first result is the following *approximate maximum principle*, where we only assume that the target set Y is closed.

THEOREM 5.1 Let $\bar{u}$ be a solution of the problem (4.3)–(4.4). Then there exists a sequence $\{\delta_n\}$ of positive numbers with $\delta_n \to 0$, a sequence $\{u^n\} \subset V$, a sequence $\{y^n\} \subset Y$ such that

(5.1) $$d(u^n, \bar{u}) + \|y_n - f(\bar{u})\| \leq \delta_n$$

and a sequence $\{(\mu_n, z_n)\} \subset \mathbf{R} \times E$ satisfying

(5.2) $$\mu_n \geq 0, \quad \|(\mu_n, z_n)\| = 1$$

and such that, for every $(\eta^n, \xi^n) \in \overline{\text{conv}}\, \partial(f_0, f)(u^n)$ and every $w_n \in K_Y(y^n)$ we have

(5.3) $$\mu_n \eta^n + \langle z_n, \xi^n - w^n \rangle \geq - \delta_n(1 + \|w^n\|).$$

Moreover, for every cluster point (μ, z) of the sequence $\{(\mu_n, z_n)\}$ in $\mathbf{R} \times E$ (E endowed with its weak topology) we have

(5.4) $$\mu \geq 0, \quad z \in N_Y(f(\bar{u})),$$

and, for every $(\eta, \xi) \in \liminf_{n \to \infty} \overline{\text{conv}}\, \partial(f_0, f)(u^n)$ we have

(5.5) $$\mu\eta + \langle z, \xi \rangle \geq 0 .$$

For a proof (which is based on Ekeland's variational principle [3]), see [10].

The result is somewhat incomplete in two senses. Firstly, it depends on the unknown sequence $\{u^n\}$ (although, as we shall see below, in some important applications sufficiently large subsets of $\liminf_{n \to \infty} \overline{\text{conv}}\, \partial(f_0, f)(u^n)$ that do not depend on $\{u^n\}$ can be identified.) Secondly, the vector (μ, z) may vanish, rendering (5.5) inoperative. In order to avoid this, we must impose additional assumptions on Y. One of the several results available is Theorem 5.2 below.

Let $\{\Delta_n;\ n = 1,2,\ldots\}$ be a sequence of sets in E. Following [5] we say that $\{\Delta_n\}$ has *finite codimension* in E if and only if there exists a subspace H with $\dim H^{\perp} < \infty$ such that

$$\Delta_H = \bigcap_{n \geq 1} \Pi_H(\overline{\text{conv}}\,(\Delta_n))$$

has nonempty interior, where Π_H is the projection of E into H.

THEOREM 5.2 Let the assumptions of Theorem 5.1 hold. Assume that for every sequence $\{u^n\}$, $u^n \in V$ with $d_n(u^n, \bar{u}) \to 0$ fast enough and every sequence $\{y^n\} \subset Y$ with $y^n \to \bar{y}$ there exists $\rho > 0$ such that the sequence

(5.6) $$\{\Delta_n\} = \{\overline{\text{conv}}\,(K_Y(y^n) \cap B(0, \rho)) - \overline{\text{conv}}\,\partial(f_0, f)(u^n)\}$$

has finite codimension in E. and is such that

(5.7) $$\bigcup_{n \geq 1} (I - \Pi_H)(\Delta_n)$$

is bounded. Then the multiplier (μ, z) in (5.5) does not vanish.

For the proof see [10], where variants of the assumptions above are also considered. We note that, for the finite dimensional problem (1.1)–(1.2), the conditions in Theorem 5.2 are automatically satisfied, and (5.5) reduces to the Kuhn–Tucker necessary conditions.

To illustrate what Theorems 5.1 and 5.2 mean we note that in the time optimal problem for a system (see §2), (5.5) reduces to

$$\langle z, \xi(\bar{t}, s, u, v, \bar{u}(s))\rangle \geq 0 \text{ a.e. in } s, \ v \in U$$

For the system generated by (2.1)–(2.2) ξ is given by (2.7) so that the maximum principle can be cast in the familiar form

$$\langle S(t, s; \bar{u})^* z, f(s, y(s, \bar{u}), \bar{u}(s))\rangle = \min_{v \in U} \langle S(t, s; \bar{u})^* z, f(s, y(s, \bar{u}), v)\rangle,$$

$S(t, s; \bar{u})^* z = z(s)$ the solution of

$$z'(s) = -(A + \partial_y(t, y(t, u), u(t))^* z(s),$$

$$z(\bar{t}) = z .$$

In this generality, however, not all information pertaining to the maximum principle (such as constancy of the Hamiltonian) is available.

§6. *Convergence of suboptimal elements.* Let $\{u^n\}$ be a suboptimal sequence for the problem (4.7)–(4.8). Information on $\{u^n\}$ (hopefully leading to convergence) is obtained by the same technical arguments (chiefly, Ekeland's variational principle) as those used for optimal elements in §5. There are three steps:

(a) *The sequence maximum principle.* This is a separate approximate maximum principle for each term u^n of the suboptimal sequence. It looks very much like the approximate maximum principle (Theorem 5.1). What is constructed here is a second sequence $\{\tilde{u}^n\}$, $\tilde{u}^n \in V_n$ and a sequence $\{y^n\} \subset Y$ such that

(6.1) $$d(\tilde{u}^n, u^n) + \|\tilde{y}^n - f_n(\tilde{u}^n)\| \leq \delta_n \to 0,$$

and a sequence $\{(\mu_n, z_n)\} \subset \mathbf{R} \times E$ such that

(6.2) $$\mu_n \geq 0, \quad \|(\mu_n, z_n)\| = 1$$

and such that, for every $(\eta^n, \xi^n) \in \overline{\text{conv}}\, \partial(f_{0n}, f_n)(u^n)$ and every $w_n \in K_Y(\tilde{y}^n)$ we have

(6.3) $$\mu_n \eta^n + \langle z_n, \xi^n - w^n \rangle \geq -\delta_n(1 + \|w^n\|).$$

(b) *The convergence principle.* The sequence maximum principle becomes a *convergence principle* if we show that the sequence $\{(\mu_n, z_n)\}$ of multipliers (or at least a subsequence thereof) is convergent to a *nonzero* element (μ, z) of $\mathbf{R} \times E$. If the convergence is weak (resp. strong) we speak of a *weak* (resp. *strong*) convergence principle. Note that, in view of (6.2), the requirement that $(\mu, z) \neq 0$ is unnecessary for a strong convergence principle. Weak convergence principles can be proved by means similar to those used to show that the multiplier (μ, z) in (5.5) does not vanish. In fact, if for some $\rho > 0$ the sequence

(6.4) $$\{\Delta_n\} = \{\overline{\text{conv}}\,(K_Y(\tilde{y}^n) \cap B(0, \rho)) - \overline{\text{conv}}\, \partial(f_{0n}, f_n)(\tilde{u}^n)\}$$

has finite codimension in E and (5.7) holds, then a weak convergence principle holds. Stronger assumptions on the sequence $\{\Delta_n\}$ in (6.4) produce strong convergence principles (see [10]).

(c) *Convergence of suboptimal sequences.* For the reasons pointed out in §3, this step is best carried out for particular instances of the general problem. As an illustration, we consider the system generated by (2.1)–(2.2). Here, (5.3) (with $w^n = 0$) is

(6.5) $$\langle S(t_n, s; \tilde{u}^n)^* z_n, \{f(s, y(s, \tilde{u}^n), v) - f(s, y(s, \tilde{u}^n), \tilde{u}(s)\}\rangle \geq -\delta_n \to 0$$

a. e. in $0 \leq t \leq \bar{t}$ for all $v \in U$. Since $d(\tilde{u}^n, u^n) \to 0$, it is enough to deal with $\{\tilde{u}^n\}$. The key to the convergence results is the existence of a subsequence of $\{\tilde{u}^n\}$ such that

$$\tilde{u}^n \to \bar{u} \text{ weakly in } L^2,$$

$$y(s, \tilde{u}^n) \to y(s, \bar{u}) \text{ strongly in } 0 \leq s \leq \bar{t},$$

$$S(t_n, s; \tilde{u}^n)^* \to S(\bar{t}, s; \bar{u})^* \text{ strongly in } 0 \leq s \leq \bar{t}.$$

This can be proved if A generates an analytic semigroup or (in a modified form) when $R(\mu; A) = (\mu I - A)^{-1}$ is compact for some μ; see [5], [6], [7]. Using this convergence property we obtain

$$\langle S(\bar{t}, s; \bar{u})^* z, f(s, y(s, \bar{u}), v) - f(s, y(s, \bar{u}), \tilde{u}^n(s))\rangle \geq -\rho_n \to 0$$

We write $\zeta(s) = S(\bar{t}, s; \bar{u})z$, $U(s) = f(s, y(s, \bar{u}), U)$, $w(s) = f(s, y(s, \bar{u}), v)$, $v^n(s) = f(s, y(s, \bar{u}), \tilde{u}^n(s))$: (6.5) becomes

$$\langle \zeta(s), w(s) - v^n(s)\rangle \geq -\rho_n = \delta. \tag{6.6}$$

Let V be an arbitrary set in E, $\zeta \in E$, $\zeta \neq 0$, $\rho > 0$. Define

$$V(\zeta, \rho) = \{v \in V; \langle \zeta, w - v\rangle \geq -\rho, w \in V\}.$$

Assuming that the vector $\zeta(s)$ in (6.6) does not vanish and

$$\operatorname{diam} U(s)(\zeta(s), \rho) \to 0 \text{ as } \rho \to 0+ \tag{6.7}$$

we have a convergence theorem for the sequence

$$\{f(s, y(s, \bar{u}), \tilde{u}^n(s))\} \tag{6.8}$$

from which convergence of $\{\tilde{u}^n\}$ can in some cases be obtained. Condition (6.7) holds, for instance, for uniformly convex sets in finite dimensional spaces, but the results are not limited to convex sets: convergence can also be obtained in some nonconvex situations (see [7], [10]).

We mention finally that convergence of suboptimal controls can be obtained by other methods as well. We mention briefly two of these methods. In [8] convergence results are obtained from geometry of Hilbert spaces, especifically from situations where weak convergence implies strong convergence (such as weak convergence from inside to the boundary of a sphere). An interesting feature of this kind of result is that it does not depend on the maximum principle but only on its consequence, the "bang-bang" principle, thus it is potentially more general that the others. On the other hand, in [11], convergence results in various norms are obtained by direct estimates for systems defined by quasilinear initial value problems of the form (2.1)-(2.2). These estimates depend on the maximum principle and are totally explicit; in particular, rates of convergence can be obtained in many cases, in general of the type

$$\|u^n - \bar{u}\| = O(\sqrt{\varepsilon_n}) .$$

In some control problems where the control set is 1-dimensional, this rates can be improved to linear, that is,

$$\|u^n - \bar{u}\| = O(\varepsilon_n) .$$

REFERENCES

[1] J-P. AUBIN and I. EKELAND, *Applied Nonlinear Analysis*, Wiley, New York, 1984

[2] F. CLARKE, *Optimization and Nonsmooth Analysis*, Wiley, New York, 1983

[3] I. EKELAND Nonconvex minimization problems, Bull. Amer. Math. Soc. 1 (NS) (1979) 443-474

[4] H. O. FATTORINI, A unified theory of necessary conditions for nonlinear nonconvex control systems, Applied Math. Optim. 15 (1987) 141-185

[5] H. O. FATTORINI, Optimal control of nonlinear systems: cónvergence of suboptimal controls, I, to appear in Proceedings of Special Session on Operator Methods in Optimal Control Problems, Annual AMS Meeting, New Orleans, January 1986.

[6] H. O. FATTORINI, Optimal control of nonlinear systems: convergence of suboptimal controls, II, to appear in Proceedings of IFIP Workshop on Control Systems Described by Partial Differential Equations, Gainesville, February 1986

[7] H. O. FATTORINI, Convergence of suboptimal controls for point targets, Proceedings of Conference on Optimal Control of Partial Differential Equations, Oberwolfach, May 1986, International Series of Numerical Mathematics vol. 78 (1987), 91-107, Birkhäuser Verlag, Basel

[8] H. O. FATTORINI, Some remarks on convergence of suboptimal controls, to appear in Proceedings of the First International Conference on Advances on Communications and Control Systems, Washington, D.C, June 1987

[9] H. O. FATTORINI, Convergence of suboptimal controls: the point target case, to appear.

[10] H. O. FATTORINI and H. FRANKOWSKA, Necessary conditions for infinite dimensional control problems, to appear.

[11] H. O. FATTORINI and H. FRANKOWSKA, Rates of convergence for suboptimal controls, to appear.

[12] H. FRANKOWSKA, The maximum principle for differential inclusions with end point constraints, SIAM J. Control 25 (1987)

[13] H. FRANKOWSKA and Cz. OLECH, R-convexity of the integral of set-valued functions, Amer. J. Math. (1982) 117-129

[14] H. FRANKOWSKA and Cz. OLECH, Boundary solutions of differential inclusions, J. Diff. Equations 44 (1982) 156-165.

[15] D. G. LUENBERGER, *Linear and nonlinear programming*, Addison-Wesley, Reading 1984

[16] J. STOER and R. A. TAPIA, The local convergence of sequential linear programming methods, to appear in Proceedings of the First International Conference on Advances on Communications and Control Systems, Washington, D. C, June 1987

AN INTRODUCTION TO THE METHODS BASED ON UNIQUENESS FOR EXACT CONTROLLABILITY OF DISTRIBUTED SYSTEMS

Jacques-Louis Lions
Collège de France, 3, rue d'Ulm, F-75005 Paris
and CNES (Centre National d'Etudes Spatiales)

INTRODUCTION

The problem of exact controllability (E.C.) for a distributed system consists in trying to steer the state *in a given finite time* T from a given initial state to another given final state.

The *state equation* is a linear Partial Differential Equation. All problems considered here are *linear*.

We act on the system through the boundary (or locally or in a pointwise fashion).

This is a classical problem. Several methods of attack are known : cf. D.L.RUSSELL [1].

We have observed in J.L. LIONS [1] that another method can be based on Uniqueness properties. We give here an *introduction* to this method ant to some applications, based on *penalty approximations*.

1. EXACT CONTROLLABILITY AND OPTIMAL CONTROL

1.1. Setting of the problem.

In $\Omega \subset \mathbb{R}^n$, bounded with smooth boundary, let us consider the operator A defined by

$$A\phi = -\frac{\partial}{\partial x_i}\left(a_{ij}(x)\frac{\partial\phi}{\partial x_j}\right) + a_i(x)\frac{\partial\phi}{\partial x_i} + a_o(x)\phi \qquad (1.1)$$

where the functions a_{ij} are supposed smooth enough, and where we have :

$$\left| \begin{array}{l} a_{ij} = a_{ji}\ ,\quad a_{ij}(x)\xi_i\xi_j \geq \alpha\,\xi_i\xi_i\ ,\quad \alpha>0\ ,\ \forall\ \xi_i \in \mathbb{R}\ , \\ a_i\ ,\ a_o \in L^{\infty}(\Omega)\,. \end{array} \right. \qquad (1.2)$$

The *state of the system* we want to exact control is given by the solution $y = y(v)$ of

$$y'' + Ay = 0 \quad \text{in} \quad \Omega\times(0,T), \tag{1.3}$$

(where we set $y'' = \frac{\partial^2 y}{\partial t^2}$), subject to the *initial conditions*

$$y(o) = y^o \ , \ y'(o) = y^1 \quad \text{in} \quad \Omega \tag{1.4}$$

($y(o)$ denotes the function $x \to y(x,o)$; similar notation for $y'(o)$), and with the *boundary condition* :

$$y = \begin{cases} v & \text{on} \quad \Sigma_o \subset \Sigma = \Gamma\times(0,T), \qquad \Gamma = \partial\Omega , \\ 0 & \text{on} \quad \Sigma\setminus\Sigma_o . \end{cases} \tag{1.5}$$

In (1.5) v denotes the *control function.* □

Remark 1.1.

The problem (1.3)(1.4)(1.5) is a *non homogeneous* boundary value problem for the *hyperbolic operator* $\frac{\partial^2}{\partial t^2} + A$. At least if the boundary Γ of Ω and the coefficients are smooth enough, there are *infinitely many ways* to make this problem precise, depending on the *function spaces* where we choose y^o , y^1 and v. Cf. J.L.LIONS and E. MAGENES [1] . □

Remark 1.2.

In (1.5) Σ_o denotes an arbitrary subset of Σ for the time being. But later on the choice of Σ_o will play a crucial role. □

The problem of *Exact Controllability* (E.C.) is now the following : let T *be given* > 0. For any couple $\{y^o,y^1\}$ (in a given function space) find if possible v (in a suitable function space) such that the corresponding solution y satisfies

$$y(T) = y'(T) = 0. \tag{1.6}$$

Remark 1.3.

Due to the *finite speed* of propagation of singularities, it is *necessary* to take T *large enough.* □

Remark 1.4.

Of course it will be necessary to make precise the choices of function spaces in E.C. □.

The problem of *Reverse* Exact Controllability (R.E.C.) is now the following : given T (large enough), given $\{z^o, z^1\}$ arbitrarily in a given function space, find if possible v such that if y denotes the solution of (1.3)(1.4)(*with* $y^o = y^1 = 0$) and (1.5) then

$$y(T) = z^o \;, \quad y'(T) = z^1. \tag{1.7}$$

Remark 1.5.

The R in R.E.C. stands for Reverse. It could as well stand for Reachability ! □

Remark 1.6.

Of course the two problems E.C. and R.E.C. are *equivalent* for operators $\frac{\partial^2}{\partial t^2} + A$ which are *reversible*.

But they are *not* equivalent for *irreversible* operators.

Just to mention a simple case, assume that

$$A_1\phi = A\phi + \int_o^t b(t,\sigma)\phi(x,\sigma)d\sigma \,, \tag{1.8}$$

$b(t,\sigma)$ given function in $\mathbb{R}^2_+$.

Then the two problems have to be considered in a separate fashion. □

1.2. Optimal control

It is quite clear (cf. details in the references given in the bibliography) that if E.C. (or R.E.C.) admits a solution, it admits *an infinity of these*.

The point in using optimal control here is to find "*the best*" *solution*. *But this notion depends on the choice of a norm for* v.
We *choose*

$$v \in L^2(\Sigma_o). \tag{1.9}$$

Remark 1.7.

We want to emphasize this point. Infinitely many other choices are possible ! Cf.J.L. LIONS [1][2][3]. □

We then introduce the sets

$$\mathcal{U}(y^o,y^1) = \{v \mid v \in L^2(\Sigma_o) \text{ such that the solution } y \text{ of (1.3)(1.4)(1.5) satisfies to (1.6)}\}. \tag{1.10}$$

(This is still a little bit "fuzzy", since we have not yet made precise the hypothesis on y^o, y^1) ;

$$\left| \begin{array}{l} R\mathcal{U}(z^o,z^1) = \{v \mid v \in L^2(\Sigma_o) \text{ such that the solution } y \text{ of} \\ (1.3),\ (1.4) \text{ with } y^o = y^1 = 0,\ (1.5) \text{ satisfies to } (1.7)\}. \end{array} \right. \tag{1.11}$$

If we assume that $\mathcal{U}(y^o,y^1)$ (resp. $R\mathcal{U}(z^o,z^1)$ is not empty) (of course it will be necessary *to prove* later on that, indeed, with suitable choices of T and of Σ_o these sets are indeed not empty !!), we then consider the following problems of *optimal control* : *minimize the functional*

$$\frac{1}{2} \int_{\Sigma_o} v^2 d\Sigma \tag{1.12}$$

on $\mathcal{U}(y^o,y^1)$ (or *on* $R\mathcal{U}(y^o,y^1)$). □

Remark 1.8.

We are going to obtain a constructive method to obtain the *optimal* control which minimizes (1.12) in $\mathcal{U}(y^o,y^1)$ or in $R\mathcal{U}(y^o,y^1)$. We obtain in this way a method to verify if $\mathcal{U}(y^o,y^1)$ (or $R\mathcal{U}(y^o,y^1)$) *is not empty.* □

2. PENALIZATION

2.1. Penalized functional

For every $\varepsilon > 0$, we set

$$J_\varepsilon(v,y) = \frac{1}{2} \int_{\Sigma_o} v^2 d\Sigma + \frac{1}{2\varepsilon} \iint (y''+Ay)^2 \tag{2.1}$$

where we set

$$\iint_{\Omega\times(0,T)} \phi \, dxdt = \iint \phi .$$

In (2.2) we assume that

$$\left| \begin{array}{l} y'' + Ay \in L^2(\Omega \times (0,T)), \\ y(o) = y^o , \quad y'(o) = y^1 , \quad y(T) = y'(T) = 0 , \\ y = \left| \begin{array}{ll} v & \text{on } \Sigma_o \\ 0 & \text{on } \Sigma \setminus \Sigma_o \end{array} \right. \end{array} \right. \tag{2.2}$$

and we consider

$$\inf J_\varepsilon(v,y) , \quad v,y \text{ subjects to } (2.2) . \tag{2.3}$$

Remark 2.1.

Under reasonable assumptions on y^o,y^1, the set of the $\{v,y\}$'s such that (2.2) holds true is *not empty*. It is only when $y''+Ay = 0$ that the question is not trivial ! □

Remark 2.2.

We consider here an approximation of E.C. We deal in the same way with R.E.C. We shall obtain in this way methods for E.C. *and* for R.E.C. □

Remark 2.3.

One has of course

$$J_\varepsilon(v,y) = \frac{1}{2}\int_{\Sigma_0} v^2 d\Sigma \text{ if } v \in \mathcal{U}(y^0,y^1) \text{ and } y = y(v). \quad \square \tag{2.4}$$

Remark 2.4.

Problem (2.3) admits a unique solution, denoted by $u_\varepsilon, y_\varepsilon$.
We want now to study the *convergence* of the method.

2.2. Convergence of the penalization approximation.

This is standard procedure in the Calculus of Variations.
One has

$$J_\varepsilon(u_\varepsilon,y_\varepsilon) \le J_\varepsilon(v,y) = \frac{1}{2}\int_{\Sigma_0} v^2 d\Sigma \quad \text{if} \quad v \in \mathcal{U}(y^0,y^1), \quad y=y(v). \tag{2.5}$$

Therefore, as $\varepsilon \to 0$,

$$\begin{aligned} &\{u_\varepsilon, \frac{1}{\sqrt{\varepsilon}}(y''_\varepsilon + Ay_\varepsilon)\} \text{ remains in a bounded set of} \\ &L^2(\Sigma_0) \times L^2(\Omega\times(0,T)). \end{aligned} \tag{2.6}$$

We can extract a subsequence, still denoted by $u_\varepsilon, y_\varepsilon$, such that

$$u_\varepsilon \to u \quad \text{in } L^2(\Sigma_0) \text{ weakly,} \tag{2.7}$$

$$y_\varepsilon \to y \quad \text{in } L^\infty(0,T;L^2(\Omega)) \text{ weak star} \tag{2.8}$$

(see below for this point),

and $y = y(u)$ satisfies (1.2)(1.3)(with $v = u$) (1.4).

[For (2.8) one observes that

$$\begin{aligned} &y''_\varepsilon + Ay_\varepsilon = \sqrt{\varepsilon}\, g_\varepsilon, \quad g_\varepsilon \text{ bounded in } L^2(\Omega\times(0,T)), \\ &y_\varepsilon(0) = y^0, \quad y'_\varepsilon(0) = y^1, \\ &y_\varepsilon = \begin{vmatrix} u & \text{on } \Sigma_0, \\ 0 & \text{on } \Sigma\setminus\Sigma_0. \end{vmatrix} \end{aligned}$$

Then (2.8) follows by using results given in J.L. LIONS [4] (and reproduced with more details in J.L. LIONS [31])].

But

$$J_\varepsilon(u_\varepsilon, y_\varepsilon) \geq \frac{1}{2}\int_{\Sigma_o} u_\varepsilon^2 \, d\Sigma$$

hence by (2.7)

$$\liminf J_\varepsilon(u_\varepsilon, y_\varepsilon) \geq \frac{1}{2}\int_{\Sigma_o} u^2 d\Sigma, \qquad u \in \mathcal{U}(y^o, y^1). \tag{2.9}$$

But by (2.5)

$$\limsup J_\varepsilon(u_\varepsilon, y_\varepsilon) \leq \inf \frac{1}{2}\int_{\Sigma_o} v^2 d\Sigma, \quad v \in \mathcal{U}(y^o, y^1).$$

Comparing with (2.9) it follows that

$$u_\varepsilon \to u \quad \text{in } L^2(\Sigma_o) \text{ strongly.}$$ □

2.3. Optimality system for (2.3)'.

We *define*

$$p_\varepsilon = -\frac{1}{\varepsilon}(y''_\varepsilon + Ay_\varepsilon). \tag{2.10}$$

The Euler equation for (2.3) is

$$\int_{\Sigma_o} u_\varepsilon v \, d\Sigma - \iint p_\varepsilon(\eta'' + A\eta) = 0 \tag{2.11}$$

$\forall \eta$ such that

$$\left| \begin{array}{l} \eta'' + A\eta \in L^2(\Omega \times (0,T)), \\ \eta(o) = \eta'(o) = \eta(T) = \eta'(T) = 0, \\ \eta = \left| \begin{array}{l} v \text{ on } \Sigma_o, \\ 0 \text{ on } \Sigma\backslash\Sigma_o. \end{array} \right. \end{array} \right. \tag{2.12}$$

It follows from (2.11)(2.12) that

$$\left| \begin{array}{l} p''_\varepsilon + A^* p_\varepsilon = 0, \\ p_\varepsilon = 0 \text{ on } \Sigma, \\ \dfrac{\partial p_\varepsilon}{\partial \nu_{A^*}} = u_\varepsilon \text{ on } \Sigma_o. \end{array} \right. \tag{2.13}$$

We have set

$$\frac{\partial}{\partial \nu_{A^*}} = a_{ij} \, \nu_i \frac{\partial}{\partial x_j}$$

and we have in fact *in the present situation* $\frac{\partial}{\partial\nu_{A^*}} = \frac{\partial}{\partial\nu_A}$. □

We now arrive *at the key point : what kind of estimate (if any) can we deduce from* (2.13) *knowing that* u_ε *is bounded in* $L^2(\Sigma_o)$?

There is a *systematic* way to study this question, as it is explained in Section 3 below. □

Remark 2.5.

We can use the same method for R.E.C. In (2.2) one has just to impose, instead of the end conditions of (2.2), the conditions :

$$y(o) = y'(o) = 0 , \quad y(T) = z^o, \quad y'(T) = z^1 .$$ □

Remark 2.6.

Let G be a Hilbert space on Σ_o. *One can replace everywhere* $L^2(\Sigma_o)$ *by* G. □

Remark 2.7.

One can even introduce *Banach* spaces G instead of $L^2(\Sigma_o)$. But we do not know applications of this remark. □

3. NEW FUNCTION SPACES

3.1. Formal definition

Let us consider the equation

$$\left| \begin{array}{l} \phi'' + A^*\phi = 0 , \\ \phi = 0 \text{ on } \Sigma, \\ \phi(o) = \phi^o , \quad \phi'(o) = \phi^1 . \end{array} \right. \tag{3.1}$$

This problem makes sense in *the reversible* case.

We then set

$$\|\{\phi^o, \phi^1\}\|_F = \left(\int_{\Sigma_o} \left(\frac{\partial\phi}{\partial\nu_{A^*}}\right)^2 d\Sigma\right)^{\frac{1}{2}} . \tag{3,2}$$

This quantity defines a *norm* iff we have the following *Uniqueness* property :

$$\left| \begin{array}{l} \phi'' + A^*\phi = 0 \quad \text{in } \Omega\times(0,T) , \\ \phi = 0 \text{ on } \Sigma, \\ \frac{\partial\phi}{\partial\nu_{A^*}} = 0 \text{ on } \Sigma_o \\ \textit{implies} \quad \phi = 0. \end{array} \right. \tag{3,3}$$

Remark 3.1

This is a Holmegren's type result. If, for instance, $A = A^* = -\Delta$, $\Sigma = \Gamma \times (0,T)$, (3.3) is true if $T >$ diameter of Ω.
Cf. other examples in J.L. LIONS [3]. □

Assuming we have uniqueness, then (3.2) does define a norm. We then *define*

$$\left| \begin{array}{l} F = \textit{Hilbert space} \text{ obtained by } \textit{completion} \text{ of smooth functions} \\ \phi^o, \phi^1, \text{ such that } \phi^o = \phi^1 = 0 \text{ on } \Gamma, \text{ for the norm (3.2).} \quad \square \end{array} \right. \tag{3.4}$$

Let us consider now the equation

$$\left| \begin{array}{l} \phi'' + A^*\phi = 0 \quad \text{in } \Omega\times(0,T) \\ \phi(T) = \phi^o, \quad \phi'(T) = \phi^1, \\ \phi = 0 \quad \text{on } \Sigma. \end{array} \right. \tag{3.5}$$

This problem makes sense in the reversible case and it is the best thing to consider in the *irreversible* case !

Assuming the Uniqueness property (3.3) to hold true, we define a (new) *norm* by

$$\left| \begin{array}{l} \|\{\phi^o,\phi^1\}\|_G = \left(\int_{\Sigma_o} \left(\frac{\partial\phi}{\partial\nu_{A^*}}\right)^2 d\Sigma \right)^{\frac{1}{2}}, \\ \phi \text{ solution of (3.5).} \end{array} \right. \tag{3.6}$$

We define G by completion as in (3.4). □

Remark 3.2.

If $A^* = A$ = operator with coefficients *independent* of t, then $G = F$.

3.2. Applications of F and of G.

If we assume that (3.3) holds true, then we can define F and G and it follows from (2.13) that, as $\varepsilon \to 0$,

$$\{p_\varepsilon(o), p'_\varepsilon(o)\} \text{ remains in a bounded set of } F \tag{3.7}$$

$$\{p_\varepsilon(T), p'_\varepsilon(T)\} \text{ remains in a bounded set of } G. \tag{3.8}$$

Then one can pass to the limit, in suitable topologies (cf. details in J.L. LIONS [3]).
Let p denote the limit of p_ε.

We find the following optimality systems :

Optimality system for $\inf \frac{1}{2}\int_{\Sigma_o} v^2 d\Sigma$, $v \in \mathcal{U}(y^o, y^1)$.

$$\left| \begin{array}{l} y'' + Ay = 0 , \\ p'' + A^* p = 0 , \\ y = \left| \begin{array}{ll} u & \text{on } \Sigma_o \\ 0 & \text{on } \Sigma \backslash \Sigma_o \end{array} \right. , \quad p = 0 \text{ on } \Sigma , \end{array} \right. \tag{3.9}$$

$$\left| \begin{array}{ll} y(o) = y^o , & y'(o) = y^1 , \\ y(T) = 0 , & y'(T) = 0 , \end{array} \right. \tag{3.10}$$

$$\frac{\partial p}{\partial \nu_{A^*}} = u \text{ on } \Sigma_o . \quad \square \tag{3.11}$$

Optimality system for $\inf \frac{1}{2}\int_{\Sigma_o} v^2 d\Sigma$, $v \in R\mathcal{U}(y^o, y^1)$.

Equations (3.9)(3.11) are unchanged. Conditions (3.10) are replaced by

$$\left| \begin{array}{l} y(o) = 0 , \; y'(o) = 0 , \\ y(T) = z^o , \; y'(T) = z^1 . \end{array} \right. \quad \square \tag{3.12}$$

Remark 3.3.

We know that

$$\{ p(o), p'(o) \} \in F , \quad \{ p(T), p'(T) \} \in G. \quad \square \tag{3.13}$$

We now derive the methods HUM (and RHUM) *from the optimality systems*.

4. H U M.

Let ϕ be the solution of (3.1) where we take $\{\phi^o, \phi^1\} \in F$. We then define ψ by

$$\left| \begin{array}{l} \psi'' + A\psi = 0 , \\ \psi(T) = \psi'(T) = 0, \\ \psi = \left| \begin{array}{ll} \dfrac{\partial \phi}{\partial \nu_A} & \text{on } \Sigma_o , \\ 0 & \text{on } \Sigma \backslash \Sigma_o . \end{array} \right. \end{array} \right. \tag{4.1}$$

This makes sense in the reversible case.

We then define (of course all this has to be made more precise. Cf. J.L. LIONS [2][3])

$$\Lambda\{\phi^o , \phi^1\} = \{\psi'(o) , -\psi(o)\} . \tag{4.2}$$

One has

$$\Lambda \in \mathcal{L}(F ; F') , \quad \Lambda^* = \Lambda. \tag{4.3}$$

Multiplying (4,1) by ϕ and integrating by parts (the integration by parts are valid by definition of the weak solution of (4.1) based on transposition method as in J.L. LIONS and E. MAGENES |1 l), one finds that

$$<\Lambda\{\phi^o, \phi^1\}, \{\phi^o, \phi^1\}> = \int_{\Sigma_o} (\frac{\partial\phi}{\partial\nu_{A^*}})^2 d\Sigma = \|\{\phi^o, \phi^1\}\|_F^2 . \tag{4.4}$$

Therefore

$$\Lambda \text{ is an isomorphism from } F \text{ onto } F'. \quad \square \tag{4.5}$$

This is the H U M method. It solves, *assuming uniqueness* (hence the terminology), the E.C. problem. Indeed if one assumes that

$$\{y^1, -y^o\} \in F' , \tag{4.6}$$

one solves

$$\Lambda\{\phi^o, \phi^1\} = \{y^1, -y^o\}. \tag{4.7}$$

One has then the optimal control given by

$$u = \frac{\partial\phi}{\partial\nu_{A^*}} \text{ on } \Sigma_o . \tag{4.8}$$

Indeed for the choice (4.8) one verifies, by comparison with the Optimality System in Section 3.2, that $\psi = y$, $\phi = p$.

5. R H U M

We start now from (3.5)(the only sensible thing to do in the irreversible case !). We then consider ψ defined by

$$\left| \begin{array}{l} \psi'' + A\psi = 0 , \\ \psi(o) = \psi'(o) = 0 , \\ \psi = \left| \begin{array}{ll} \frac{\partial\phi}{\partial\nu_{A^*}} & \text{on } \Sigma_o \\ 0 & \text{on } \Sigma\backslash\Sigma_o . \end{array} \right. \end{array} \right. \tag{5.1}$$

We then define the operator μ by

$$\mu\{\phi^o, \phi^1\} = \{-\psi(T), \psi(T)\} . \tag{5.2}$$

One has

$$\mu \in \mathcal{L}(G ; G') , \quad \mu^* = \mu \tag{5.3}$$

and

$$<\mu\{\phi^o, \phi^1\}, \{\phi^o, \phi^1\}> = \int_{\Sigma_o} (\frac{\partial\phi}{\partial\nu_{A^*}})^2 d\Sigma = \|\{\phi^o, \phi^1\}\|_G^2 \tag{5.4}$$

Therefore μ *is an isomorphism from* G *onto* G'. $\square$

This gives the R H U M method. *Assuming Uniqueness*, one supposes that

$$\{z^1, -z^o\} \in G'. \tag{5.5}$$

One then uniquely solves in G the equations

$$\mu\{\phi^o, \phi^1\} = \{-z^1, z^o\}. \tag{5.6}$$

If one takes

$$u = \frac{\partial\phi}{\partial\nu_{A^*}} \text{ on } \Sigma_o \tag{5.7}$$

one solves REC.

Indeed with the choice (5.7) one has $\psi=y$, $\phi=p$ in the Optimality system (3.9)(3.11)(3.12). ||

6. CONCLUSION

Of course we have only presented the beginning of the theory ! We refer to J.L.LIONS [2][3] for further details.

Applications to plate theory are given in J. LAGNESE and J.L. LIONS [1].

Applications to distributed control and to *simultaneous* control of several systems are given in A. HARAUX [1] and E. ZUAZUA [2].

Applications to other models of elasticity are given in E. ZUAZUA [1] (who also wrote the notes of Volume 1 of J.L. LIONS [3]). For the case with constant coefficients $(A = -\Delta)$ the space F has been characterized for the first time by L.F. HO [1]. Another approach giving more complete information on F is given in C. BARDOS, G. LEBEAU, J. RAUCH [1]. The case of variable coefficients, and also for other boundary conditions, has been studied by V. KOMORNIK [1].

Irreversible situations, using RHUM, are studied in J.L. LIONS [3], Vol. 2. Of special interest here is the case of elasticity with memory. We refer here to G.LEUGERING [1] and to the Bibliography therein. Some remarks are made in the A. [3], Volume 2. The case of domains Ω with corners or when dealing with mixed boundary conditions (say of Dirichlet type on one part of the boundary, Neumann type on the other part) is studied in P. GRISVARD [1][2].

Asymptotic problems connected with E.C. (or R.E.C.) are studied in the A., Volume 2 and in D. CIORANESCU [1](cf. also the Bibliography therein).

Applications to the problems treated, by other methods, in W. LITTMAN and L. MARKUS [1] are considered in E. ZUAZUA [3].

REFERENCES

BARDOS, C., LEBEAU, G. and RAUCH, J. [1] Appendix in LIONS, J.L. [3], Vol. 1.

CIORANESCU, D. [1] Book in preparation.

GRISVARD, P. [1] Contrôlabilité exacte dans les polygones et polyèdres. C.Rendus Académie des Sci. Paris, 304, (1987), pp. 367-370.

[2] To appear.

HARAUX, A. [1] To appear J.M.P.A., 1988.

HO, L.F. [1] Observabilité frontière de l'équation des ondes. C.R. Acad. Sci. Paris, 302, (1986), pp. 443-446.

KOMORNIK, V. [1] Contrôlabilité exacte en un temps minimal. C.R. Acad. Sci., Paris 304, (1987), pp. 223-225.

[2] Exact Controllability in Short Time for the Wave Equation. To appear.

LAGNESE, G. and LIONS, J.L. [1] Modelling, Analysis and Control of Thin Plates. Collection R.M.A., Masson, Paris, 1988.

LEUGERING, G. [1] Cf. These Proceedings and the Bibliography therein.

LIONS, J.L. [1] Contrôlabilité exacte des systèmes distribués. C.R. Acad. Sci. Paris, 302 (1986), pp. 471-475.

[2] Exact Controllability, Stabilization and Perturbations for Distributed Systems. J. Von Neumann Lecture, Boston, July, 1986, SIAM Review, 1988.

[3] Contrôlabilité exacte, Stabilisation et Perturbations des systèmes distribués. Collection R.M.A., Masson, Paris :
Vol. 1 (1988), Contrôlabilité exacte, Notes written by E. ZUAZUA.
Vol. 2 (1988), Perturbations
Vol. 3 under preparation. Stabilization.

[4] Contrôle optimal des systèmes distribués singuliers. Dunod, Paris, 1983.

LIONS, J.L. and MAGENES, E. [1] Problèmes aux limites non homogenes et applications. Vol. 1 and 2, Dunod, Paris, 1968.

LITTMAN, W. and MARKUS, L. [1] Boundary Control Theory for Vibrating Beams and Equations of Scole. To appear.

RUSSELL, D.L. [1] Controllability and Stabilizability Theory for Linear Partial Differential Equations. Recent Progress and Open Questions. SIAM Review 20, (1978), pp. 639-739.

ZUAZUA, E. [1] Contrôlabilité exacte d'un modèle de plaques vibrantes en un temps arbitrairement petit. C.R. Acad. Sci. Paris, 304, Série I, 1987, p.173-176.

[2] To appear.

[3] To appear.

STABILITY OF WAVE EQUATIONS WITH NONLINEAR DAMPING
IN THE DIRICHLET AND NEUMANN BOUNDARY CONDITIONS *

I. Lasiecka
Applied Mathematics Department
University of Virginia
Thornton Hall
Charlottesville, Virginia 22903

1. Introduction

Let Ω be an open, bounded domain in R^n with a smooth boundary $\Gamma = \Gamma_0 \cup \Gamma_1$, where Γ_0 and Γ_1 are disjoint portions of the boundary. Let $\gamma(u)$ be a monotonne increasing possibly multivalued function defined on R^1 such that $0 \subset \gamma(0)$. Let $A_D u = \Delta u$ for $u \in H_0^1(\Omega) \cap H^2(\Omega)$. Consider the following second-order hyperbolic problems:

$$\begin{cases} y_{tt}(x_1 t) = \Delta y(x,t) & x \in \Omega; \quad t > 0 ; \\ y(x,0) = y_0(x) & x \in \Omega ; \\ y_t(x,0) = y_1(x) & \\ y(x,t) = 0 & x \in \Gamma_0 ; \quad t>0 ; \end{cases} \tag{1.1}$$

with either Dirichlet

$$y(x,t) + \gamma\left(\frac{\partial}{\partial \eta} A_D^{-1} y_t(x,t)\right) \ni 0 \quad x \in \Gamma_1; \ t > 0; \tag{1.1.D}$$

or Neumann

$$\frac{\partial y}{\partial \eta}(x,t) + \gamma(y_t| \, y_t(x,t)) \ni 0 \quad x \in \Gamma_1; \ t > 0 \tag{1.1.N}$$

boundary conditions on Γ_1 .

(*) Research partially supported by the National Science Foundation under Grant DMS-8301668 and the Air Force Office of Scientific Research under Grant AFOSR-84-0365.

Our aim is to prove that under suitable conditions imposed on the nonlinear term γ, systems (1.1.D) (resp. (1.1.N)) are asymptotically stable (when $t \to \infty$) in the strong topology of $L_2(\Omega) \times H^{-1}(\Omega)$ (resp. $H^1(\Omega) \times L^2(\Omega)$).

In the linear case, when $\gamma(y) = y$, it is known [see [Ch.1], [L-1], [L-T.1]) that the solutions (y, y_t) of the wave equations with linear dissipative boundary conditions decay to zero in the strong topology of the underlined spaces. If, in addition, certain geometric conditions are imposed on the domain Ω, then the solutions $(y\ ,y_t)$ decay to zero exponentially. In the nonlinear case, problem (1.1.N) defined on the one-dimensional domain Ω was treated in [Ch.2]. In this special case, Chen established the asymptotic behavior of the solutions when $t \to \infty$. Techniques of [Ch.2], based on the methods of characteristics, can not be generalized to the higher dimensions of Ω. Subsequently, in [L-2], in order to establish the asymptotic behavior of solutions to (1.1.N) defined on arbitrary bounded domains in R^n, different approach based on stability theory for monotone generators was used. Still, techniques of [L.2] could not handle the Dirichlet boundary conditions. In the present paper, we propose to provide a more general abstract treatment which would allow to incorporate, as the special cases, both Neuman and Dirichlet problems.

The outline of the paper is as follows: in section 2 we shall formulate and prove the results in the abstract framework, while in section 3 we shall show how to apply these results to the specific situation in question of problems (1.1.N) and (1.1.D). Finally, in section 4 we prove the main stability results for problems (1.1.N) and (1.1.D).

2. Abstract results

Let V, H and U, U_0 be Hilbert spaces such that $V \subset H \subset V'$ and $U \subset U_0 \subset U'$. Let A be a maximal monotone and hemicontinuous operator from V to V'. Let $\Phi: U_0 \to R^1$ be a proper convex, semicontinuous function with $\partial\Phi: U \to U'$ - its subgradient. Finally, let be given a linear operator $B: U' \to V'$ with $B^*: V \to U$, where $(Bu, v)_H = \langle u, B^* v\rangle_{U_0}$ for all $u \in U'$; $v \in V$.

Consider the nonlinear operator $D: V \to V'$ given by:

$$Dy = (A + B\partial\Phi B^*)y; \quad y \in V . \tag{2.1}$$

We shall prove that under certain conditions imposed on the operator B, D is maximal monotone on H.

In fact, let us assume

$$B^* \in \mathcal{L}(V;U) \text{ and } B^* \textit{ is surjective from } V \textit{ onto } U. \tag{H-1}$$

Theorem 2.1

Assume (H-1). Then the operator D is defined by (2.1) is maximal monotone on H.

proof:

The proof of the Theorem 2.1 is based on the following Lemma

Lemma 2.2

Assume (H-1). Then for any $v^o \in V$, $z \in \partial(\Phi B^*)v^o$ iff $z \in B\partial(\Phi)B^*v^o$.

Proof of Lemma 2.2:

Notice first, that since $\Phi: U \to R^1$ is lower semicontinuous and proper convex, the same is true for $\Phi B^*: V \to R^1$. Hence

$$\partial(\Phi B^*): V \to V' \text{ is maximal monotone .} \tag{2.2}$$

We prove next that

$$z \in B\partial(\Phi)B^*v^o \Rightarrow z \in \partial(\Phi B^*)v^o. \tag{2.3}$$

Let $z \in B\partial(\Phi)B^*v^o$. Then $z = Bf$ where $f \in \partial(\Phi)B^*v^o \in U'$ and

$$\Phi(B^*v^o) - \Phi(u) \geq \langle f, B^*v^o - u\rangle_{U_0} \text{ for all } u \in U. \tag{2.4}$$

Since by (H-1) for any $v \in V$; $u \equiv B^*v \subset U$, (2.4) yields

$$\Phi(B^*v^o) - \Phi(B^*v) \geq \langle f, B^*v^o - B^*v\rangle_{U_0} \text{ for all } v \in V, \text{ or equivalently}$$

$$\Phi(B^*v^o) - \Phi(B^*v) \geq (Bf, v^0 - v)_H \text{ for all } v \in V. \tag{2.5}$$

From the definition of subgradient and from (2.5) we infer

$$z = Bf \in \partial(\Phi B^*)v^0.$$

Conversly,

$$z \in \partial(\Phi B^*)v^o \Rightarrow z \in B(\partial\Phi)B^* v^o. \tag{2.6}$$

To prove (2.6), we first show that for any $z \in \partial(\Phi B^*)v^o$, there exists $f \in U'$ such that

$$z = Bf\,. \tag{2.7}$$

Proof of (2.7):

Notice first that

$$\partial(\Phi B^*)v^o \subset N(B^*)' = \overline{R(B)}. \tag{2.8}$$

In fact; let $v = v^o + h;\ h \in V$. Then for $z \in \partial(\Phi B^*)v^o$, $(z,h)_H \le \Phi(B^*)v^o - \Phi(B^*)(v^o - h)$ for all $h \in V$. Let $\bar{h} \in N(B^*)$. Then $(z,\bar{h}) \le \Phi(B^*)v^o - \Phi(B^*)v^o = 0$. Also $-\bar{h} \in N(B^*)$, hence $-(z,\bar{h}) \le 0$ which implies $(z,\bar{h}) = 0$ for all $\bar{h} \in N(B^*)$, proving (2.8). From (2.8) we obtain $z = \lim_{n \to \infty} Bf_n$, $f_n \in U'$. We shall prove that

$$|f_n|_{U'} \le C|z|_{V'}. \tag{2.9}$$

In fact, since

$$|Bf_n|_{U'} \le 2|z|_{V'} \quad n > N_o,$$

$$<f_n, B^*h>_{U_o} = (Bf_n, h)_H \le 2|z|_{V'}|h|_V. \tag{2.10}$$

On the other hand

$$|f_n|_{U'} = \sup_{\psi_n \in U} \frac{<f_n, \psi>_{U_o}}{|\psi_n|_U}. \tag{2.11}$$

By the surjectivity of B^* (hypothesis (H-1)), for any $\psi_n \in U$ there exists $h_n \in V$ such that $\psi_n = B^* h_n$ and

$$|h_n|_V \le C|\psi_n|_U\,. \tag{2.12}$$

(2.10), (2.11) and (2.12) then yield

$$|f_n|_{U'} = \sup_{\psi_n \in U} \frac{<f_n, B^* h_n>_{U_0}}{|\psi_n|_U} \le 2|z|_{V'}|h_n|_V \le \frac{2C|z|_{V'}|\psi_n|_U}{|\psi_n|_U} = 2C|z|_{V'}$$

which proves (2.9).

From (2.9) $f_n \xrightarrow{\omega} f \in U'$. Since $B: U' \to V'$ is continuous; $Bf_n \xrightarrow{\omega} Bf$ *in* V'. On the other hand $z = \lim_{n \to \infty} Bf_n$, hence $z = Bf$ proving (2.7).

Now we are in a position to complete the proof of (2.6). Let $z \in \partial(\Phi B^*)v^0$. By the definition of the subgradient, and by (2.7) for some $f \in U'$

$$(\Phi B^*)v^o - (\Phi B^*)v \ge (z, v^o - v)_H = (Bf, v^o - v)_H \text{ for } all\ v \in V; \tag{2.13}$$

We need to prove that $f \in \partial\Phi(B^* v^o)$, which is equivalent to showing that $f \in U'$ satisfies:

$$\Phi(B^* v^o) - \Phi u \ge <f, B^* v^o - u>_{U_0} \text{ for } all\ u \in U. \tag{2.14}$$

By using once more surjectivity of B^*, for any $u \in U$ we can find $v \in V$; $B^* v = u$. Thus from (2.13) (which holds for all $v \in V$) we obtain

$$\Phi(B^*)v^o - \Phi u \ge <f, B^* v^o - u>_{U_0} \text{ for } all\ u \in U,$$

which is equivalent to (2.14). Proof of the Lemma 2.2 is thus completed. □

<u>Corallary 2.3</u>

Assume (H-1). Then $B\partial\Phi B^*$ is maximal monotone $V \to V'$.

The proof of Corollary 2.3 follows from Lemma 2.2 and from the fact that ΦB^* is proper convex and lower sem-icontinuous on V.

To complete the proof of Theorem 2.1 it is enough to notice that D is the sum of maximal monotone operators with A being hemicontinuous $V \to V'$. Thus the assertion of Theorem 2.1 follows from Corollary 2.7 in [B-1].

Corollary 2.4

Assume (H-1) and moreover assume that

A *is coercive* $V \to V'$. (*ie* $(Av,v)_H \geq c|v|_v^2$) (H-2)

Then

$A + B\,\partial(\Phi)B^* + I$ is surjective from $V \to V'$

and $(A + B\,\partial\Phi B^* + I)^{-1}$ takes bounded sets in V' into bounded sets in V.

Corollary 2.4 follows directly from Theorem 2.1 and from the standard properties of maximal monotone operators (see [B-1]).

We shall apply the results of Theorem 2.1 to the following second order inclusion

$$\begin{cases} y_{tt} + (A + B\,\partial(\Phi)B^*)y_t \ni 0 \\ y(0) = y_0 \in D(A^{1/2});\ \ y_t(0) = y_1 \in H \end{cases} \tag{2.15}$$

where $D(A^{1/2})$ stands for the domain of $A^{1/2}$, when A is considered as unbounded operator defined on H.

Here we additionally assume

that A is a linear, selfadjoint, positive operator on H (hence $A^{1/2}$ *is well defined)* and

$$|x|_{D(A^{1/2})} \equiv |A^{1/2}x|_H \tag{H-3}$$

The wellposedness of (2.15) follows from

Theorem 2.5

Assume $(H\!-\!3)$ and $(H\!-\!1)$ *with* $V = D(A^{1/2})$.
Let $C : D(A^{1/2})\,x\,H \to D(A^{1/2})\,x\,H$ be defined as:

$$C \equiv \begin{bmatrix} 0 & -I \\ A & B\partial(\Phi)B^* \end{bmatrix}$$

Then C generates a nonlinear semigroup of contractions on $\overline{D(C)}$. For all $(y_0, y_1) \in D(C)$, we have $(y(t), y_t(t)) \in D(C)$ where y, y_t satisfy (2.15).

Proof

C is monotone. In fact; for all $x_1, x_2 \in D(C)$

$$\left[C \begin{bmatrix} x_1 \\ x_2 \end{bmatrix}, \begin{matrix} x_1 \\ x_2 \end{matrix} \right]_{D(A^{1/2}) x H} = -(x_2, x_1)_{D(A)^{1/2}} + (Ax_1 + B\partial(\Phi)B^* x_2, x_2)_H = -(A^{1/2} x_2, A^{1/2} x_1)_H$$

$$+ (Ax_1, x_2)_H + \langle \partial\Phi B^* x_2, B^* x_2)_{U_0} \geq 0 \, .$$

To prove that C is maximal monotone, it is enough to show that

$$Range\ (C+I) = D(A^{1/2})\, xH \tag{2.16}$$

proof of 2.16:
Let $(f_1, f_2) \in D(A^{1/2})\, xH$. Consider

$$\begin{cases} -x_2 + x_1 = f_1 \\ Ax_1 + B\partial\Phi B^* x_2 + x_2 \ni f_2, \end{cases}$$

$$Ax_1 + B\partial\Phi B^*(x_1 - f_1) + x_1 - f_1 \ni f_2.$$

Let $z \equiv x_1 - f_1$. Hence

$$Az + B\partial(\Phi)B^* z + z \ni f_2 - Af_1 \, . \tag{2.17}$$

Since A is coercive and continuous on $D(A^{1/2})$, (ie: $(Az, z)_H = | z|^2_{D(A^{1/2})}$ and $(Az, v)_H = | z|_{D(A^{1/2})} | v|_{D(A^{1/2})}$), we apply Corollary 2.4 to (2.17). In fact, with $(f_1, f_2) \in D(A^{1/2}) xH$, we have $f_2 - Af_1 \in D(A^{1/2})'$ and by Corollary 2.4, there exists solution to (2.17) z such that $z \in D(A^{1/2})$. Thus $x_1 \in D(A^{1/2})$ and $x_2 \in D(A^{1/2})$ which proves (2.16), hence the first statement of Theorem 2.5. The second statement follows from standard semigroup theory. □

3. Abstract formulation of (1.1.D) and (1.1.N).

In this section we shall reformulate problems (1.1.D) and (1.1.N) in the abstract nonlinear semigroup framework.

(i) Dirichlet boundary conditions

Let $A_D: L_2(\Omega) \to L_2(\Omega)$ be defined as $A_D u \equiv \Delta u$; $u \in \mathcal{D}(A_D) \equiv H^2(\Omega) \cap H_0^1(\Omega)$. If A_D is considered as acting on $\mathcal{D}(A_D^{1/2})'$, then A_D is continuous and coercive from $L_2(\Omega) \to \mathcal{D}(A_D)'$. In that case, $D(A_D^{1/2}) = L_2(\Omega)$. Thus we set $H \equiv \mathcal{D}(A_D^{1/2})'$; $V \equiv D(A_D^{1/2}) = L_2(\Omega)$; $V' \equiv \mathcal{D}(A_D)'$. It is immediate to check that the operator A_D satisfies (H-3).

Next, let us define $U_0 \equiv L_2(\Gamma_1)$; $U \equiv H^{1/2}(\Gamma_1)$; $U' \equiv H^{-1/2}(\Gamma_1)$. We construct a function $\Phi: L_2(\Gamma_1) \to R_1$ by the following formula

$$\Phi(u) \equiv \begin{cases} \int_\Gamma j(u)d\Gamma_1 & \text{if } j(u) \in L_1(\Gamma_1) \\ \infty & \textit{otherwise} \end{cases} \tag{3.1}$$

where $\gamma \equiv \partial j$.

It is well known ([B-2]), that Φ is proper convex, lower semicontinuous on $L_2(\Gamma_1)$. Moreover, $f \in \partial\Phi(v)$ iff $f(x) \in \gamma(v(x))$.

Finally we define the operator $B: H^{1/2}(\Gamma_1) \to \mathcal{D}(A)'$; $Bu \equiv A_D D u$ where

$$D \in \mathcal{L}(H^{-1/2}(\Gamma_1) \to L_2(\Omega)) \textit{ is given by} \tag{3.2}$$

$$\Delta Dg = 0 \text{ and } Dg|_{\Gamma_1} = g;\ Dg|_{\Gamma_0} = 0.$$

One can check, by using Greens Formula that $B^* \in \mathcal{L}(L_2(\Omega) \to H^{1/2}(\Gamma))$ is given by

$$B^{*}v = D^{*}v = \frac{\partial}{\partial\eta} A_D^{-1} v; \quad v \in L_2(\Omega); \tag{3.3}$$

where $(Dg,v)_{L_2(\Omega)} = \langle g, D^{*}v \rangle_{L_2(\Gamma_1)}$ and $(Bu,v)_{(A_D^{1/2})'} = \langle u, B^{*}v \rangle_{L^2(\Gamma_1)}$.

With the above notation, the semigroup representation of (1.1.D) is the following:

$$\begin{cases} y_{tt} + (A_D + A_D D(\partial\Phi) D^{*}) y_t \ni 0; \\ y(0) = y_0 \in L_2(\Omega), \quad y_t(0) = y_1 \in \mathcal{D}(A_D^{1/2})' = H^{-1}(\Omega)^{*}. \end{cases} \tag{3.4}$$

One can verify (details are in [L-3]) that B^{*} is continuous and surjective from $L_2(\Omega)$ onto $H^{1/2}(\Gamma_1)$, hence hypothesis (H-1) holds. Moreover, operator C defined in Theorem 2.5 with $A = A_D$, B, $\partial\Phi$, H, V, U^0, V as above satisfies (see [L-3])

$$\overline{D(C)} = D(A_D^{1/2}) \times H = L_2(\Omega) \times \mathcal{D}(A_D^{1/2})' \tag{3.5}$$

Therefore, we are in a position to apply to the problem (3.4) Theorem 2.5.

Theorem 3.1

For all $(y_0, y_1) \in L_2(\Omega) \times H^{-1}(\Omega)$ *, there exists unique strongly continuous solution to (1.1.D) $(y, y_t) \in L_2(\Omega) \times H^{-1}(\Omega)$. Moreover, the operator C defined in Theorem 2.5 generates a strongly continuous semi-group of contractions on $L_2(\Omega) \times H^{-1}(\Omega)$.

(ii) Neumann boundary conditions

Nonhomogenous Neumann boundary conditions (in contrast with Dirichlet) give raise to variational problems, thus they can be treated by using variational techniques. Below, however, we shall demonstrate how the results on wellposedness in the Neumann case can be also obtained from our abstract treatment in section 2. To this end we define:

$A_N : L_2(\Omega) \to L_2(\Omega)$ *given by*

---

* Here we recall that $|y|_{D(A_D^{1/2})'} \simeq |y|_{H^{-1}(\Omega)}$

$$A_N u \equiv \Delta u;\ u \in \mathcal{D}(A_N) \equiv \{u \in H^2(\Omega);\ \frac{\partial u}{\partial \eta}|_{\Gamma_1} = 0;\ u|_{\Gamma_0} = 0\},$$

$$H \equiv L_2(\Omega);\ V \equiv H_{\Gamma_0}(\Omega);\ V' = (H^1_{\Gamma_0}(\Omega))'$$

where

$$H^1_{\Gamma_0}(\Omega) \equiv \{u \in H^1(\Omega);\ u|_{\Gamma_0} = 0\}.$$

$$U_0 \equiv L_2(\Gamma_1);\ U \equiv H^{1/2}(\Gamma_1);\ U' = H^{-1/2}(\Gamma_1).$$

The operator $B : H^{-1/2}(\Gamma_1) \to (H^1_{\Gamma_0}(\Omega))'$ is defined as:

$Bu \equiv A_N N u$ where

$$N \in \mathcal{L}(H^{-1/2}(\Gamma_1) \to H^1_{\Gamma_0}(\Omega)) \text{ is given by} \tag{3.6}$$

$$\Delta N g = 0;\ \frac{\partial}{\partial \nu} N g|_{\Gamma_1} = g;\ Ng|_{\Gamma_0} = 0.$$

One can check that $B^* \in \mathcal{L}(H^{-1/2}(\Gamma_1) \to (H^1_{\Gamma_0}(\Omega))')$ is given by

$$B^* v = N^* A_N v = v|_{\Gamma_1}, \tag{3.7}$$

where $(Ng, v)_{L_2(\Omega)} = \langle g, N^* v\rangle_{L_2(\Gamma_1)}$ and

$$(Bu, v)_{L_2(\Omega)} = \langle u, B^* v\rangle_{L_2(\Gamma_1)}$$

Now, the semigroup representation of (1.1.N) takes the form

$$\begin{cases} y_{tt} + A_N N \partial(\Phi) N^* A_N y_t \ni 0 \\ y(0) = y0 \in H^1_{\Gamma_0}(\Omega);\; y_t(0) = y_1 \in L_2(\Omega) \end{cases} \tag{3.8}$$

where Φ is defined by (3.1). Similarily as in the Dirichlet case, one verifies that:

$$A_N : H^1_{\Gamma_0}(\Omega) \to (H^1_{\Gamma_0}(\Omega))' \text{ is selfadjoint, continuous and coercive} \cdot \tag{3.9}$$

$$\mathcal{D}(A_N^{1/2}) = D(A_N^{1/2}) = H^1_{\Gamma_0}(\Omega) = V; \tag{3.10}$$

$$B^* \text{ is continuous and surjective from } H^1_{\Gamma_0}(\Omega) \text{ onto } H^{1/2}(\Gamma_1); \tag{3.11}$$

$$\overline{D(C)} = D(A_N^{1/2}) \times H = H^1_{\Gamma_0}(\Omega) \times L^2(\Omega); \text{ where the operator C is defined in Theorem 2.5.} \tag{3.12}$$

Since hypothesis (H-1) and (H-3) are verified, we can apply Theorem (2.5) to the problem (1.1.N). This yields

Theorem 3.2

For all $y_0 \in H^1_{\Gamma_0}(\Omega); y_1 \in L_2(\Omega)$; there exists unique strongly continuous solution to (1.1.N) $(y, y_t) \in H^1_{\Gamma_0}(\Omega) \times L_2(\Omega)$. Moreover, the operator C defined in Theorem 2.5 generates strongly continuous semi-group at contractions on $H^1_{\Gamma_0}(\Omega) \times L_2(\Omega)$.

4. Asymptotic stability

In order to obtain asymptotic stability results for the dynamics in question, we need to impose further restrictions on the nonlinear term γ and on the domain Ω. We shall assume

$$\begin{cases} (i)\;\; z \cdot u > 0 \;\text{ for } all\; z \in \gamma(u);\; u \neq 0 \\ (ii)\;\; \gamma(0) = 0. \end{cases} \tag{A-1}$$

$\partial\Phi u \subset$ *bounded set in* $H^{-1/2+\varepsilon}(\Gamma_1)$ for $u \in$ *bounded set in* $H^{1/2}(\Gamma_1)$. (A-2)

$(x-x^0)\vec{n} \leq 0$ *on* Γ_0 *where* $\vec{n}$ *is outward normal to* Γ_0 and $x^0 \in R^n$. (A-3)

Our main result is the following

Theorem 4.1

Assume (A-1) - (A-3). Then the slution (y, y_t) of (1.1.D) (resp. (1.1.N)) decays to zero in the strong topology of $L_2(\Omega) \times H^{-1}(\Omega)$ *(resp.* $H^1_{\Gamma_0}(\Omega) \times L_2(\Omega)$) for all initial conditions in the same space.

Remark 1

It can be shown (see [L-2]) that (A-2) holds in the following situations:

(*i*) $dim\ \Omega = 1$;

(*ii*) $dim\ \Omega = 2$ and $|z| \leq C|x|^p + D$ for *any* $z \in \gamma(x)$, *any* $p \geq 0$;

(*iii*) $dim\ \Omega = 3$ and $|z| \leq C|x|^p + D$ for *any* $z \in \gamma(x)$; $p < 3$.

Remark 2

It is well known that assumption (A-3) holds for "star shaped" domains.

Remark 3

The Neumann problem (1.1.N), with $\gamma(0) \neq 0$ has been considered in [L-2]. In that case, it was shown in [L-2], that all limit solutions approach asymptotically the limit set which is contained in the ball B(0, R) where R depends only on the geometry of the domain Ω but not on the norm of initial conditions.

Proof of Theorem 4.1

Let C_D (resp. C_N) denote the operator C defined in Theorem 2.5 correspondng to the Dirichlett (resp. Neumann) boundary cnditions. We shall prove first

Lemma 4.2

Assume (A-1) and (A-2). Then the resolvents $R(\eta, C_D)$ (resp $R(\eta, C_N)$) are compact on $L^2(\Omega)\times H^{-1}(\Omega)$ (resp. $H^1_{\Gamma_0}(\Omega)\times L_2(\Omega)$).

Proof of Lemma 4.2 for the Neumann case is given in [L-2]. We shall then consider only the Dirichlet case. Without loss of generality, w assume η=1. Let $x = R(1,C_D)f$ and $f=(f_1, f_2)\in$ bounded set in $L_2(\Omega)\times H^{-1}(\Omega)$. Then

$$\begin{cases} a)\ -x_2+x_1=f_1 \\ b)\ A_Dx_1+A_DD\partial(\Phi)D^*x_2+x_2-f_2\ni 0 \end{cases} \tag{4.1}$$

or

$$A_Dx_2+A_DD\partial\Phi D^*x_2-f_2+A_Df_1\ni 0. \tag{4.2}$$

Since $A_Df_1-f_2\in\mathcal{D}(A_D)'=V'$,from Corollary 2.5, it follows that for (f_1, f_2) in a bounded set in $L_2(\Omega)\times H^{-1}(\Omega)$, x_2 belong to a bounded set in $V=L^2(\Omega)$. Hence, by (4.1a), x_1 are also in a bounded set in $L_2(\Omega)$. On the other hand, (4.1.b) is equivalent to:

$$\begin{cases} \Delta x_1=x_2-f_2 \\ x_1|_{\Gamma_0}=0 \\ x_1|_{\Gamma_1}\in-\partial\Phi(D^*x_2)\,. \end{cases} \tag{4.3}$$

Since $D^*\in\mathcal{L}(L_2(\Omega)\to H^{1/2}(\Gamma_1))$, by using assumption (A-2) we infer that $\partial\Phi(D^*x_2)\subset$ bounded set in $H^{-1/2+\varepsilon}(\Gamma_1)$, thus $x_1|_{\Gamma_1}\in H^{-1/2+\varepsilon}(\Gamma_1)$. By standard elliptic reglarity, we conclude (using again $x_2-f_2\in H^{-1}(\Omega)$) that $x_1\in$ bounded set in $H^{\varepsilon}(\Omega)$. Thus, we have obtained that for any $(f_1,f_2)\in$ bounded set in $L_2(\Omega)\times H^{-1}(\Omega)$, $(x_1, x_2)\in$ bounded set in $H^{\varepsilon}(\Omega)\times L^2(\Omega)$. The conclusion follows then by compact imbedding of $H^{\varepsilon}(\Omega)\times L^2(\Omega)$ in $L_2(\Omega)\times H^{-1}(\Omega)$. □

To continue with the proof of Theorem 4.1, we recall that by virtue of Theorems 3.1 and 3.2, operators C_D (resp C_N) are maximal dissipative on $L_2(\Omega)\times H^{-1}(\Omega)$ (resp $H^1_{\Gamma_0}(\Omega)\times L_2(\Omega)$). Moreover by Lemma 4.2, the resolvents $R(\eta,C_D)$ and $R(\eta,C_N)$ are compact. Therefore we are in a position to apply Theorem 1 in [D-S] which asserts the existence of ω-limit sets such that with $\tilde{y}(0)\equiv(y_0,y_1)$

$$e^{C_D t}\tilde{y}(0)\underset{t\to\infty}{\to}\omega_D \;\; in \; L_2(\Omega)\times H^{-1}(\Omega) \tag{4.4}$$

$$e^{C_N t}\tilde{y}(0)\underset{t\to\infty}{\to}\omega_N \;\; in \; H^1_{\Gamma_0}(\Omega)\times H^{-1}(\Omega). \tag{4.5}$$

Moreover, the corresponding semigroups restricted to $\overline{co}\,\omega_D$ (*resp* $\overline{co}\,\omega_N$) are affine groups of isometries.

It is convenient to introduce the following notation

$$\mathcal{A}_D\equiv\begin{bmatrix}0 & -I\\ A_D & 0\end{bmatrix};\quad \mathcal{B}_D\equiv\begin{bmatrix}0\\ A_D D\end{bmatrix};$$

$$\mathcal{A}_N\equiv\begin{bmatrix}0 & -I\\ A_N & 0\end{bmatrix};\quad \mathcal{B}_N\equiv\begin{bmatrix}0\\ A_N N\end{bmatrix}.$$

Then

$$C_D=\mathcal{A}_D+\mathcal{B}_D(\partial\Phi)\,\mathcal{B}^*_D\;;$$

$$C_N=\mathcal{A}_N+\mathcal{B}_N(\partial\Phi)\mathcal{B}^*_N\,.$$

and (1.1.D) and (1.1.N) are equivalent to:

$$\begin{cases}\tilde{y}_t+\mathcal{A}_D\tilde{y}+\mathcal{B}_D(\partial\Phi)\mathcal{B}^*_D\tilde{y}\ni 0\\ \qquad\qquad\qquad\qquad\qquad ;\\ \tilde{y}(0)=(y_0,y_1)\in L_2(\Omega)\times H^{-1}(\Omega)\end{cases} \tag{4.6}$$

and

$$\begin{cases} \tilde{y}_t + \mathcal{A}_N\tilde{y} + \mathcal{B}_N\partial\Phi\mathcal{B}_N^*\tilde{y} \ni 0 \\ \tilde{y}(0) = (y_0, y_1) \in H^1_{\Gamma_0}(\Omega) \times L_2(\Omega). \end{cases} \tag{4.7}$$

Since $e^{C_D t}$ (resp $e^{C_N t}$) are isometries for $\tilde{y}(0) \in \overline{co}\,\omega_D$ (resp $\overline{co}\,\omega_N$), after multiplying (4.6) (resp 4.7) by $\tilde{y}(t)$, intergrating by parts and evoking assumption (A-1) we obtain

$$\begin{cases} \mathcal{B}_D^*\tilde{y}(t) = 0 \\ |\tilde{y}(t)|_{L_2(\Omega)\times H^{-1}(\Omega)} = |\tilde{y}(0)|_{L_2(\Omega)\times H^{-1}(\Omega)} \\ \text{for } all\ \tilde{y}(0) \in \overline{co}\,\omega_D \text{ and } t\geq 0. \end{cases} \tag{4.8}$$

$$\begin{cases} \mathcal{B}_N^*\,\tilde{y}(t) = 0 \\ |\tilde{y}(t)|_{H^1_{\Gamma_0}(\Omega)\times L_2(\Omega)} = |\tilde{y}(0)|_{H^1_{\Gamma_0}(\Omega)\times L^2(\Omega)} \\ \text{for } all\ \tilde{y}(0) \in \overline{co}\,\omega_N. \end{cases} \tag{4.9}$$

The equivalent pde versions of conditions (4.8) and (4.9) are: for $y_0, y_1 \in \overline{co}\,\omega_D$ we have

$$\begin{cases} y_{tt} = \Delta y;\ \ y(0) = y_0;\ y_t(0) = y_1 \quad , \\ y|_{\Gamma_0} = 0;\ \ y|_{\Gamma_1} + \gamma(0) \ni 0 \ on\ \Gamma_1 \ , \\ \frac{\partial}{\partial\eta} A_D^{-1} y_t|_{\Gamma_1} = 0 \ , \\ |y(t)|^2_{L_2(\Omega)} + |y_t(t)|^2_{H_{-1}(\Omega)} = |y_0|^2_{L_2(\Omega)} + |y_1|^2_{H^{-1}(\Omega)}. \end{cases} \tag{4.10}$$

For $y_0, y_1 \in \overline{co}\,\omega_N$, we have

$$
\begin{cases}
y_{tt} = \Delta y;\ y(0) = y_0;\ y_t(0) = y_1 , \\
y|_{\Gamma_0} = 0;\ \frac{\partial}{\partial \eta} y + \gamma(0) \ni 0 \ on\ \Gamma_1 , \\
y_t|_{\Gamma_1} = 0 , \\
|y(t)|^2_{H^1_{\Gamma_0}(\Omega)} + |y_t(t)|^2_{L_2(\Omega)} = |y_0|^2_{H^1_{\Gamma_0}(\Omega)} + |y_1|^2_{L_2(\Omega)} .
\end{cases}
\tag{4.11}
$$

Relations (4.10) and (4.11) provide us with a characterazation of elements of ω-limit sets. The proof of Theorem 4.1 will be completed as soon as we prove

<u>Lemma 4.3</u>

Assume (A-1) (ii) and (A-3). Let $\left(y(t), y_t(t)\right)$ satisfy (4.10) (resp (4.11)). Then $y_0 = y_1 = 0$. (i.e. $\omega_D = \{0\}$ and $\omega_N = \{0\}$)

<u>Proof of Lemma 4.3</u> (sketch)

(i) <u>Dirichlet case:</u> Let $p(t) \equiv A_D^{-1} y_t$. Then

$$
\begin{cases}
p_{tt} = \Delta p;\ p(0) = p_0 = A_D^{-1} y_1;\ p_t(0) = p_1 = y_0 , \\
p|_{\Gamma} = 0;\ \frac{\partial}{\partial \eta} p|_{\Gamma_1} = 0.
\end{cases}
\tag{4.12}
$$

By employing multipliers methods used in [L-4], [L-L-T] we obtain the following inequality:

$$
(T - C_\Omega) E_0 - \int_{\Sigma_0} |\nabla p|^2 \vec{h}\cdot\vec{n}\, d\Sigma_o \leq \int_{\Sigma_1} \left|\frac{\partial}{\partial \eta} p\right|^2 d\Sigma_1
\tag{4.13}
$$

$$
where\ E_0 \equiv |p_0|^2_{H_0^1(\Omega)} + |p_1|^2_{L_2(\Omega)}
$$

$\vec{h}(x) \equiv x - x^0; \quad \Sigma_1 \equiv \Gamma_1 x(0,T); \quad \Sigma_0 = \Gamma_0 x(0,T).$

Using assumption (A-3) (ii) and boundary conditions in (4.12) we obtain from (4.13): $E_0 = 0$; hence $A_D^{-1} y_1 = y_0 = 0$.

(ii) Neumann case

Let $p(t) \equiv y_t(t)$, where y (t) satisfies (4.11).

Then p(t) satisfies

$$\begin{cases} p_{tt} = \Delta p;\ p(0) = p_0 = y_1;\ p_t(0) = p_1 = A_N y_0. \\ p|_{\Gamma} = 0;\ \frac{\partial}{\partial \eta} p|_{\Gamma_1} = 0 \end{cases} \tag{4.14}$$

From inequality (4.13) applied to (4.14) with "smooth" data (i.e: $y_1 \in H^1_{\Gamma_0}(\Omega)$; $y_0 \in \mathcal{D}(A_N)$) we obtain, as in the Dirichlet case, that $p_0 = p_1 = 0$, hence $y_0 = y_1 = 0$. Standard density argument completes the proof for the case of $y_1 \in L_2(\Omega)$ and $y_0 \in H^1_{\Gamma_0}(\Omega)$. Thus $\omega_N = 0$. □

The proof of Theorem 4.1 is thus completed □

References

[Ch-1] G. Chen. A Note on boundary stabilization of the wave equation. SIAM J. Contr. Optims. 19: 106-113.

[Ch-2] G. Chen, H. K. Wang. Asymptotic behavior of solutions of the one dimensional wave equations with nonlinear dissipetive boundary conditions. Manuscript.

[B-1] H. Brezis. Maximal monotone operators. Lecture Notes, North Holland, Amsterdam 1973.

[B-2] H. Brezis. Problems unilateraux J. Math. pure et appl. 51. 1972, p. 1-168.

[L-1] J. Lagnese. Decay of solutions of wave equations in a bounded region with boundary dissipation. J. Diff. Equations 50; 2: 163-182.

[L-2] I. Lasiecka. Strong stabilization of a nonlinear wave equation with dissipation on the boundary and related problems. Paper dedicated to A. V. Balakrishnan on the occasion of 60th birthday Recent Advances in Communication and Control Theory. R. E. Kalman, ed al (1981) New York, Optimization Software Inc. Publication Division.

[L-3] I. Lasiecka. Asymptotic stabilization of wave and plate equations with nonlinear dissipation on the boundary. Manuscript in preparation.

[L-4] J. L. Lions. Controllabilite exacte de systemes distribues: remarque sur la theorie generale et les applications. Proc. Seventh International Confer. on Analysis and Optimization of Systems Antibes, France, June 25 --27 (1986) pg.1-13 Lecture Notes, Springer Verlag.

[L-L-T] I. Lasiecka, J. L . Lions, R. Triggiani. Nonhomogeneous boundary value problems for second-order hyperbolic operators J. Math pures et appliques 65; 149-192 (1986).

PART II

INVITED PAPERS

DUALITY METHODS FOR NON QUALIFIED DISTRIBUTED CONTROL PROBLEMS

Frederic Abergel and Roger Temam

402 McAllister Bldg.
Penn State University
State College, PA 16802

Laboratoire d'Analyse Numérique
Université Paris 11
91405-ORSAY Cédex

Introduction

It is well-known that one often meets serious difficulties when studying a problem of optimal control with pointwise constraints, for a system governed by a partial differential equation.
In this article, we want to expose some results concerning quadratic cost problems, with an elliptic state equation, in the case of pointwise constraints on the control and on the state. We shall consider either unilateral or bilateral constraints.
For such problems, it is usually easy to obtain the existence and uniqueness of an optimal control: this is a standard result for the minimization of a coercive ℓ.s.c. strictly convex functional over a Hilbert space.
Nevertheless, we are interested in the more demanding problem of finding some necessary and sufficient conditions for optimality, in short, a S.O.C. (System of Optimality Conditions).
In classical situations, such a S.O.C. obtains by (sub-)differentiating the functional that we want to minimize, and introducing the adjoint state as a Lagrange multiplier. Unfortunately, this approach does not suit the kind of constraints we consider, for it requires a non-void interior type assumption that is generally invalid when pointwise constraints are met. Our approach is based on some classical duality methods of Convex Analysis, and applies whenever the state equation is linear. Let us now recall the basic facts we need [7].
Let A, B be two separable Hilbert spaces, Λ a continuous linear operator from A to B, and F (resp. G) a convex function from A (resp. B) into $\overline{\mathbb{R}}$; we study the problem

$$\operatorname*{Inf}_{a\in A}\{F(a) + G(\Lambda a)\} \qquad (\mathcal{Q})$$

and introduce its dual problem

$$\operatorname*{Sup}_{b\in B}\{-F^*[\Lambda^* b] - G^*(-b)\} \qquad (\mathcal{Q}^*)$$

Here, Λ^* is the transposed operator, and F^* (resp. G^*) is the conjugate function of F (rep. G).
Under certain lower semi-continuity assumptions, we have:

$$\operatorname{Inf}(\mathcal{Q}) = \operatorname{Sup}(\mathcal{Q}^*) \qquad (0.1)$$

and, if $\bar{a}$ (resp. $\bar{b}$) is a solution of (Q) (resp. (Q^*)):

$$F(\bar{a}) + F^*(\Lambda^*\bar{b}) = \langle \Lambda\bar{a},\bar{b}\rangle \tag{0.2}$$

$$G(\Lambda\bar{a}) + G^*(-\bar{b}) = \langle \Lambda\bar{a},-\bar{b}\rangle \tag{0.3}$$

Here, $\{(0.2),(0.3)\}$ is the S.O.C. we are looking for. In the problems we study, we can easily formulate (Q), and compute (Q^*); the difficulty lies precisely in studying the latter. Typically, we have the following situation:

(Q) with L^∞ constraints $\longleftrightarrow$ (Q^*) coercive in a L^1 norm

It is well-known that optimization problems are not well posed in L^1; hence, we shall proceed as follows:

1) Extend (Q^*) to a larger space, involving bounded measures (a nice substitute for L^1 in such situations)
2) Prove the existence of generalized solutions of (Q^*) in that space.
3) Extend the formal S.O.C. to the actual solutions of (Q) and (Q^*).

In Section I, we study a "model" problem of distributed control, and show how our method works, while in section II, we give various examples of optimal control problems that can be handled similarly.

I A distributed control problem

Here is the problem we are interested in:

$$\text{P} \quad (\text{I.1}) \quad \begin{cases} \text{To find } (z,v) \text{ in } L^2(\Omega)\times L^2(\Omega) \text{ minimizing the cost function} \\ J(z,v) = \dfrac{1}{2\eta}\displaystyle\int_\Omega v^2dx + \frac{1}{2}\int_\Omega |z-z_d|^2dx \\ \text{and subject to: } (-\Delta z+z) = v \text{ in } \Omega,\ z=0 \text{ on } \Gamma. \end{cases}$$

Moreover, we assume that the state of the system is subject to:

$$|z| \le \alpha \quad \text{a.e. in } \Omega \tag{I.a}$$

or

$$z \le \alpha \quad \text{a.e. in } \Omega \tag{I.b}$$

Ω is a bounded open set of $\mathbb{R}^N$ $(N \ge 1)$, with a smooth boundary Γ. If $N = 3$, P corresponds to the optimal heating of Ω: v is the volumic heating, z_d is the desired temperature, (I.a),(I.b) are technological constraints. We first consider the constraints (I.a).
In order to study problem P, we consider it as a problem of convex optimization, in the Hilbert space $X = H^2(\Omega) \cap H_0^1(\Omega)$; we recall that X is the natural space of solutions for the Dirichlet homogeneous boundary-value

problem

$$\begin{cases} -\Delta z + z = v & \text{in } L^2(\Omega) \\ z = 0 & \text{on } \Gamma \end{cases} \tag{I.2}$$

(see [8]).
We rewrite P as follows:

$$(\mathcal{P}) \qquad \operatorname*{Inf}_{z\in X}\{F(z) + G(\Omega z)\} \tag{I.3}$$

where: $\Lambda z = (z, -\Delta z + z)$, $F \equiv 0$, and

$$G(P_1, P_2) = \begin{cases} \frac{1}{2\eta}\int_\Omega |P_2|^2 dx + \frac{1}{2}\int_\Omega |P_1 - z_d|^2 dx & \text{if } |P_1| \le \alpha \\ +\infty & \text{otherwise} \end{cases}$$

The dual problem $(\mathcal{P}^*)$ is computed as in [7], and we find:

$$(\mathcal{P}^*) \qquad \operatorname*{Sup}_{P\in X}\left\{-\frac{\eta}{2}\int_\Omega |P|^2 dx + \frac{1}{2}\int_\Omega |z_d|^2 dx - \int_\Omega \Psi_\alpha(-\Delta p + p + z_d)dx\right\} \tag{I.4}$$

where we have set: $\Psi_\alpha(s) = \begin{cases} \frac{s^2}{2} & \text{if } |s| \le \alpha \\ \alpha\left(|s| - \frac{\alpha}{2}\right) & \text{if } |s| \ge \alpha. \end{cases}$

As announced above, we do not know whether $(\mathcal{P}^*)$ has a solution in X, due to the lack of coercivity; considering the definition of Ψ_α, it seems natural to extend $(\mathcal{P}^*)$ to the space

$$BL_0(\Omega) = \{u \in L^2(\Omega), (-\Delta u + u) \in M_1(\Omega),\ u = 0 \text{ on } \Gamma\}$$

(BL stands for "bounded Laplacian," and "0" refers to the Dirichlet boundary condition). The study of the extended problem is mainly based on the notion of convex function of a measure, defined and studied in [5] [6]. Let us recall some basic facts about this notion: let h be a convex function from $\mathbb{R}$ to $\overline{\mathbb{R}}$, with $h(0) = 0$, $h \ge 0$, and such that:

$$h(s) \le C(1+|s|), \tag{I.5}$$

and let us furthermore assume that its conjugate function h^*, defined by $h^*(s) = \operatorname*{Sup}_{t\in\mathbb{R}} (s.t - h(t))$ is bounded on its domain. We then have

<u>Proposition I.1</u>: Let μ be a bounded measure on Ω, $\mu = gdx + \theta_s|\mu_s|$ ([3]); $h(\mu)$ is the bounded measure defined by:

$$h(\mu) = (h\circ g)dx + h_\infty(\theta_s) \cdot |\mu_s| \tag{I.6}$$

where h_∞ is the asymptotic function of h:

$$h_\infty(s) = \lim_{t\to+\infty} \frac{h(t\cdot s)}{t}$$

Moreover, we have the duality formula, for φ in $C_0(\Omega)$:

$$\langle h(\mu),\varphi\rangle = \mathrm{Sup}\left[\int_\Omega \varphi\cdot f\cdot d\mu - \int_\Omega \varphi\cdot h^*(f)dx\right]$$

the supremum being taken for $f \in C_0(\Omega)$, $h^*(f) \in L^1(\Omega)$.
We can now define the generalized problem (Q^*), simply by extending the functional in $(\mathcal{P}^*)$ to $BL_0(\Omega)$, thanks to Proposition I.1. Moreover, the existence of solutions of (Q^*) in $BL_0(\Omega)$ is (almost) trivial: one just has to check some lower semicontinuity properties [5] [6].
The last difficulty is to establish the S.O.C; we easily determine its formal expression, by assuming there exists a solution of $(\mathcal{P}^*)$ in X:

$$\text{(S.O.C)} \quad \begin{cases} \bar{q} = -\frac{1}{\eta}(-\Delta\bar{z}+\bar{z}) & \text{(I.7)} \\ \int_\Omega \Psi_\alpha(-\Delta\bar{q}+\bar{q}+z_d) = \int_\Omega (-\Delta\bar{q}+\bar{q}+z_d)\bar{z} - \frac{1}{2}\int_\Omega |\bar{z}|^2 dx & \text{(I.8)} \end{cases}$$

It is obvious from (I.8) that there is a problem left, namely, the meaning of "$\int_\Omega (-\Delta q+q+z_d)\cdot z$", when q is in $BL_0(\Omega)$, and z in X. The following proposition takes care of that problem:

<u>Proposition I.2</u>: Let q (resp. $\bar{z}$) be feasible for (Q^*) (resp. $(\mathcal{P})$); the distribution "$(-\Delta q+q)\cdot z$", defined by

$$\langle(-\Delta q+q)\cdot z,\varphi\rangle = \langle q,(-\Delta(z\varphi)+z\varphi)\rangle \tag{I.9}$$

is a bounded measure on Ω, and the Green's formula

$$\int_\Omega (-\Delta q+q)\cdot z = \int_\Omega q\cdot(-\Delta z+z) \tag{I.10}$$

is valid.

<u>Remark I.1</u>: The right-hand side of (I.10) is well defined, thanks to the assumption "$q \in L^2(\Omega)$".

Proposition I.2 appears to be the key result for the study of P, allowing us to extend the general framework of the introduction to our particular problem; we do obtain (I.7) (I.8) as a S.O.C. for $(\mathcal{P})$, (I.8) being understood as in Proposition I.2.

Remark I.2: One can easily re-write (I.8), so as to enlighten the role of the singular part of the measure "$-\Delta\bar{q}+\bar{q}+z_d$", see also [4].

As for the unilateral case (I.b), although the idea is basically the same, the technique becomes more complicated: the interested reader is referred to [2].

II Some other examples

In this section, we give a brief survey of several situations where the duality methods provide a satisfactory S.O.C.

a) "Opposite constraints" problem

The situation is now as described on the figure; we consider the same state equation and cost function, and we want:

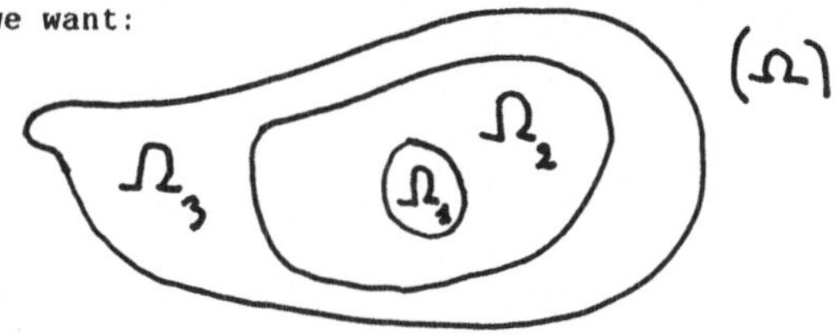

$z \geq \beta$ in Ω_1

$z \leq \alpha$ in Ω_3

and no constraints in Ω_2.

Such a situation corresponds, for instance, to the problem of finding the optimal burning of a region Ω_1, without damaging too much its neighborhood.

This problem is studied and solved, in the case $\Omega \subset \mathbb{R}^3$, with the same techniques as in Section I.

b) Boundary control problem(s):

One can consider several boundary control problems, and try to use the duality methods. Nevertheless, some new technical difficulties may appear, related to some regularity (or non-regularity) results for non homogeneous boundary problems, see [1].

Remark II.2: If one is interested in parabolic state equation, the method formally applies, but some important difficulties arise, particularly in trying to approximate the feasible states properly.

Bibliography

[1] Abergel, F. A non well-posed problem in convex optimal control, to appear in Appl. Math. Optim.

[2] Abergel, F. - Temam, R. "Optimality conditions for some non well-posed distributed control problems", submitted to S.I.A.M. J. Control and Optimization.

[3] Bourbaki, N. (1965) Integration, 2nd edn, Hermann, Paris.

[4] Casas, E. (1986) Control of an elliptic problem with pointwise state constraints, S.I.A.M. J. Control and Optimization, Vol. 24, #6.

[5] Demengel, F. - Temam, R. (1984) Convex function of a measure and applications, Indiana University Math J., 33.

[6] Demengel, F. - Temam, R. (to appear) Convex function of a measure and applications: the unbounded case.

[7] Ekeland, I. - Temam, R. (1972) Analyse convexe et problemes variationnels, Dunod, Paris.

[8] Lions, J.L. - Magenes, E. (1968) Problemes anx limites non homogenes et applications, Vol. 1, Dunod, Paris.

IDENTIFICATION OF OPERATORS IN SYSTEMS GOVERNED BY EVOLUTION EQUATIONS ON BANACH SPACE

N. U. Ahmed

Department of Electrical Engineering and Department of Mathematics
University of Ottawa, Ottawa, Ontario, Canada K1N 6N5

INTRODUCTION

In this paper we are mainly concerned with the questions of identification of operators and existence of optimal relaxed controls for systems governed by evolution equations on Banach space. We summarize here some of the results recently obtained by the author. These results are broad extensions of the previous results of the author presented at the 1986 annual meeting of the AMS held in New Orleans, Louisiana[1]. The contents of the paper are organized into four major sections:

(A) Cauchy Problems Involving C_0 - Semigroups.

(B) Initial Boundary Value Problems Based on Analytic Semigroups.

(C) Semilinear Initial Boundary Value Problems (Relaxed Controls).

(D) Nonlinear Problems Involving Monotone and Accretive Operators.

(A) CAUCHY PROBLEMS INVOLVING C_0 - SEMIGROUPS

Let X be a Banach space, $M \geq 1$ and $w \in R$. Define

$$G(M,w) \equiv \{ \text{ the class of infinitesimal generators of } C_0 \text{ s.g.'s } \{T(t), t \geq 0\} \text{ in } X \text{ with } \| T(t)\|_{\mathcal{L}(X)} \leq M\exp(wt), t \geq 0 \}.$$

Consider the problem (P1):

Let Q be a metric space and $A : Q \to G(M,w)$ and consider the system,

$$\dot{x} = A(\xi)x, \quad x(0) = x_0 \in X, \quad \xi \in Q, \quad t \in I = (0,T], \tag{1}$$

with the cost functional given by,

$$J(\xi) \equiv \int_I g(t, x_\xi(t))dt, \tag{2}$$

where x_ξ is the mild solution of the Cauchy problem corresponding to $\xi \in Q$. The problem is to find a $q^0 \in Q$ such that $J(q^0) \leq J(q)$ for all $q \in Q$. The solution to this problem is given in the following theorem.

Theorem 1

Let $Q \equiv (Q, \rho)$ be a compact metric space and $A : Q \rightarrow G(M, w)$ such that

(a1) for each λ, with $\mathrm{Re}\lambda > w$,

$$R(\lambda, A(q^n)) \xrightarrow{\tau_{so}} R(\lambda, A(q^0)) \tag{3}$$

whenever $q^n \xrightarrow{\rho} q^0$, where $R(\lambda, A) \equiv (\lambda I - A)^{-1}$ denotes the resolvent of A and τ_{so} denotes the strong operator topology on $\mathcal{L}(X)$,

(a2) $g : I \times X \rightarrow \bar{R}$, $t \rightarrow g(t, x)$ is measurable for each $x \in X$, $x \rightarrow g(t, x)$ is lower semicontinuous (l.s.c.) on X for almost all(a.a.) $t \in I$ and $g(t, x) > -\infty$ for all $t \in I$ and $x \in X$ with $\| x \| < \infty$.

Then there exists a $q^0 \in Q$ such that $J(q^0) \leq J(q)$ for all $q \in Q$.

Proof

The proof follows from Trotter-Kato theorem [Theorem 4.2, Pazy, p85] and Fatou's lemma. By Trotter-Kato we have $x^n(t) \rightarrow x^0(t)$ in X as $n \rightarrow \infty$, with x^n and x^0 being the solutions of equation(1) corresponding to q^n and q^0 respectively. Fatou's lemma and the hypothesis on g imply lower semicontinuity of J on Q with $J(q) > -\infty$ for all $q \in Q$. Hence the conclusion follows from compactness of Q. ■

Consider the problem(P2):

$$\begin{aligned} &\dot{x} = A(q)x + Bx, \quad x(0) = x_0 \in X, \qquad t \in I, \\ \text{with} \quad &J(q, B) \equiv \int_I g(t, x_{q,B}(t))dt = \min. \end{aligned} \tag{4}$$

for $q \in Q$ and $B \in D \subset \mathcal{L}(X)$. The solution is given by the folowing theorem.

Theorem 2

Consider the problem(P2) and suppose the following assumptions hold:

(a1) g and A are as in theorem 1,

(a2) Q is a compact metric space and D is a sequentially compact subset of $\mathcal{L}_s(X) \equiv (\mathcal{L}(X), \tau_{so})$.

Then there exists a pair $(q^0, B^0) \in Q \times D$ such that

$$J(q^0, B^0) \leq J(q, B) \quad \text{for all} \quad (q, B) \in Q \times D.$$

Proof

The proof follows from Trotter-Kato theorem and the perturbation theory for semigroups where the generator of the semigroup is perturbed by a bounded linear operator in X. ■

We can prove similar results for a class of relatively bounded perturbations of the generators of dissipative semigroups.

Definition 3

A densely defined linear operator A, with the domain and range, $D(A)$ and $R(A)$ in X, is said to be *dissipative* if

$$\mathrm{Re}(Ax, x^*)_{X,X^*} \leq 0 \quad \text{for all} \quad x^* \in \nu(x) \text{ and } x \in D(A)$$

where $\nu(x) \equiv \{x^* \in X^* : (x, x^*)_{X,X^*} = \| x\|^2 = \| x^* \|^2_{X^*}\}$ is the duality map. A dissipative operator A is said to be *m-dissipative* if the range $R(I - A) = X$. This implies that $R(\lambda I - A) = X$ for each $\lambda > 0$.

Definition 4(Relative bounds)

A linear operator B, with $D(B)$, $R(B) \subset X$, is said to be *relatively bounded* with respect to A if (i) $D(A) \subset D(B)$ and (ii) there exist finite non-negative numbers α, β such that

$$\| Bx \| \leq \alpha \| Ax \| + \beta \| x \| \quad \text{for all } x \in D(A).$$

Consider the system

$$\dot{x} = Ax + Bx, \quad x(0) = x_0 \in X \tag{5}$$

with $A \in G(1,0)$ fixed. For $0 \leq \alpha < 1$ and $\beta \geq 0$ fixed, define the set

$$\mathcal{A}_{\alpha,\beta} \equiv \{B : D(A) \subset D(B),\ B \text{ dissipative},\ \| Bx \| \leq \alpha \| Ax \| + \beta \| x \| \quad \text{for } x \in D(A)\}. \tag{6}$$

Our problem(P3) is: find $B^0 \in \mathcal{A}_{\alpha,\beta}$ such that

$$J(B^0) \equiv \int_I g(t, x_{B^0}(t))dt \leq \int_I g(t, x_B(t))dt \equiv J(B), \tag{7}$$

for all $B \in \mathcal{A}_{\alpha,\beta}$, where x_B is the mild solution of the C.P.(5). For the solution of this problem we introduce the vector space $[D(A)] \equiv \{D(A)$ furnished with the topology induced by the graph norm $\| x \|_{D(A)} = \| Ax \| + \| x \|$, $x \in D(A)\}$. Since A is a closed operator $[D(A)]$ is a Banach space. Define $\mathcal{L}_s([D(A)], X) \equiv \{\mathcal{L}([D(A)], X)$ *furnished with the strong operator topology* $\tau_{so}\}$ and similarly $\mathcal{L}_w([D(A)], X)$ for the weak operator topology τ_{wo}.

Note that $\mathcal{A}_{\alpha,\beta} \subset \mathcal{L}([D(A)], X)$ and for X^* strictly convex $\mathcal{A}_{\alpha,\beta}$ is sequentially closed in $\mathcal{L}_s([D(A)], X)$ but not necessarily sequentially compact. If X is reflexive, then $\mathcal{A}_{\alpha,\beta}$ is sequentially compact in the weak operator topology. The solution of problem(P3) is given in the following theorem.

Theorem 5

Let X be a Banach space with strictly convex dual X^* and $A \in G(1,0)$ and $x_0 \in X$ fixed. Suppose $\mathcal{A}^0_{\alpha,\beta} \subset \mathcal{A}_{\alpha,\beta} \subset \mathcal{L}_s([D(A)], X)$ is sequentially compact and g is as in theorem 1. Then there exists a $B^0 \in \mathcal{A}^0_{\alpha,\beta}$ such that $J(B^0) \leq J(B)$ for all $B \in \mathcal{A}^0_{\alpha,\beta}$.

Proof

The proof is based on the following facts: Lumer-Phillip's theorem which states that, for a densely defined linear operator A in X, $A \in G(1,0)$ if and only if A is m-dissipative; and the stability theorem which states that if $A \in G(1,0)$ and $B \in \mathcal{A}_{\alpha,\beta}$ then $A + B \in G(1,0)$; and a modified version of Trotter-Kato theorem. ■

Now we consider the general problem(P4): Let $\mathcal{G} \equiv C_0 \times F \times \mathcal{A}^0_{\alpha,\beta}$, with $C_0 \subset X$, $F \subset L_1(I, X)$ and $\mathcal{A}^0_{\alpha,\beta}$, as defined earlier, be the admissible set of control parameters for the system,

$$\begin{cases} \dot{x} = Ax + Bx + f, & t \in I \\ x(0) = \xi, \quad A \in G(1,0) \text{ fixed}, \end{cases} \tag{8}$$

with $(\xi, f, B) \in \mathcal{G}$. The problem is to find $(\xi^0, f^0, B^0) \in \mathcal{G}$ such that

$$J(\xi^0, f^0, B^0) \equiv \int_I g(t, x_{\xi^0,f^0,B^0}(t))dt \quad \leq \quad \int_I g(t, x_{\xi,f,B}(t))dt \equiv J(\xi, f, B)$$

for all $(\xi, f, B) \in \mathcal{G}$. The existence result is given in the following theorem without proof. Detailed proof will appear elsewhere.

Theorem 6

Suppose X is reflexive, C_0 is a weakly compact subset of X, F is a weakly sequentially compact subset of $L_1(I, X)$, g satisfies the basic hypotheses of theorem 1 with lower semicontinuity replaced by weak lower semicontinuity and $\mathcal{A}^0_{\alpha,\beta}$ is a sequentially compact subset of $\mathcal{L}_s([D(A)], X)$ satisfing further: If $B_n \xrightarrow{\tau_{so}} B_0$ in $\mathcal{L}_s([D(A)], X)$ then $B^*_n|_{D(A^*)} \xrightarrow{\tau_{so}} B^*_0|_{D(A^*)}$ and there exists a number $\eta = \eta(\alpha, \beta) < \infty$, such that $sup_n\{\| B^*_n\xi^* \|_{X^*}, \| B^*_0\xi^* \|_{X^*}\} \leq \eta \| \xi^* \|_{D(A^*)}$ for all $\xi^* \in D(A^*)$. Then there exists a $(\xi^0, f^0, B^0) \in \mathcal{G}$ such that $J(\xi^0, f^0, B^0) \leq J(\xi, f, B)$ for all $(\xi, f, B) \in \mathcal{G}$. ■

The following result gives conditions for continuous dependence of the extremal on the objective functional.

Theorem 7

Consider the system(5) with $A \in G(1,0)$, $x_0 \in X$ fixed and $B \in \mathcal{A}^0_{\alpha,\beta}$ with the objective functional

$$J_n(B) = \int_I g_n(t, x_B(t))dt.$$

Suppose $\mathcal{A}^0_{\alpha,\beta}$ satisfies the assumptions of theorem 5 and $\{g_n\} \subset C(I \times X, R_+)$, $R_+ = (-\infty, +\infty]$, satisfying

(i) there exist a, r, $0 < a < \infty$, $0 \leq r < \infty$ such that $|g_n(t,x)| \leq a(1+ \| x \|^r)$ for all $n \in N_+$, $x \in X$

(ii) there exists a $g_0 \in C(I \times X, R_+)$ such that $g_n(t,x) \to g_0(t,x)$ uniformly on bounded subsets of $I \times X$.

Then, if B_n minimizes J_n and B_0 minimizes $J_0(B) \equiv \int_I g_0(t, x_B(t))dt$, $J_n(B_n) \to J_0(B_0)$ as $n \to \infty$.

Proof

The proof follows from theorem 5 and Lebesgue dominated convergence theorem. ∎

(B) INITIAL BOUNDARY VALUE PROBLEMS BASED ON ANALYTIC SEMIGROUP

Consider the system,

$$\begin{aligned} \frac{\partial \varphi}{\partial t} + \mathcal{A}\varphi &= f \quad \text{in} \quad I \times \Omega \\ (B_0 + \tilde{B})\varphi &= g \quad \text{in} \quad I \times \partial\Omega \\ \varphi(0, \cdot) &= \varphi_0 \quad \text{in} \quad \Omega, \end{aligned} \tag{9}$$

with Ω an open bounded connected subset of R^n, $I = (0, T]$, $\mathcal{A}$ and B_0 are fixed spatial and boundary operators, $\tilde{B}$ is an unkown boundary operator, f, g, φ_0 are given. We assume that $D(\mathcal{A}) \subset D(B_0) \subset D(\tilde{B})$ and that $\mathcal{B}_a$ is a suitable class of boundary operators such that $D(B_0) \subset D(\tilde{B})$ for all $\tilde{B} \in \mathcal{B}_a$. Our problem (P5) is to find a $\tilde{B}^0 \in \mathcal{B}_a$ such that, for all $\tilde{B} \in \mathcal{B}_a$,

$$\begin{aligned} J(\tilde{B}^0) &\equiv \int_{I\times\Omega} \ell_i(t,x;\varphi^0(t,x))dxdt + \int_{I\times\partial\Omega} \ell_b(t,x;\varphi^0(t,x))dxdt \\ &\leq \int_{I\times\Omega} \ell_i(t,x;\varphi(t,x))dxdt + \int_{I\times\partial\Omega} \ell_b(t,x;\varphi(t,x))dxdt \equiv J(\tilde{B}) \end{aligned} \tag{10}$$

where φ^0 is the solution of (9) corresponding to $\tilde{B}^0$ and φ is that corresponding to $\tilde{B}$ respectively.

First we give a semigroup formulation of the problem (9) and (10). Let X and Y denote suitable Banach spaces of functions or generalized functions on Ω and $\partial\Omega$ respectively and W is another Banach space with the injection $W \subset X$ continuous. We shall use the following basic assumptions.

(A1) $\mathcal{A} \in \mathcal{L}(W, X)$, $B_0 \in \mathcal{L}(W, Y)$, $W_0 \equiv \mathrm{Ker} B_0$, $W_1 \equiv \mathrm{Ker}\mathcal{A}$,

(A2) $B_0 : W \to Y$ is surjective and $(B_0|_{W_1})^{-1} \equiv R \in \ Iso(Y, W_1)$,

(A3) $A \equiv (\mathcal{A}\,|_{W_0}) \in \ Iso(W_0, X)$.

Clearly defining $P \equiv A^{-1}\mathcal{A}$ we have $P^2 = P$, that is, P is a projection. Under the assumptions

(A1)-(A3) we have $W = W_0 \oplus W_1$ (topological direct sum). Assuming $-A$ to be the generator of an analytic semigroup $S(t), 0 \leq t < \infty$, in X one can justify the use of the variation of constants formula to write φ as the solution of the integral evolution equation,

$$\varphi(t) = \tilde{\eta}(t) + \int_0^t L(t-\theta)\tilde{f}(\theta, \varphi(\theta))d\theta, \quad t \in I, \tag{11}$$

written in the abstract form as, $\varphi = \tilde{\eta} + L\tilde{F}\varphi$, where L denotes the linear Volterra integral operator,

$$(Lz)(t) \equiv \int_0^t S(t-\theta)z_1(\theta)d\theta + \int_0^t AS(t-\theta)Rz_2(\theta)d\theta, \quad t \in I,$$

$z \equiv (z_1, z_2) \in L_p(I, Z), Z \equiv X \times Y$, and $(\tilde{F}\varphi)(t) \equiv \tilde{f}(t, \varphi(t))$ the Nemytskii operator with $\tilde{f} = (f, \tilde{g}), \tilde{g} = g - \tilde{B}\varphi$ and $\tilde{\eta}(t) \equiv S(t)\varphi_0$. In general the nonlinear integral equation (11) represents a large class of semilinear initial boundary value problems. For the linear problem we set $f(t,\varphi) \equiv f(t)$, $g(t,\varphi) \equiv g(t)$ and $-\tilde{B} = \Gamma$ giving the linear Volterra integral equation

$$\varphi(t) = \eta(t) + \int_0^t AS(t-\theta)R\Gamma\varphi(\theta)d\theta, \quad t \in I \tag{12}$$

where

$$\eta(t) \equiv S(t)\varphi_0 + \int_0^t S(t-\theta)f(\theta)d\theta + \int_0^t AS(t-\theta)Rg(\theta)d\theta.$$

Throughout the rest of this section we assume that there exists a Banach space E such that the embeddings $W \subsetneq E \subsetneq X$ are continuous and dense, A is the generator of an analytic s.g. $S(t)$, $0 \leq t < \infty$, in E and $\mathcal{B}_a \subset \mathcal{L}(E, Y)$.

For the existence of solution of the integral evolution equation (12) we have the following result.

Theorem 8

Suppose the following assumptions hold:

(a1) $\mathcal{B}_a \subset \{\Gamma \in \mathcal{L}(E,Y) : \| \Gamma \|_{\mathcal{L}(E,Y)} \leq k < \infty\}$,

(a2) there exist constants $c > 0$ and $\beta \in (0,1]$ such that $\| AS(t)R \|_{\mathcal{L}(Y,E)} \leq c/t^{1-\beta}$ for $t > 0$,

(a3) $\varphi_0 \in E$ and for some $p > (1/\beta)$, $f \in L_p(I, E)$, $g \in L_p(I, Y)$.

Then for each $\Gamma \in \mathcal{B}_a$, the integral evolution equation has a unique solution $\varphi \in C(\bar{I}, E) \equiv C(E)$.

Proof

Define the operator G by

$$(G\varphi)(t) \equiv \eta(t) + \int_0^t AS(t-\theta)R\Gamma\varphi(\theta)d\theta, \quad t \in I, \quad \varphi \in C(E).$$

One can verify that $G : C(E) \to C(E)$ and that, for some positive integer $n_0 = n_0(k, c, T, \beta)$, G^{n_0} is a contraction in $C(E)$. Hence the conclusion follows from Banach fixed point theorem. ■

Note that problem (P5) can be reformulated as the following abstract problem: find $\Gamma_0 \in \mathcal{B}_a \subset \mathcal{L}(E,Y)$ such that $J(\Gamma_0) \equiv \int_I h(t, \varphi_{\Gamma_0}(t))dt$ is minimum where φ_{Γ_0} is the solution of the integral evolution equation (12) for $\Gamma = \Gamma_0$.

Theorem 9

Suppose (i) $\mathcal{B}_a$ is a sequentially compact subset of $\mathcal{L}_s(E,Y) \equiv (\mathcal{L}(E,Y), \tau_{so})$ and (ii) $t \to h(t,e)$ is measurable for each $e \in E$, $e \to h(t,e)$ is lower semicontinuous on E for a.a. $t \in \bar{I}$, and $h(t,e) > -\infty$ for $t \in \bar{I}$ and $| e |_E < \infty$.
Then there exists a $\Gamma_0 \in \mathcal{B}_a$ such that $J(\Gamma_0) = \text{Inf}\{J(\Gamma), \Gamma \in \mathcal{B}_a\}$.

Remark 10

If in theorem 9 $\mathcal{B}_a$ is merely a sequentially compact subset of $\mathcal{L}_w(E,Y) \equiv (\mathcal{L}(E,Y), \tau_{w_0})$ then the conclusion remains valid provided lower semicontinuity (l.s.c.) is replaced by weak lower semicontinuity (w.l.s.c.).

Next we present the necessary conditions of optimality for the problem (P5) with a quadratic cost functional given by,

$$\begin{cases} h(t,\xi) \equiv \frac{1}{2}\langle Q(C\xi - z_0(t)), C\xi - z_0(t)\rangle_{F^*,F} \\ J(\Gamma) \equiv \int_I h(t, \varphi_\Gamma(t))dt, \quad \Gamma \in \mathcal{B}_a, \end{cases} \tag{13}$$

where $C \in \mathcal{L}(E,F)$ is the output operator with F being a reflexive Banach space, $Q \in \mathcal{L}^+(F,F^*)$, $Q = Q^*$ and $z_0 \in C(F)$ is the measured output.

Theorem 11 (Necessary Conditions of Optimality)

Suppose $\mathcal{B}_a$ is a closed bounded convex subset of $\mathcal{L}(E,Y)$. Then, in order that $\Gamma_0 \in \mathcal{B}_a$ be optimal for the problem (12), (13), it is necessary that there exists a $\tilde{\psi}^0 \in C(Y_w^*) \equiv \{$ the space of Y^*- valued w^*- continuous functions on $\bar{I}\}$ such that

(i)

$$\int_I \langle(\Gamma - \Gamma_0)\varphi^0(t), \tilde{\psi}^0(t)\rangle_{Y,Y^*} dt \;\; \geq \;\; 0 \;\; \text{for all } \Gamma \in \mathcal{B}_a \tag{14}$$

with

$$\tilde{\psi}^0(t) \equiv \int_t^T R^* S^*(\theta - t) A^* \psi^0(\theta) d\theta, \quad t \in \bar{I}$$

where $\psi^0 \in L_\infty(E^*) \equiv L_\infty(I, E^*)$, is the unique solution of the adjoint integral evolution equation,

(ii)

$$\begin{cases} \psi^0(t) = w^0(t) + \int_t^T \Gamma_0^* R^* S^*(\theta - t) A^* \psi^0(\theta) d\theta \\ w^0(t) \equiv C^* Q(C\varphi^0(t) - z_0(t)), \quad t \in I \end{cases} \tag{15}$$

and

(iii) φ^0 is the solution of the integral equation (12) corresponding to $\Gamma = \Gamma_0$.

(C) SEMILINEAR INITIAL BOUNDARY VALUE PROBLEMS (RELAXED CONTROLS)

Consider the problem (P6) of relaxed controls for the semilinear initial boundary value problem (9) with nonlinear $\tilde{B}$ given by $-\tilde{B} \equiv b(\mu, t, e)$ with μ representing the control. Writing this as an integral equation we obtain the following control problem:

$$\begin{cases} \varphi(t) = \eta(t) + \int_0^t AS(t-\theta)Rb(\mu_\theta, \theta, \varphi(\theta))d\theta \\ J(\mu) \equiv \int_I \ell(\mu_\theta, \theta, \varphi(\theta))d\theta = \min. \end{cases} \tag{16}$$

where $b: M \times I \times E \to Y$, $\ell: M \times I \times E \to R$ and, in particular, $\beta(\mu, t, e) = \int_U \beta(u, t, e)\mu(du)$, $\beta = b/\ell$, U a compact Polish space, $\mu \in M \equiv M(U)$ the space of probability measures on the Borel subsets of U, and $\mathcal{M} \equiv \mathcal{M}(I, M)$ is the space of M-valued functions on I furnished with L.C. Yong's fine topology.

We present the existence of optimal relaxed controls in the following theorem.

Theorem 12 (Existence of Optimal Relaxed Controls)

Consider the control problem (16) and suppose b and ℓ satisfy the following assumptions:

(b1) $b: U \times I \times E \to Y$ is continuous,

(b2) there exists a constant $k_1 = k_1(U)$, $0 < k_1 < \infty$, such that

$$\| b(u,t,e) \|_Y \leq k_1(1 + | e |_E)$$

$$\| b(u,t,e_1) - b(u,t,e_2) \|_Y \leq k_1 | e_1 - e_2 |_E$$

(ℓ1) $\ell \geq 0$, and continuous on $U \times I \times E$ and $\ell(u,t,e_n) \to \ell(u,t,e_0)$ uniformly with respect to $(u,t) \in U \times I$ whenever $e_n \to e_0$ in E.

(ℓ2) there exist constants $k_2, r > 0$ such that

$$| \ell(u,t,e) | \leq k_2(1 + | e |_E^r) \quad \text{for all} \quad (u,t) \in U \times I.$$

Then there exists an optimal control $\mu^0 \in \mathcal{M}$, that is, $J(\mu^0) \leq J(\mu)$ for all $\mu \in \mathcal{M}$. ■

For the necessary conditions of optimality we have the following result.

Theorem 13

Suppose b and ℓ satisfy the assumptions of theorem 12 and further they are once Gateaux differentiable in $e \in E$ for all $(u,t) \in U \times I$. Then, in order that $\mu^0 \in \mathcal{M}$ be an optimal control, it is necessary that there exists a $\tilde{\psi}^0 \in C(Y_w^*)$ such that

(i)

$$\int_I \left\{ \langle \tilde{\psi}^0(t), b(\mu_t - \mu_t^0, t, \varphi^0(t)) \rangle_{Y^*,Y} + \ell(\mu_t - \mu_t^0, t, \varphi^0(t)) \right\} \geq 0 \qquad (17)$$

for all $\mu \in \mathcal{M}$ with $\tilde{\psi}^0$ as given in equation (14) where $\psi^0 \in L_\infty(E^*)$, is the unique solution of the adjoint integral equation,

(ii)

$$\psi^0(t) = e_0^*(t) + \int_t^T L_0^*(\theta,t)\psi^0(\theta)d\theta, \quad t \in I \qquad (18)$$

with

$$e_0^*(t) \equiv \ell_e(\mu_t^0, t, \varphi^0(t))$$

$$L_0(\theta,t) \equiv AS(\theta - t)Rb_e(\mu_t^0, t, \varphi^0(t)), \quad 0 \leq t < \theta \leq T$$

where ℓ_e and b_e are the Gateaux differentials of ℓ and b in e on E,

and

(iii) φ^0 is the solution of

$$\varphi^0(t) = \eta(t) + \int_0^t AS(t-\theta)Rb(\mu_\theta^0, \theta, \varphi^0(\theta))d\theta, \quad t \in I.$$

(D)NONLINEAR PROBLEMS INVOLVING MONOTONE AND ACCRETIVE OPERATORS

We have results on existence and necessary conditions of optimality for nonlinear (degenerate) evolution equations of the form $\frac{d}{dt}(E\varphi) + A\varphi = f$ where A is a nonlinear monotone hemicontinuous bounded coercive operator from $L_p(I,V)$ to $L_{p'}(I,V^*)$, $(1/p) + (1/p') = 1$, with the injections $V \subseteq H \subseteq V^*$ continuous and dense and $E \in \mathcal{L}^+(H)$ to be identified. We have also results for systems, $\frac{d\varphi}{dt} + A(q)\varphi = 0$, with $A : Q \to M_{ac}(X) \equiv$ the class of accretive operators in X. Here nonlinear semigroup theory plays a central role. Further we have also results for quasilinear equations of Kato type, $\dot{x} + A(q,t,x)x = f(t,x)$, where q is to be identified from a compact metric space Q.

Due to limited space we shall present here only an existence result for systems involving monotone operators. Consider the system $\frac{d}{dt}(E\varphi) + A\varphi = f$ and suppose the nonlinear operator A satisfies the following properties:

(A1) there exist a constant $c \in (0,\infty), h \in L_1(I)$ and $1 < p < \infty$ such that

$$\langle A(t)v, v \rangle_{V^*,V} \geq h(t) + c \parallel v \parallel_V^p \quad \text{for } t \in I, v \in V$$

(A2)

$$\langle A(t)u - A(t)v, u - v\rangle_{V^*,V} \geq 0 \text{ for } t \in I, u, v \in V$$

(A3)

$$\| A(t)v \|_{V^*} \leq c(1 + \| v \|_V^{p-1}) \text{ for } t \in I, v \in V$$

and

(A4) whenever $v_n \to v_0$ weakly in $L_p(I,V) \equiv Y$

$$\lim_{n\to\infty} \int_I \langle A(t)v_n - A(t)v_0, v_n(t) - v_0(t)\rangle_{V^*,V} dt = 0.$$

Lemma 14

Suppose A satisfies (A1)-(A4) with $2 \leq p < \infty$, $f \in L_{p'}(I,V^*)$ and $\varphi_0 \in H$. Then for each $E \in \mathcal{L}^+(H)$, the Cauchy problem,

$$\frac{d}{dt}(E\varphi) + A\varphi = f, \quad \varphi(0) = \varphi_0, \tag{19}$$

has a unique weak solution $\varphi \in L_p(I,V)$ with $E^{\frac{1}{2}}\varphi \in C(\bar{I},H)$ in the sense that

$$-\int_I (\varphi, \frac{d}{dt}(E\psi))dt + \int_I \langle A\varphi, \psi\rangle_{V^*,V} dt = (E^{\frac{1}{2}}\varphi_0, E^{\frac{1}{2}}\psi(0)) + \int_I \langle f, \psi\rangle_{V^*,V} dt \tag{20}$$

for all $\psi \in L_p(I,V)$ such that $\frac{d}{dt}\psi \in L_2(I,H)$ and $\psi(T) = 0$. ■

Lemma 15

Suppose the assumptions of lemma 14 hold. Let $\mathcal{B}$ be a bounded subset of $\mathcal{L}^+(H)$ and let φ_E denote the solution of the C.P.(19) corresponding to $E \in \mathcal{B}$. Then

(i) $\{\varphi_E, E \in \mathcal{B}\}$ is a bounded subset of $L_p(I,V)$,

(ii) $\{E^{\frac{1}{2}}\varphi_E, E \in \mathcal{B}\}$ is a bounded subset of $L_\infty(I,H) \cap C(\bar{I},H)$,

(iii) $\{\frac{d}{dt}(E\varphi_E), E \in \mathcal{B}\}$ is a bounded subset of $L_{p'}(I,V^*)$. ■

Lemma 16

Suppose lemma 15 holds and A is strictly monotone. Then $E \to \varphi_E$ is continuous in the sense that whenever $E_n \xrightarrow{\tau_{so}} E_0$ in $\mathcal{L}^+(H)$, $\varphi_{E_n} \to \varphi_{E_0}$ weakly in $L_p(I,V)$. ■

By virtue of the above lemmas we have the following result.

Theorem 17

Suppose $\mathcal{B} \subset (\mathcal{L}^+(H), \tau_{so})$ is sequentially compact and $F : L_p(I, V) \to (-\infty, +\infty]$ is weakly lower semicontinuous. Then there exists an $E_0 \in \mathcal{B}$ such that

$$J(E_0) \equiv F(\varphi_{E_0}) \leq F(\varphi_E) \equiv J(E) \quad \text{for all} \quad E \in \mathcal{B}$$

where φ_E and φ_{E_0} are the solutions of (19) corresponding to E and E_0 respectively. ■

Remark 18

Note that theorem 17 can not be improved unless lemma 16 is improved.

Remark 19

The proofs of the existence results presented here crucially depend on the topology assumed for the space of bounded and unbounded operators and operator valued functions. In the course of this work it was discovered that there is a serious need for development of functional analysis dealing with operator valued functions.

REFERENCES

[1] N. U. Ahmed, *Identification of Operators in Differential Equations on Banach Space*, in "Operator Methods for Optimal Control Problems", Lecture Notes in Pure and Applied Mathematics, Vol. 108, (Edited by Sung J. Lee), Marcel Dekker Inc, New York, 1987.

[2] A. Pazy, *Semigroups of Linear Operators and Applications to Partial Differential Equations*, Springer-Verlag, New York Berlin Heidelberg Tokyo, 1983.

OPTIMAL CONTROL OF STATE-CONSTRAINED UNSTABLE SYSTEMS OF ELLIPTIC TYPE

J. Frédéric Bonnans and **E. Casas**[#]

INRIA - Domaine de Voluceau
BP.105, 78153 Le Chesnay
FRANCE

Facultad de Ciencias
39005 Santander
SPAIN

Abstract There has been since a few years an interest in the derivation of the optimality system of control problems involving an ill-posed state equation. Recently the authors have obtained new results concerning state-constrained problems of such type. We review and somewhat extend these results, and make a comparison with the case of problems without state constraints or with well-posed state equations.

I Introduction

We are concerned with the following control system :

$$\begin{aligned} -\Delta y + \phi(y) &= u \text{ in } \Omega, \\ y &= 0 \text{ on } \Gamma. \end{aligned} \qquad (1.1)$$

Here Ω is a smooth open bounded subset of $\mathbb{R}^n$ with $n \leq 3$, Γ is the boundary of Ω, Δ is the Laplace operator, ϕ is a C^1 mapping from $\mathbb{R}$ into $\mathbb{R}$ and we call u (resp. y) the control (resp. the state). If ϕ is non-decreasing, (1.1) has for each u in $L^2(\Omega)$ a unique solution in $Y = H^2(\Omega) \cap H_0^1(\Omega)$. Reminding the fact that

$$H^2(\Omega) \subset C(\Omega) \text{ for } n \leq 3, \qquad (1.2)$$

[#]The work of this author was supported in part by CAICYT, Madrid.

a standard application of the implicit function theorem ensures that the mapping $u \to y$ is C^1 from $L^2(\Omega)$ onto Y (see for instance [1], [2]). However, if ϕ is not nondecreasing, system (1.1) may have zero or several solutions (see some examples in [10]).

Let K be a non-empty, closed and convex subset of $L^2(\Omega)$, $\sigma \geq 2$ and y_d in $L^\sigma(\Omega)$ be given, and let $J : L^\sigma(\Omega) \times L^2(\Omega) \to \mathbb{R}$ be the functional

$$J(y,u) = \frac{1}{\sigma} \int_\Omega | y(x)-y_d(x)|^\sigma dx + \frac{N}{2} \int_\Omega u^2(x)dx. \tag{1.3}$$

Let B be a closed convex subset of a Banach space Z with $\overset{o}{B} \neq \emptyset$ and let $T:Y \to \mathbb{R}^m$ and $L:Y \to Z$ be linear continuous mappings. Let a be given in $\mathbb{R}^m$. We will consider the following control problem :

(P) $\qquad \min J(y,u)$ s.t. $u \in K$, $y \in Y$, $Ty = a$, $Ly \in B$.

Such problems, but without state constraints, have been studied by Lions ([12], ch.3) for the particular case $\phi(s) = - s^3$. These results have been generalized by Komornik [11] for a C^1 mapping ϕ. We will see that those results are a special case of the results given here. Lions ([12], ch.3, $9) also considered the case when the control is distributed and there are some constraints on the state, but not on the control. In that case the control may be thought as a function of the state and the results are not comparable to those given here. The optimal control of state-constrained systems has been widely studied for convex problems with linear state equations ([7], [8], [13], [14]) and more recently for nonconvex problems with nonlinear (but well-posed) equations by the authors ([1], [2]). We will see that the results given here are similar to those of [1] and [2], at least for unqualified problems.

II The main results

We first state a result concerning the existence of solutions for (P). This allows to check that the hypothesis are convenient. Then we state the main results concerning the optimality conditions, and discuss them.

Theorem 1 (see [5]) We suppose that (P) is feasible and that

(i) either $N > 0$ or K is bounded in $L^2(\Omega)$,

(ii) We may write $\phi(s) = \phi_1(s) + \phi_2(s)$ with ϕ_i continuous, $i = 1,2$, ϕ_1 non decreasing and, for some $C > 0$:

$$|\phi_2(s)| \leq C(1 + |s|^{\sigma/2}).$$

Then (P) has at least one solution. □

The proof of this Theorem lies on classical tricks (pass to the limit in a minimizing sequence). Hypothesis (ii) above means that the nonmonotone part of ϕ must be dominated in some sense by the criterion.

We denote the subdifferential of a convex mapping by ∂.

Theorem 2 Let $(\bar{y},\bar{u})$ be a solution of (P). Then there exists $\bar{p}$ in $L^2(\Omega)$, $\bar{\lambda}$ in $\mathbb{R}^m$, $\bar{\mu}$ in Z' and $\bar{\alpha} \geq 0$ such that

$$\bar{\alpha} + \|\bar{p}\| + \|\bar{\lambda}\| + \|\bar{\mu}\|_{Z'} > 0, \tag{2.1}$$

$$- \Delta\bar{p} + \phi'(\bar{y})\bar{p} = \bar{\alpha}\,|\bar{y}-y_d|^{\sigma-2}(\bar{y}-y_d) + T^*\bar{\lambda} + L^*\bar{\mu} \text{ in } Y', \tag{2.2}$$

$$\langle \bar{\mu}, z - L\bar{y}\rangle \leq 0, \quad \forall z \in B, \tag{2.3}$$

$$\int_\Omega (\bar{p} + \bar{\alpha} N \bar{u})(v - \bar{u}) \geq 0, \quad \forall v \in K. \quad \square \tag{2.4}$$

In fact (2.2) must be understood in the adjoint sense, that is

$$\int_\Omega \bar{p}\,(-\Delta y + \phi'(\bar{y})y)\, d_n = \bar{\alpha}\int_\Omega |\bar{y}-y_d|^{\sigma-2}(\bar{y}-y_d)y + \langle\bar{\lambda},Ty\rangle + \langle\bar{\mu},Ly\rangle, \qquad \forall y \in Y.$$

As Δ maps Y onto $L^2(\Omega)$, the above equation makes sense.

When such results are stated in a non-qualified form (i.e. $\bar{\alpha}$ may be 0) one has to take care that the statements are not trivial. In fact (2.2)-(2.4) is trivially satisfied with $\bar{\alpha}$, $\bar{\lambda}$, $\bar{\mu}$, $\bar{p}$ equal to 0. However (2.1) makes the statement non trivial. Some additional hypothesis allow to strenghten (2.1). For instance suppose that

$$TY = \mathbb{R}^m \text{ and } \overline{LY} = Z. \tag{2.5}$$

This is equivalent to T^* and L^* injective. Then obviously (2.1) is equivalent to the stronger statement

$$\bar{\alpha} + \|\bar{p}\| + \|T^* \bar{\lambda}\| + \|L^* \bar{\mu}\| > 0. \tag{2.6}$$

Suppose in addition that $T^* \bar{\lambda} + L^* \bar{\mu} = 0$ and $\bar{\mu} \in \partial I_B(L\bar{y})$ implies $\bar{\lambda} = 0$ and $\bar{\mu} = 0$. This will be the case if

$$L^* \partial I_B(L\bar{y}) \cap R(T^*) = 0 \tag{2.7}$$

However if (2.5) holds, as $\overset{o}{B} \neq \emptyset$ we may apply the chain rule to $\partial(I_B oL)$: hence (2.7) is equivalent to

$$\partial(I_B oL)\bar{y} \cap R(T^*) = 0. \tag{2.8}$$

If (2.5) and (2.8) hold, then under the hypothesis of Theorem 1, if $\bar{\alpha} = 0$ and $\bar{p} = 0$ then $T^*\bar{\lambda} + L^* \bar{\mu} = 0$, hence $\bar{\lambda} = 0$ and $\bar{\mu} = 0$, in contradiction with (2.1). This proves then

Corollary 1 Under the hypothesis of Theorem 2, if (2.5) and (2.8) hold, then (2.2)-(2.4) hold with

$$\bar{\alpha} + \|\bar{p}\|_{L^2(\Omega)} > 0. \quad \square \tag{2.1)'$$

Conversely it is trivial to prove using the above arguments that Corollary 1 implies Theorem 2. A proof of Corollary 1 is given in [5] in the special case $T : C_0(\Omega) \to \mathbb{R}^m$ and $L : C_0(\Omega) \to Z$. Let us sketch the idea of the proof of Theorem 2 (details will appear in [6]). We pass to the limit in the optimality system of the approximated problem

$$(P_\varepsilon)\quad \begin{array}{l} \min\limits_{y,u,w} \; J(y,u) + \frac{1}{2\varepsilon}\int_\Omega (w+\phi(y))^2dx + \frac{1}{2\varepsilon}\|Ty-a\|^2 + \\ \qquad + \frac{1}{2}\int(u-\bar{u})^2dx + \frac{1}{2}\int_\Omega (w+\phi(\bar{y}))^2 + \frac{1}{2}\|Ly-L\bar{y}\|^2, \\ \text{s.t.}\, u \in K,\ w \in L^2(\Omega),\ y \in Y,\ Ly \in B \text{ and} \\ -\Delta y = u+w \text{ in } \Omega, \\ \quad y = 0 \text{ ou } \Gamma. \end{array}$$

Problem (P_ε) involves two controls u and w and a well-posed state equation. Its optimality system can be formulated using the method of [1].

After a normalization of the multipliers appearing in the optimality system of (P_ε) the problem is to avoid that all approximate multipliers vanish when $\varepsilon \to 0$. For this the following Lemma, which has its own interest, is used :

Lemma 1 (see [5]) Let Z be a Banach space and B be a convex subset of Z (not necessarily closed) with non-empty interior. Let $\{(z_n, \eta_n)\}$ be a sequence in $Z \times Z'$ such that $z_n \in B$, $z_n \to z$ and $\eta_n \in \partial I_B(z_n)$. If $\liminf \|\eta_n\| > 0$, then 0 is not a weak star limit-point of $\{\eta_n\}$. □

We use Lemma 1 above with $Ly_\varepsilon = z_\varepsilon$. It is in order to obtain the <u>strong</u> convergence Ly_ε that we added the term $\frac{1}{2}\|Ly-L\bar{y}\|^2$ in the criterion of problem (P_ε) : the convergence of the infimum of (P_ε) towards the infimum of (P), which is easy to obtain, implies then that $Ly_\varepsilon \to L\bar{y}$ in Z. The terms of the criterion of (P_ε) involving $\bar{u}$ and $\phi(\bar{y})$ enforce the convergence and imply that $(u_\varepsilon, y_\varepsilon)$ does not converge towards another solution of (P).

Of course, the results of Theorem 2 are still valid if there is no constraint of the form $Ty=a$ and/or $Ly \in B$. Then one has simply to delete the terms refering to $(T,\bar{\lambda})$ and/or $(L,\bar{\mu})$ in the conclusion of Theorem 2. In this case, our hypothesis essentially reduce to those of [12], ch.3 and [11]. In fact, in these two references the unqualified form of the optimality systems is obtained as a step in the proof and an additional hypothesis on K allows to reject the possibility that $\bar{\alpha}$ be null. This qualification result can be obtained also for some state constrained systems ; see the examples in [5].

For state constrained problems with a well posed state equation the results of [2] are similar to the results obtained here. However the well-posedness of the system can be used to get qualification results in some cases (see [2]). Also, as the equation of the adjoint state (2.2) is well posed, (2.1) can obviously be replaced by the weaker statement

$$\bar{\alpha} + \|\bar{\lambda}\| + \|\bar{\mu}\|_Z > 0. \tag{2.1)''$$

A last case is when ϕ is affine. Then (P) is a convex problem. Assuming that no equality constraint is present and the following qualification hypothesis holds :

There exists u_0 in K such that Ly_{u_0} is in $\overset{\circ}{B}$,

the conclusion is obtained in a qualified form (i.e. with $\bar{\alpha} = 1$). This can be proved using the standard results of subdifferential calculus.

III Applications

In this section we consider two different control problems and we derive the optimality system for each of them by using Theorem 2.

First problem

(P) $\min J(y,u)$ s.t. (1.1), $u \in K$, $y \in Y$, $y(x_i) = a_i$,
$1 \leq i \leq m$, $\int_\Omega \|\nabla y(x)\| \, dx \leq r$.

Here $r > 0$, $\{x_i\}_{i=1}^m$ are given in Ω, $Ty = \{y(x_i)\}_{i=1}^m$ and $Z = \{\nabla y : y \in W_0^{1,1}(\Omega)\}$ endowed with the norm

$$\|z\|_Z = \int_\Omega \|z(x)\| \, dx.$$

Because of the Poincaré inequality in $W_0^{1,1}(\Omega)$, Z is a Banach space and B is the closed ball in Z of center 0 and radius r ; define L by $Ly = \nabla y$. It is easy to verify that hypothesis (2.5) is satisfied. Hence from Corollary 1 it follows

Theorem 3 Let $(\bar{y},\bar{u}) \in Y \times K$ be a solution of (P). Then there exist a real number $\bar{\alpha} \geq 0$ and elements $\bar{\lambda}$ in R^m, $\vec{f} \in L^\infty(\Omega)^n$ and $\bar{p} \in W_0^{1,s}(\Omega)$ for all $s < n/(n-1)$ satisfying

$$\bar{\alpha} + \|\bar{p}\|_{W_0^{1,s}(\Omega)} + \|\bar{\lambda}\| + \|\operatorname{div} \vec{f}\| > 0, \tag{3.1}$$

$$\begin{aligned} &- \Delta\bar{y} + \phi(\bar{y}) = \bar{u} \text{ in } \Omega, \\ &\bar{y} = 0 \text{ on } \Gamma, \end{aligned} \tag{3.2}$$

$$\begin{aligned} &-\Delta\bar{p} + \phi'(\bar{y})\bar{p} = \bar{\alpha}|\bar{y}-y_d|^{\sigma-2}(\bar{y}-y_d) + \sum_{i=1}^{m} \bar{\lambda}_i\delta_{[x_i]} - \operatorname{div} \vec{f} \text{ in } \Omega, \\ &\bar{p} = 0 \text{ on } \Gamma, \end{aligned} \tag{3.3}$$

$$\int_\Omega \vec{f}(z - \nabla\bar{y})dx \leq 0 \ \forall z \in B, \tag{3.4}$$

$$\int_\Omega (\bar{p} + \bar{\alpha}N\bar{u})(v - \bar{u})dx \geq 0, \ \forall v \in K \tag{3.5}$$

Second Problem

Here we assume that $T : C_0(\Omega) \to R^m$ and $L : C_0(\Omega) \to Z$ are linear and continuous mappings. Some examples corresponding to this case can be found in [5]. The interest of this particular case of Theorem 2 is that $T^*\bar{\lambda}$ and $L^*\bar{\mu}$ belong to $C_0(\Omega)'$ and there (as in previous problem) $\bar{p}$ is in $W_0^{1,s}(\Omega)$ and hence the adjoint state equation can be rigurously written in the following way :

Theorem 4 Let $(\bar{y},\bar{u}) \in Y \times K$ be a solution of (P). Then there exist a real number $\bar{\alpha}$ and elements $\bar{\lambda}$ in R^m, $\bar{\mu}$ in Z' and $\bar{p}$ in $W_0^{1,s}(\Omega)$ for all $s < n/(n-1)$ satisfying (2.1), (2.3), (2.4) and

$$- \Delta\bar{p} + \phi'(\bar{y})\bar{p} = \alpha|\bar{y} - y_d|^{\sigma-2}(\bar{y} - y_d) + T^*\bar{\lambda} + L^*\bar{\mu} \text{ in } \Omega,$$

$$\bar{p} = 0 \text{ on } \Gamma. \tag{3.6}$$

REFERENCES

[1] J.F. BONNANS and E. CASAS, Contrôle de systèmes non linéaires comportant des contraintes distribuées sur l'état, Rapport de Recherche n°300, INRIA, 1984.

[2] J.F. BONNANS and E. CASAS, Contrôle de systèmes elliptiques semilinéaires comportant des contraintes distribuées sur l'état, Collège de France Seminar, 1984. To appear in "Nonlinear partial differential equations and their applications" vol. VIII, H. BREZIS & J.L. LIONS eds, 69-86, Pitman, Boston.

[3] J.F. BONNANS and E. CASAS, Quelques méthodes pour le contrôle optimal de problèmes comportant des contraintes sur l'état, Anal. Stiinficice Univ. "Al. I. Cuza" din Iasi 32. S.Ia, Mathematica, 58-62, 1986.

[4] J.F. BONNANS and E. CASAS, On the choice of the function spaces for some state-constrained control problems. Numer. Funct. Anal & Optimiz. 7(4), 333-348, 1984-1985.

[5] J.F. BONNANS and E. CASAS, Optimal control of semilinear multistate systems with state constraints, INRIA report n°722, 1987.

[6] J.F. BONNANS and E. CASAS, To appear.

[7] E. CASAS, Quelques problèmes de contrôle avec contraintes sur l'état. C.R. Acad. Paris. 296 série I, 509-512, 1983.

[8] E. CASAS, Control of an elliptic problem with pointwise state constraints. SIAM J. on Control and Optimization 24, 1309-1318, 1986.

[9] F.H. CLARKE, Optimization and nonsmooth analysis, Wiley-Interscience, New York, 1983.

[10] M.G. CRANDALL and P. RABINOWITZ, Bifurcation perturbation of simple eigenvalues and linearized stability. Arch. Rat. Mech. Anal. 53, 161-180.

[11] V. KOMORNIK, On the control of strongly nonlinear systems I, Studia Sci. Math. Hungar. (to appear), 1987.

[12] J.L. LIONS, Contrôle de systèmes distribués singuliers, Dunod, Paris, 1983.

[13] U. MACKENROTH, Convex parabolic boundary control problems with pointwise state constraints. J. Math. Anal. Appl. 87, 256-277, 1982.

[14] U. MACKENROTH, On some elliptic optimal control problems with state constraints, Optimisation 17, 595-607, 1986.

OPTIMAL CONTROL OF QUASILINEAR ELLIPTIC EQUATIONS

Eduardo Casas and Luis A. Fernández
Departamento de Matemáticas, Estadística y Computación
39005 Santander, SPAIN

1.- INTRODUCTION

In this paper, we study some optimal control problems of systems governed by quasilinear strongly elliptic equations. Our main interest consists in the derivation of optimality conditions.

Let Ω be an open and bounded subset of $\mathbb{R}^N$ with Lipschitz continuous boundary Γ (Nečas (10)). Let us consider the following differential operator

$$Ay = -\operatorname{div} a(x,\nabla y)$$

where

$$a(x,\eta) = (a_1(x,\eta),\ldots,a_N(x,\eta))$$

We will assume the conditions (see Tolksdorf (11))

$$\begin{array}{l} a_j(\cdot,\eta) \text{ is a measurable function on } \Omega \\ a_j(x,\cdot) \text{ belongs to } C^1(\mathbb{R}^N) \quad j=1,\ldots,N \end{array} \tag{1.1}$$

$$\sum_{i,j=1}^{N} \frac{\partial a_j}{\partial \eta_i}(x,\eta)\rho_i\rho_j \geq C_1(1+|\eta|)^{\alpha-2}|\rho|^2 \tag{1.2}$$

$$\sum_{i,j=1}^{N} \left|\frac{\partial a_j}{\partial \eta_i}(x,\eta)\right| \leq C_2(1+|\eta|)^{\alpha-2} \tag{1.3}$$

$$a_j(x,0) = 0 \qquad j=1,\ldots,N \tag{1.4}$$

for some $\alpha\in(1,+\infty)$, some strictly positive constants C_1 and C_2, all $x\in\Omega$ and all $\eta,\rho\in\mathbb{R}^N$.

Now we consider the Dirichlet problem associated to operator A

$$\begin{array}{ll} Ay = u & \text{in } \Omega \\ y = 0 & \text{on } \Gamma \end{array} \tag{1.5}$$

In the sequel $W^{-1,\beta}(\Omega)$ will denote the dual of the Sobolev space $W_0^{1,\alpha}(\Omega)$, where $1<\alpha,\beta<+\infty$ and $\frac{1}{\alpha}+\frac{1}{\beta}=1$, and $D(\Omega)$ the space of infinitely

---

This research was partially supported by CAICYT (Madrid)

differentiable functions with a compact support in Ω.

It is well known that problem (1.5) has a unique weak solution $y(u)\in W_0^{1,\alpha}(\Omega)$ for each $u\in W^{-1,\beta}(\Omega)$ (J.L. Lions (7)).

There exists a vast litterature on the control of nonlinear elliptic equations. Here we mention the control of elliptic variational inequalities (Barbu (1), Mignot (8), Mignot and Puel (9), Bermúdez and Saguez (2), Friedman (6)) and the state constrained control problems governed by semilinear elliptic equations (Bonnans and Casas (3),(4)). However we do not know results about the control of quasilinear elliptic equations.

Our aim is to consider a control problem associated to the system (1.5). In these optimal control problems the main question we must investigate is the differentiability of the state respect to the control.

We will distinguish two cases, depending on the polynomial growth order α of operator coefficients. In case $\alpha\geq 2$, we prove the previous differentiability in the Gâteaux sense, introducing some function spaces naturally associated with the state equation. In case $1<\alpha<2$, we approximate the initial control problem by a family of problems corresponding to the case $\alpha=2$.

Let us point out that operator

$$Ay = -\operatorname{div}\left((b(x) + |\nabla y|)^{\alpha-2}\nabla y\right)$$

satisfies the hypotheses (1.1) to (1.4) if $0<b_0\leq b(x)\leq b_1<+\infty$. However inequality (1.2) is not verified for $b(x)\equiv 0$ and so this case is not included in the present work.

The details of the proofs and some regularity results for the optimal control and state will be given in a paper to appear.

2.- THE CONTROL PROBLEM (P_α). CASE $\alpha\geq 2$

Given $y\in W^{1,\alpha}(\Omega)$, let $H_0^y(\Omega)$ be the Hilbert space completed of $D(\Omega)$ respect to the norm

$$||w|| = \left(\int_\Omega (1 + |\nabla y|)^{\alpha-2}|\nabla w|^2 dx\right)^{1/2}$$

Clearly, we have the continuous injections

$$W_0^{1,\alpha}(\Omega) \subset H_0^y(\Omega) \subset H_0^1(\Omega)$$

More general spaces of this type have been considered by Coffman

(5) and Trudinger (12).

The main result of this section is the next

Theorem 2.1.- Let us consider the functional

$$F : L^2(\Omega) \longrightarrow H_0^1(\Omega)$$

defined by $F(u) = y(u)$. Then F is Gâteaux differentiable when $H_0^1(\Omega)$ is endowed with the weak topology. Moreover if $DF(u).v=z$ then z belongs to $H_0^{y(u)}(\Omega)$ and it is the unique solution in this space of problem

$$- \operatorname{div} \left(\frac{\partial a}{\partial \eta} (x, \nabla y(u)) \nabla z\right) = v \quad \text{in } \Omega$$
$$z = 0 \quad \text{on } \Gamma$$

Sketch of proof. Given $u, v \in L^2(\Omega)$ and $t>0$, we consider the following problems

$$- \operatorname{div} a(x, \nabla y) = u \quad \text{in } \Omega$$
$$y = 0 \quad \text{on } \Gamma$$

$$- \operatorname{div} a(x, \nabla y_t) = u+tv \quad \text{in } \Omega$$
$$y_t = 0 \quad \text{on } \Gamma$$

Substrating these equations, multiplying by $y_t - y$ and integrating in Ω, we obtain

$$\int_\Omega \frac{a(x,\nabla y_t) - a(x,\nabla y)}{t} . \frac{\nabla y_t - \nabla y}{t} dx = \int_\Omega v . \frac{y_t - y}{t} dx$$

Taking $z_t = \dfrac{y_t - y}{t}$ and using the Mean Value Theorem

$$\int_\Omega \nabla z_t^T \frac{\partial a}{\partial \eta} (x, w_t) \nabla z_t \, dx = \int_\Omega v z_t \, dx$$

with $w_t(x) = \nabla y(x) + \theta_t(x)(\nabla y_t(x) - \nabla y(x))$, $0<\theta_t(x)<1$.

From the hypothesis (1.2), we deduce that $\{z_t\}_{t>0}$ is bounded in $H_0^y(\Omega)$ and finally we pass to the limit in the equation

$$\int_\Omega \nabla\psi^T \frac{\partial a}{\partial \eta}(x, v_t) \nabla z_t \, dx = \int_\Omega v\psi \, dx \qquad \forall \psi \in D(\Omega)$$

where $v_t(x) = \nabla y(x) + \theta_t(x;\psi)(\nabla y_t(x) - \nabla y(x))$, $0<\theta_t(x;\psi)<1$.

Now let us introduce the control problem associated to the system

(1.5). Let K be a nonempty, convex and closed subset of $L^2(\Omega)$ and let

$$J : L^2(\Omega) \longrightarrow \mathbb{R}$$

be the functional defined by

$$J(v) = \frac{1}{2} \int_\Omega |y(v) - y_d|^2 dx + \frac{r}{2} \int_\Omega |v|^2 dx$$

where y_d is a fixed element of $L^2(\Omega)$ and r is a non-negative constant.

Then we consider the control problem

$$(P_\alpha) \quad \begin{array}{l} \text{Minimize } J(v) \\ \text{Subject to } v \in K \end{array}$$

As usual we derive existence of solution and also we get the first order optimality system by using the previous theorem.

Theorem 2.2.- Assume that

Either K is bounded in $L^2(\Omega)$ or $r>0$

Then there exists (at least) one solution of (P_α). Moreover if $\bar{u}$ is a solution of (P_α), then there exist elements $\bar{y} \in W_0^{1,\alpha}(\Omega)$ and $\bar{p} \in H_0^{\bar{y}}(\Omega)$ such that

$$\begin{aligned} -\operatorname{div} a(x, \nabla\bar{y}) &= \bar{u} \quad \text{in } \Omega \\ \bar{y} &= 0 \quad \text{on } \Gamma \end{aligned}$$

$$\begin{aligned} -\operatorname{div}\left[\left(\frac{\partial a}{\partial \eta}(x, \nabla\bar{y})\right)^T \nabla\bar{p}\right] &= \bar{y} - y_d \quad \text{in } \Omega \\ \bar{p} &= 0 \quad \text{on } \Gamma \end{aligned}$$

$$\int_\Omega (\bar{p} + r.\bar{u})(v - \bar{u})\, dx \geq 0 \quad \forall v \in K$$

3.- THE CONTROL PROBLEM (P_α). CASE $1<\alpha<2$

In this section, we consider the same control problem (P_α) as in section 2 with slight variations. These are motivated by the fact that $W_0^{1,\alpha}(\Omega)$ is not included in $L^2(\Omega)$ in some cases (for example, if N=3 and α is close to 1). Therefore we formulate (P_α) in the following way

$$(p_\alpha) \quad \begin{array}{l} \text{Minimize } J(v) \\ \text{Subject to } v \in K \end{array}$$

with K a non empty, convex and closed subset of $L^\beta(\Omega)$ and $J : L^\beta(\Omega) \longrightarrow \mathbb{R}$ given by

$$J(v) = \frac{1}{\alpha}\int_\Omega |y(v)-y_d|^\alpha dx + \frac{r}{\beta}\int_\Omega |v|^\beta dx$$

where y_d is a fixed element of $L^\alpha(\Omega)$ and $r \geq 0$.

As in the preceding case, it is easily followed

Theorem 3.1.- Assume that

Either K is bounded in $L^\beta(\Omega)$ or $r>0$

Then there exists (at least) one solution of (P_α).

We do not know if J es differentiable, therefore in order to derive the optimality conditions, for each $\varepsilon>0$ we consider the perturbed operator

$$A_\varepsilon y = -\varepsilon\Delta y - \operatorname{div} a(x,\nabla y)$$

that satisfies the hypotheses (1.1) to (1,4) for $\alpha=2$. Hence, given $v\in H^{-1}(\Omega) = (H_0^1(\Omega))'$ there exists a unique weak solution $y_\varepsilon(v)\in H_0^1(\Omega)$ of problem

$$\begin{aligned} A_\varepsilon y &= v \quad \text{in } \Omega \\ y &= 0 \quad \text{on } \Gamma \end{aligned}$$

The next theorem shows that $\{A_\varepsilon\}_{\varepsilon>0}$ is an approximating family of operator A.

Theorem 3.2.- Assume that $v_\varepsilon \longrightarrow u$ weakly in $L^\beta(\Omega)$ as $\varepsilon \to 0$. Then $y_\varepsilon(v_\varepsilon) \longrightarrow y(u)$ strongly in $W_0^{1,\alpha}(\Omega)$ as $\varepsilon \to 0$.

It is not difficult to deduce this theorem from the M-property (7) and the following lemma

Lemma.- There exist positive constants k_0, k_1 and k_2 depending only on N, α, c_1 and c_2 such that

a) $\displaystyle\sum_{j=1}^N (a_j(x,\eta)-a_j(x,\eta'))(\eta_j-\eta_j') \geq k_1(1+|\eta|+|\eta'|)^{\alpha-2}|\eta-\eta'|^2$

b) $\displaystyle\sum_{j=1}^N |a_j(x,\eta)| \leq k_2(1+|\eta|)^{\alpha-2}|\eta|$

c) $\displaystyle\int_\Omega (a(x,\nabla y)-a(x,\nabla y'))(\nabla y-\nabla y')dx \geq$

$$\geq k_0 ||\nabla y-\nabla y'||^2_{L^\alpha(\Omega)} \; ||1+|\nabla y|+|\nabla y'|\;||^{\alpha-2}_{L^\alpha(\Omega)}$$

Proof.- For a) and b) see Tolksdorf (11)

c) is a simple consequence of Hölder's inequality that we apply with $s = \frac{2}{2-\alpha} > 1$ and $t = \frac{2}{\alpha} > 1$. Then

$$\int_\Omega |\nabla y - \nabla y'|^\alpha dx \leq \left(\int_\Omega (1+|\nabla y|+|\nabla y'|)^\alpha dx\right)^{\frac{2-\alpha}{2}} \left(\int_\Omega \frac{|\nabla y - \nabla y'|^2}{(1+|\nabla y|+|\nabla y'|)^{2-\alpha}} dx\right)^{\frac{\alpha}{2}}$$

Together with a) this completes the proof.

Now let $\bar{u}$ be a fixed solution of (P_α). We introduce the cost functional

$$J_\varepsilon(v) = \frac{1}{\alpha}\int_\Omega |y_\varepsilon(v) - y_d|^\alpha dx + \frac{r}{\beta}\int_\Omega |v|^\beta dx + \frac{1}{\beta}\int_\Omega |v-\bar{u}|^\beta dx$$

and the corresponding control problem

$$(P_\varepsilon) \quad \begin{cases} \text{Minimize } J_\varepsilon(v) \\ \text{Subject to } v \in K \end{cases}$$

Since J_ε is differentiable we can derive the following result analogue to theorem 2.2.

<u>Theorem 3.3</u>.- For each $\varepsilon > 0$, there exists (at least) one solution of (P_ε).

Moreover, if u_ε is a solution of (P_ε), then there exist elements y_ε and p_ε in $H_0^1(\Omega)$ such that

$$\begin{aligned} -\operatorname{div}(\varepsilon \nabla y_\varepsilon + a(x,\nabla y_\varepsilon)) &= u_\varepsilon \quad && \text{in } \Omega \\ y_\varepsilon &= 0 && \text{on } \Gamma \end{aligned}$$

$$\begin{aligned} -\operatorname{div}\left[\left(\varepsilon I + \frac{\partial a}{\partial \eta}(x,\nabla y_\varepsilon)\right)^T \nabla p_\varepsilon\right] &= |y_\varepsilon - y_d|^{\alpha-2}(y_\varepsilon - y_d) \quad && \text{in } \Omega \\ p_\varepsilon &= 0 && \text{on } \Gamma \end{aligned}$$

$$\int_\Omega (p_\varepsilon + r|u_\varepsilon|^{\beta-2}u_\varepsilon + |u_\varepsilon - \bar{u}|^{\beta-2}(u_\varepsilon - \bar{u}))(v-u_\varepsilon)\, dx \geq 0 \quad \forall v \in K$$

where I denotes the identity matrix NxN.

Our purpose is to pass to the limit in the optimality conditions of the perturbed problems. Continuing with the previous notations, from the definition of J_ε we deduce that $\{u_\varepsilon\}_{\varepsilon>0}$ is bounded in $L^\beta(\Omega)$. Now using theorem 3.2 it follows that there exist $u\in L^\beta(\Omega)$ and $y(u)\in W_0^{1,\alpha}(\Omega)$ such that as $\varepsilon\to 0$

$$u_\varepsilon \longrightarrow u \quad \text{weakly in } L^\beta(\Omega)$$

$$y_\varepsilon \longrightarrow y(u) \quad \text{in } W_0^{1,\alpha}(\Omega)$$

Once again thanks to theorem 3.2 we get

$$y_\varepsilon(\bar{u}) \longrightarrow \bar{y} \quad \text{in } W_0^{1,\alpha}(\Omega)$$

where $\bar{y}$ denotes the state associated to $\bar{u}$.

From the above results, the lower semicontinuity of J_ε and the relation

$$J_\varepsilon(u_\varepsilon) \leq J_\varepsilon(\bar{u}) \quad \text{for all } \varepsilon>0$$

we deduce that $u=\bar{u}$ and therefore

$$u_\varepsilon \longrightarrow \bar{u} \quad \text{weakly in } L^\beta(\Omega)$$

$$y_\varepsilon \longrightarrow \bar{y} \quad \text{in } W_0^{1,\alpha}(\Omega)$$

as $\varepsilon\to 0$.

Now from the definition of J_ε, we obtain the strong convergence of $\{u_\varepsilon\}_{\varepsilon>0}$ to $\bar{u}$.

Finally, using hypothesis (1.2) we may conclude that $\{p_\varepsilon\}_{\varepsilon>0}$ is bounded in $W_0^{1,\alpha}(\Omega)$ and hence there exist a sequence $\varepsilon_n\to 0$ and $\bar{p}\in W_0^{1,\alpha}(\Omega)$ such that

$$p_{\varepsilon_n} \longrightarrow \bar{p} \quad \text{weakly in } W_0^{1,\alpha}(\Omega)$$

Summarizing, we have

Theorem 3.4.- There exist elements $\bar{y}$ and $\bar{p}$ in $W_0^{1,\alpha}(\Omega)$ satisfying

$$-\operatorname{div} a(x,\nabla\bar{y}) = \bar{u} \quad \text{in } \Omega$$

$$\bar{y} = 0 \quad \text{on } \Gamma$$

$$- \operatorname{div} \left(\left(\frac{\partial a}{\partial \eta} (x,\nabla\bar{y})\right)^T \nabla\bar{p} \right) = |\bar{y}-y_d|^{\alpha-2}(\bar{y}-y_d) \quad \text{in } \Omega$$

$$\bar{p} = 0 \quad \text{on } \Gamma$$

$$\int_\Omega (\bar{p}+r|\bar{u}|^{\beta-2}\,\bar{u})(v-\bar{u})\,dx \geq 0 \quad \forall v \in K$$

Moreover

$$\int_\Omega (\nabla\bar{p})^T \frac{\partial a}{\partial \eta}(x,\nabla\bar{y})^T \nabla\bar{p}\,dx \leq \int_\Omega |\bar{y}-y_d|^{\alpha-2}(\bar{y}-y_d)\bar{p}\,dx$$

REFERENCES

1. V. BARBU, "Optimal Control of variational inequalities", Pitman, London, 1984.

2. A. BERMUDEZ and C. SAGUEZ, "Optimal Control of a Signorini problem", SIAM J. Control and Optimiz. 25, pp. 576-582, 1987.

3. J.F. BONNANS and E. CASAS, "Contrôle de systèmes elliptiques semi-linéaires comportant des contraintes distribuées sus L'état", Collège de France Seminar 1984. To appear in "Nonlinear partial differential equations and their applications", H. Brezis & J.L. Lions eds., Pitman.

4. J.F. BONNANS and E. CASAS, "Optimal control of state-constrained unstable systems of elliptic type", in these proceedings.

5. C.V. COFFMAN, R. DUFFIN and V.J. MIZEL, "Positivity of weak solutions of non-uniformly elliptic equations", Ann. di Mat. pura e appl. 104, pp. 209-238, 1975.

6. A. FRIEDMAN, "Optimal control for variational inequalities", SIAM J. Control and Optimiz. 24, pp. 439-451, 1986.

7. J. L. LIONS, "Quelques méthodes de résolution des problèmes aux limites non linéaires", Dunod, Paris, 1969.

8. F. MIGNOT, "Contrôle dans les Inéquations Variationelles Elliptiques" J. Functional Analysis 22, pp. 130-185, 1976.

9. F. MIGNOT and J.P. PUEL, "Optimal control in some variational inequalities" SIAM J. Control and Optimiz. 22, pp. 466-476, 1984.

10. J. NEČAS, "Les méthodes directes en théorie des équations elliptiques", Masson, Paris, 1967.

11. P. TOLKSDORF, "Regularity for a more general class of quasilinear elliptic equations", J. Differential Equations 51, pp. 126-150, 1984.

12. N. S. TRUDINGER, "Maximum principles for linear, non-uniformly elliptic operators with measurable coefficients", Math. Z. 156, pp. 291-301, 1977.

SOME RESULTS ON LINEAR QUADRATIC PERIODIC CONTROL WITHOUT DETECTABILITY

Giuseppe Da Prato
Scuola Normale Superiore
56100 PISA, Italy

INTRODUCTION

Consider a dynamical system governed by the following state equation:

$$(1) \qquad y'(t) = A(t)y(t) + f(t) + B(t)u(t) \qquad ; \quad t \in \mathbb{R}$$

where $A(t):D(A(t)) \subset H \to H$, $B(t):U \to H$ are linear operators (with $A(t)$ generally unbounded) and H is a Hilbert space, f is a mapping from $\mathbb{R}$ to H and u (the control) is a mapping from $\mathbb{R}$ to U.

We shall assume A, B, f 2π-peridic. Under these assumptions, a natural optimal control problem is the following: (see |1|, |2|, |3|).

Minimize

$$(2) \qquad J(u) = \int_0^{2\pi} [\langle R(t)y(t),y(t)\rangle + \langle N(t)u(t),u(t)\rangle]\, dt$$

over all 2π-periodic controls u, where y is a 2π-periodic corresponding solution of (1), $R(t)$, $N(t)$ are positive operators respectively in H and U which moreover are 2π-periodic. In this connection, we remark, in fact that the usual cost function

$$(3) \qquad F(u) = \int_0^{\infty} [\langle R(t)y(t),y(t)\rangle + \langle N(t)u(t),u(t)\rangle]\, dt$$

may be identically equal to $+\infty$, as simple examples show. This fact further motivates consideration of the periodic problem (1).

We shall look for a feedback optimal control (for the cost functional $J(u)$) of the form:

$$(4) \qquad u^*(t) = -N^{-1}B^*(\bar{Q}y^* + r) \qquad .$$

Here, $\bar{Q}$ is a 2π-periodic solution of the Riccati equation

$$(5) \qquad Q' + A^*Q + QA - QBN^{-1}B^*Q + R = 0 \ ;$$

y^* is a periodic solution to the closed loop equation

(6) $$y' = Ly - BN^{-1}B^*r + f \quad ;$$

r is a periodic solution to the dual equation

(7) $$r' + L^*y + \bar{Q}f = 0 \; ;$$

and L is the feedback operator

(8) $$L = A - BN^{-1}B^*\bar{Q} \quad .$$

In [2], [3] the existence of a 2π-periodic solution of eq. (5) was proved. In order to prove the existence of periodic solutions to equation (6) and (7), we need the condition that L be non resonant; i.e. that 1 is not an eigenvalue of $U_L(2\pi,0)$, where $U_L(t,s)$ is the evolution operator relative to L. In the papers [2], [3] we assumed that $(A, \sqrt{R})$ is detectable, because in this case it follows that L is stable and this in turn implies that L is non resonant.

In this paper we shall prove some spectral properties of $U_L(2\pi,0)$ asserting in particular that if A is non resonant then L is also non resonant.

After introducing some notations, in Section 1, we then state the optimal control problem in Section 2. In Section 3 we prove new spectral properties of $U_L(2\pi,0)$ and finally we study the above control problem via Dynamic Programming in Section 4.

1. NOTATIONS

If Z is a Banach space we shall denote by $C_\#(Z)$ the Banach space of all 2π-periodic and continuous mappings $\mathbb{R} \to Z$, endowed with the sup norm. $L^2_\#(Z)$ will represent the Banach space of all the mappings $\mathbb{R} \to Z$ which are measurable, 2π-periodic and locally square Bochner integrable. For any $\alpha \in \,]0,1[$ we set $C^\alpha_\#(Z) = \{ f \in C_\#(Z);$ f is α-hölder continuous$\}$.

If W is another Banach space, we set:

$$C^s_\#(\mathcal{L}(Z;W)) = \{\gamma : \mathbb{R} \to \mathcal{L}(Z;W); \gamma(\cdot)z \in C_\#(W) \quad \text{for any} \quad z \in Z\}.$$

We shall consider two complex Hilbert spaces D and H with

$D \subset H$ and a mapping $A: \mathbb{R} \to \mathcal{L}(D;H)$ such that:

$(H)_A$

i) D is dense in H and the embedding of D in H is continuous and compact

ii) There exists $\alpha \in]0,1[$ such that $A \in C^{\alpha}_{\#}(\mathcal{L}(D;H))$

iii) $A(t)$ generates an analytic semi-group in H, $t \in \mathbb{R}$

From $(H)_A$ it follows ([4]) that there exists an evolution operator $U_A(t,s)$ relative to A; moreover $U_A(t,s)$ is compact if $t > s$. Consider now the equation:

$$(1.1) \qquad y'(t) = A(t)y(t) + f(t) \quad ; \quad f \in L^2_{\#}(H) .$$

A continuous function $y: \mathbb{R} \to H$ is said to be a mild solution to equation (1.1) if for any $t \geq s$ we have:

$$(1.2) \qquad y(t) = U_A(t,s)y(s) + \int_s^t U_A(t,\sigma)f(\sigma)\, d\sigma .$$

The following result is well-known:

<u>Proposition 1.1.</u> Assume $(H)_A$. If 1 is not eigenvalue of $U_A(2\pi,0)$, then there exists a unique 2π-periodic mild solution y of equation (1.1) which is given by:

$$(1.3) \qquad y(t) = U_A(t,0)(1 - U_A(2\pi,0))^{-1} \int_0^{2\pi} U_A(2\pi,s)f(s)ds + \int_0^t U_A(t,s)f(s)\, ds .$$

If 1 is not an eigenvalue of $U_A(2\pi,0)$, we say that A is <u>non resonant</u>.

2. SETTING OF THE PROBLEM

We consider a dynamical system:

$$(2.1) \qquad y'(t) = A(t)y(t) + f(t) + B(t)u(t)$$

where A verifies $(H)_A$. We assume:

$(H)_1$ $\qquad f \in L^2_\#(H)$, $\qquad B \in C^s_\#(\mathcal{L}(u;H))$,

where U is another Hilbert space (the space of controls).
If A is non resonant, then by Proposition 1.1, equation (2.1) has a unique mild solution $y \in C_\#(H)$.

We introduce now the cost functional. Let $R: \mathbb{R} \to \Sigma(H)$, $N: \mathbb{R} \to \Sigma(U)$ such that

$(H)_2$ $\qquad R \in C^s_\#(\Sigma(H))$, $R(t) \geq 0$; $N \in C^s_\#(\Sigma(U))$, $N(t) \geq \varepsilon > 0$, $t \in \mathbb{R}$

for some $\varepsilon > 0$. By $\Sigma(H)$ (resp. $\Sigma(U)$), we mean the real Banach space of hermitian operators in H (resp. U).

For any $u \in L^2_\#(U)$ we consider the set Λ_u (possibly empty) of all 2π-periodic mild solutions of equation (2.1). If A is non resonant, then Λ_u consists just of one element. The set U_{ad} of the admissible control is defined by

(2.2) $$U_{ad} = \{u \in L^2_\#(U);\ \Lambda_u \neq \emptyset\} .$$

For any $u \in U_{ad}$ and any $y \in \Lambda_u$ we set:

(2.3) $$J(u,y) = \int_0^{2\pi} [\langle Ry,y\rangle + \langle Nu,u\rangle]\, dt .$$

A couple (u^*,y^*) with $u^* \in U_{ad}$ and $y^* \in \Lambda_{u^*}$ is said to be optimal if

(2.4) $$J(u^*,y^*) \leq J(u,y) \qquad \forall u \in U_{ad}, \quad \forall y \in \Lambda_u .$$

3. PERIODIC SOLUTIONS OF RICCATI EQUATIONS

We assume here $(H)_A$, $(H)_1$ and $(H)_2$. Consider the Riccati equation:

(3.1) $$Q' + A^*Q + QA - QKQ + R = 0$$

where $K = BN^{-1}B^*$. $Q \in C^s_\#(\Sigma(H))$ is called a 2π-periodic solution to equation (3.1) if for any $t > s$ and any $x \in H$ we have:

$$Q(s)x = U_A^*(t,s)Q(t)U_A(t,s)x + $$
$$+ \int_s^t U_A^*(\sigma,s)\ [R - QKQ]\ (\sigma)U_A(\sigma,s)x\ d\sigma \quad .$$

In [2], [3] conditions are given which assure the existence of a 2π-periodic solution of (3.1).

In this section we assume that a 2π-periodic solution $\bar{Q}$ of (3.1) exists and we then study some properties of the feedback operator $L = A - K\bar{Q}$. It is easy to prove (by perturbation arguments) that there exists the evolution operator $U_L(t,s)$ relative to L.

We are here interested in giving a condition such that L is non resonant. The key hypothesis is the following:

$$(3.2) \qquad x \neq 0\ ,\ U_A(2\pi,0)x = x \ \Rightarrow\ R(t)U_L(t,0)x \neq 0 \quad \text{for some } t \geq 0 \quad .$$

<u>Proposition 3.1</u>. L is non resonant if and only if hypothesis (3.2) holds. In particular if A is non resonant then L is non resonant.

<u>Proof</u>. Assume first that L is non resonant and, by contradiction, that (3.2) does not hold. Then there exists $x \neq 0$ such that

$$(3.3) \qquad U_A(2\pi,0)x = x\ ,\ R(t)U_L(t,0)x = 0 \qquad \forall t \geq 0 \quad .$$

It follows

$$\bar{Q}'(t)U_L(t,0)x + A^*(t)\bar{Q}(t)U_L(t,0)x + \bar{Q}(t)A(t)U_L(t,0)x + $$
$$- \bar{Q}(t)K(t)\bar{Q}(t)U_L(t,0)x = 0$$

from which

$$\frac{d}{dt}\langle\bar{Q}(t)U_A(t,0)x,U_A(t,0)x\rangle = \|\sqrt{K}(t)\bar{Q}(t)U_A(t,0)x\|^2 \quad .$$

By integrating from 0 to 2π we get $\int_0^{2\pi} \|K(t)\bar{Q}(t)U_A(t,0)x\|^2 = 0$ which implies $K(t)\bar{Q}(t)U_A(t,0)x = 0$. It follows

$$\frac{d}{dt}U_A(t,0)x = A(t)U_A(t,0)x = L(t)U_A(t,0)x$$

so that $U_A(2\pi,0)x = U_L(2\pi,0)x = x$ which contradicts the hypothesis that L is non resonant.

Assume now (3.2) and by contradiction that L is resonant. Then there exists $x \neq 0$ such that $U_L(2\pi,0)x = x$. We have:

$$\frac{d}{dt} \langle \bar{Q}(t)U_L(t,0)x, U_L(t,0)x\rangle = - \| \sqrt{K}(t)\bar{Q}(t)U_L(t,0)x\|^2 + - \| \sqrt{R}(t)U_L(t,0)x\|^2 .$$

It follows, by integrating from 0 to 2π, $R(t)U_L(t,0)x$ which contradicts (3.2). ■

4. DYNAMIC PROGRAMMING

We assume here $(H)_A$, $(H)_1$ and $(H)_2$. Moreover we suppose that there exists a 2π-periodic solution $\bar{Q}$ of Riccati equation (3.1) and that $L = A - BN^{-1}B^*\bar{Q}$ is non resonant.

Theorem 4.1 There exists an optimal couple (y^*,u^*), where y^* is the unique 2π-periodic solution to the closed loop equation

(4.1) $$y' = Ly - BN^{-1}B^*r + f ,$$

u^* is given by

(4.2) $$u^*(t) = -N^{-1}B^*(\bar{Q}y + r)(t) ,$$

and r is the unique 2π-periodic solution to the equation

(4.3) $$r' + L^*r + \bar{Q}f = 0 .$$

Finally, the optimal cost $J(u^*,y^*)$ is given by

(4.4) $$J^* \doteq J(u^*,y^*) = \int_0^{2\pi} [\, 2 \langle r,f\rangle - \| N^{-\frac{1}{2}}B^*r\|^2 \,]\, ds .$$

Proof. Following [2], [3] we fix any admissible control u and an element $y \in \Lambda_u$. Then we compute

$$\frac{d}{dt} \langle \bar{Q}(t)y(t),y(t)\rangle + 2 \langle r(t),(t)\rangle .$$

Integrating the result between 0 and 2π we find the identity

$$(4.5) \qquad J(u,y) = J^* + \int_0^{2\pi} \| N^{-\frac{1}{2}} B^* \bar{Q} y + N^{-\frac{1}{2}} B^* r + N^{\frac{1}{2}} u \|^2 \, ds \quad .$$

Thus $J(u,y) \geq J^*$. Let now y^* be the 2π-periodic of (4.1) and let u^* be defined by (4.2). Since $y^{*\prime} = Ay^* + f - Bu^*$, we have $u^* \in U_{ad}$ and $y^* \in \Lambda_{u^*}$; moreover $J(u^*,y^*) = J^*$ and the conclusion follows.■

Example 4.2 Consider a system

$$(4.6) \qquad \begin{cases} y' = \Delta y + \phi(t) y + f(t,x) + u(t,x) \; ; \; t \in \mathbb{R}, \; x \in \Omega \\ \\ y(t,x) = 0 \; , \; t \in \mathbb{R} \; , \; x \in \partial\Omega \end{cases}$$

where Ω is a bounded subset of $\mathbb{R}^n$ with regular boundary $\partial\Omega$, $y_0 \in L^2(\Omega)$, $f,u \in L^2([0,T] \times \Omega)$ are 2π-periodic in t and $\phi \in C^{\alpha}_{\#}(\mathbb{R})$ for some $\alpha \in]0,1[$.

We set $H = U = L^2(\Omega)$, $D = H^2(\Omega) \cap H^1_0(\Omega)$, $A(t) = \Delta + \phi(t)$, $B(t) = I$.

We denote by $\{-\lambda_k\}$, $k \in \mathbb{N}$ the eigenvalues of $-\Delta$ (with Dirichlet boundary conditions) and by $\{\eta_k\}$ the correspond eigenvectors. Then $U_A(2\pi,0)$ is compact and its eigenvalues are given by $\exp(-2\pi\lambda_k + \Phi(2\pi))$ where $\Phi(t) = \int_0^t \phi(s)\,ds$.

Under the above assumptions (A,B) is stabilizable, thus ([2]) there exists a 2π-periodic solution of the Riccati equation.

We consider two cases:

a) $\qquad \lambda_k \neq \frac{1}{2\pi} \Phi(2\pi) \quad , \quad$ for any $k \in \mathbb{N}$.

Then $L = A - BN^{-1}B^*\bar{Q}$ is non resonant (Corollary 3.2) so that Theorem 4.1 applies for any choice of R and N.

b) $\qquad \lambda_k = \frac{1}{2\pi} \Phi(2\pi) \quad , \quad$ for some $k \in \mathbb{N}$.

Since $U_A(t,0)\eta_k = \exp(-t\lambda_k + \Phi(t))\,\eta_k$, we have to check (again by Corollary 3.2) if $R(t)\eta_k$ is not identically 0. In this case we can still aplly Theorem 4.1.

REFERENCES

1. BITTANTI, S., LOCATELLI, A., MAFFEZZONI, C., Periodic optimization under small perturbations, Periodic Optimization, vol. II, A. Marzollo ed., SPRINGER-VERLAG, New York, 1972, 183-231

2. DA PRATO, G., Synthesis of optimal control for an infinite dimensional periodic problem, SIAM J. Control and Optimization, vol. 25, 3, (1987), 706-714

3. DA PRATO, G., ICHIKAWA, A., Quadratic control for linear periodic system, Applied Math. and Optimiz. (to appear)

4. H. TANABE, Equations of Evolutions, Pitman, London, 1979

FURTHER DEVELOPMENTS IN THE APPLICATION OF MIN MAX DIFFERENTIABILITY TO SHAPE SENSITIVITY ANALYSIS.*

M.C.Delfour
Centre de recherches mathématiques and
Département de mathématiques et de statistique,
Université de Montréal, C.P. 6128, Succ. A,
Montréal, Québec, Canada, H3C 3J7

J.-P. Zolésio
Laboratoire de Physique Mathématique,
Université des Sciences et Techniques
du Languedoc, Pl. Eugène Bataillon,
34060 - Montpellier Cédex, France

ABSTRACT. The object of this paper is to show how "Shape gradients" can be computed without studying or characterizing the "Shape or Material derivative" of the state. To do that, we show how to parametrize the problem and the function spaces involved. The final expression is obtained by using theorems on the differentiability of a Min Max with respect to a parameter. Our approach gives a precise mathematical meaning to many formal or quick computations found in the literature. Several examples are given including non-differentiable cost functions and state contraints.

1. INTRODUCTION

In Shape Optimization problems, the optimization variable is no longer a function but the shape of a domain and the state variable is usually the solution of a boundary value problem over that domain. For Shape Sensitivity Analysis, it is thus necessary to introduce an appropriate notion of "directional derivative". Many techniques are available for this purpose, but, in our opinion, the method based on a virtual "Velocity Fields of Deformations" (cf. J. CEA [2,3], J.P. ZOLESIO [1,2,3]) seems to offer both flexibility and accuracy within an elegant framework. This method has also often been referred to as the "Speed Method". The use of "velocity" is more accurate and the addition of "virtual" is necessary to avoid confusion with the physical velocity in Continuum Mechanics problems.

The final computations of the "Shape gradient" are formally very similar to what is available in "Control Sensitivity Analysis". This has been at the origin of many formal or quick computations in the literature. For instance, illustrative examples can be found in the book of HAUG, CHOI and KOMKOV [1], the paper of DEMS and MROZ [1], the proceedings of the NATO-ASI by HAUG and CEA [1] and the recent paper on quick computations by J. CEA [1]. Yet, in many cases, no direct mathematical justification is available without a detailed study of the derivative of the state with respect to the domain variation. In this paper, we show how this critical step can be avoided by using theorems on the differentiability of a Min Max.

The study of the derivative of the state is avoided by introducing a Lagrangian formulation of the problem where the state equation is viewed as the constraint. Within that framework the cost function is equal to a Min Max over appropriate function spaces. By techniques which are specific to "Shape Analysis", the computation of the Shape Gradient can be reduced to the derivative of the Min Max with respect to a parameter $t \geq 0$ which plays the role of a "virtual time". The mathematical justification is then provided by using a theorem on the differentiability of a Min Max with respect to a real parameter. To our knowledge two such theorems are directly applicable to our problem. One which does not assume the existence of a

* This research was supported in part by the National Sciences and Engineering Council of Canada Operating Grant A-8730 and a FCAR Grant from the "Ministère de l'Education du Québec".

saddle point (cf. DELFOUR and ZOLESIO [1,2,3]) and one which takes advantage of the existence of a saddle point (cf. CORREA and SEEGER [1]).

Again this approach provides complete mathematical justification to most classical problems where the state is the minimizing element of a quadratic energy functional and the cost function is differentiable with respect to the state. It also extends to classes of convex non-differentiable cost functions and non-linear energy functionals. Some problems with state constraints can also be covered (cf. A. SOUISSI [1] for problems involving Stokes' equation with an equality constraint on the divergent of the velocity).

It is to be emphasized that the approach followed in this paper is not the only way to justify the Lagrangian formulation. A penalization approach has also been introduced by DELFOUR and ZOLESIO [4,5] to provide appropriate justification when the state is the solution of a non-linear equation or a variational inequality.

Notation. $\mathbb{R}$ will denote the field of real numbers, $\mathbb{R}^+$ the subset of positive or zero reals and $\mathbb{R}^n$ ($n \geq 1$, an integer) the n-fold Cartesian product of $\mathbb{R}$. The inner product and norm in $\mathbb{R}^n$ will be defined as

$$x \cdot y = \Sigma_{i=1,n}\, x_i\, y_i \qquad |x| = (x \cdot x)^{1/2}.$$

The dual operator of a continuous linear operator $A : X \to Y$ will be denoted by A^*. The identity matrix in R^n will be written I_d. The composition of two applications f and g will be denoted by $f \circ g$.

2. A SIMPLE ILLUSTRATIVE EXAMPLE

2.1. Problem formulation

Consider the following simple example. Let Ω be a bounded open domain in $\mathbb{R}^n$ with a smooth boundary Γ. Let $y = y(\Omega)$ be the solution of the variational problem

$$\mathrm{Inf}\{E(\Omega, \varphi) \;:\; \varphi \in H^1(\Omega)\} \tag{1}$$

where

$$E(\Omega, \varphi) = 1/2 \int_\Omega [|\nabla\varphi|^2 + |\varphi|^2 - 2\, f\, \varphi]\, dx \tag{2}$$

for some fixed function f in $H^1(\mathbb{R}^n)$. We associate with y a cost function

$$J(\Omega) = F(\Omega, y(\Omega)). \tag{3}$$

For instance we can choose the standard cost function

$$F(\Omega, y) = 1/2 \int_\Omega (y - Y_d)^2\, dx, \quad Y_d \in H^1(\mathbb{R}^n). \tag{4}$$

2.2. The Velocity Field Method

Recall the notion of a shape derivative. Let $V(t, x)$, $t \geq 0$, $x \in \mathbb{R}^n$, be a **Velocity Field of Deformation.** Under the action of V, the points of Ω are transported onto a new domain $\Omega_t = T_t(\Omega)$, where the transformation $T_t : \mathbb{R}^n \to \mathbb{R}^n$ is generated by the solutions of the equation

$$(\partial/\partial t)\, T_t(x) = V(t, T_t(x)), \quad t \geq 0, \quad T_0(x) = x \tag{5}$$

(cf. J.P. ZOLESIO [3]). Let y_t be the solution of problem (1) on the transformed domain Ω_t

$$\mathrm{Inf}\{E(\Omega_t, \varphi) \;:\; \varphi \in H^1(\Omega_t)\} \tag{1_t}$$

and associate with y_t the cost function

$$J(\Omega_t) = F(\Omega_t, y_t). \tag{3_t}$$

The Shape Derivative of J at Ω in the direction V is defined as

$$dJ(\Omega; V) = \lim_{t \to 0^+} [J(\Omega_t) - J(\Omega)]/t.$$

Traditional methods involve the computation of the **Shape Derivative** (or **Partial Derivative**) Y' or the **Material Derivative** y. The shape derivative is defined as

$$Y'(x) = \lim_{t \to 0^+} [Y(t,x) - Y(0,x)]/t \tag{6}$$

where for some $\tau > 0$ $Y(t,x)$ is an appropriate extension of $y_t(x)$ to $[0, \tau] \times D$ for some fixed domain D containing all perturbations Ω_t of Ω, $0 \le t \le \tau$. The material derivative is defined as

$$\dot{y} = \lim_{t \to 0+} [y^t - y]/t \tag{7}$$

in an appropriate function space on Ω, where y^t is the transported solution (from Ω_t to Ω)

$$y^t = y_t \circ T_t. \tag{8}$$

In classical examples Y' is the solution of a boundary value problem which depends on y and the normal component of the velocity field on the boundary Γ. However, in general, the material derivative is also the solution of a boundary value problem on Ω, but it depends on the velocity field in the whole domain. In fact the two derivatives are related through the formula

$$Y' = \dot{y} - \nabla y \bullet V \tag{9}$$

where V is the velocity field at 0, $x \to V(0,x)$. In general Y' is "rougher" than the material derivative.

The next step consists in differentiating $J(\Omega_t)$ using the material derivative or Y'. Then an appropriate adjoint variable p is introduced to eliminate those derivatives and obtain a final expression which depends on Ω, y, p and V. The adjoint variable p is the solution of a boundary value problem which is dual to the corresponding boundary value problem for the material derivative or Y'.

The final expression can then be used for Shape Sensitivity Analysis or as a necessary condition characterizing an eventual minimizing domain Ω^*.

2.3. A classical approach to the computation of the "Shape Gradient"

We turn to the simple example in section 6.1 to illustrate an approach which has been widely used and works well for classical sufficiently smooth problems.

The first step is the computation of the partial derivative Y' of the state y at time $t = 0$. To do that, consider the parametrized problem (1_t) - (3_t) for $t \ge 0$. Denote by y_t the solution of the minimization problem (1_t), where the energy functional is given by

$$E(\Omega_t, \varphi) = 1/2 \int_{\Omega_t} \{|\nabla\varphi|^2 + |\varphi|^2 - 2f\varphi\}\, dx \tag{10}$$

It is readily seen that y_t is the unique solution in $H^1(\Omega_t)$ to the variational equation

$$dE(\Omega_t, y_t; \psi) = 0 \;,\; \forall \psi \in H^1(\Omega_t)\,, \tag{11}$$

where

$$dE(\Omega_t, \varphi; \psi) = \int_{\Omega_t} \{\nabla\varphi \bullet \nabla\psi + \varphi\psi - f\psi\}\, dx\,. \tag{12}$$

Assume that $y_t(x)$, $x \in \Omega_t$, has an appropriately smooth extension $Y(t, x)$ to a large enough neighborhood of the domain Ω. Then the partial derivative Y' of Y at $t = 0$ is the unique solution of the variational equation

$$\int_\Omega \{\nabla Y' \bullet \nabla\psi + Y'\psi\}\, dx + \int_\Gamma \{\nabla y \bullet \nabla\psi + y\psi - f\psi\}\, V(0) \bullet n d\Gamma = 0 \tag{13}$$

for all ψ in $H^{3/2+\varepsilon}(\Omega)$, $\varepsilon > 0$, $V(0) = V(0,\cdot)$, n the outer unit normal to Ω and $y = y_0 = Y(0,\cdot)$ the unique solution in $H^1(\Omega)$ to the variational equation

$$\int_\Omega \{\nabla y \bullet \nabla\psi + y\psi - f\psi\}\, dx = 0 \;,\; \forall \psi \in H^1(\Omega)\,. \tag{14}$$

It is to be emphasized that y must be smooth enough to obtain the solution Y' of (13) in $H^1(\Omega)$. This illustrates the fact that Y' is "rougher" than the state y as indicated in the previous section.

The second step is the computation of the shape derivative

$$dJ(\Omega; V(0)) = \lim_{t \to 0^+} [J(\Omega_t) - J(\Omega)]/t \tag{15}$$

of the parametrized cost function

$$J(\Omega_t) = 1/2 \int_{\Omega_t} |y_t - Y_d|^2 \, dx \ , \ Y_d \in H^1(\mathbb{R}^n) . \tag{16}$$

Again by assuming the existence of a smooth extension $Y(t,\cdot)$ of y_t, we obtain

$$dJ(\Omega;V(0)) = \int_\Omega Y'(y - Y_d) \, dx + \int_\Gamma 1/2 \, |y - Y_d|^2 \, V(0) \cdot n \, d\Gamma \tag{17}$$

where Y' is the solution of the variational equation (13).

The last step is to explicitly show the dependence of $dJ(\Omega; V)$ on V and obtain an expression for the Shape Gradient. This requires the introduction of the **adjoint variable** p which is the solution in $H^1(\Omega)$ to the variational equation

$$\int_\Omega \{\nabla\varphi \cdot \nabla p + \varphi p\} \, dx + \int_\Omega \varphi(y - Y_d) \, dx = 0 \ , \ \forall \varphi \in H^1(\Omega) . \tag{18}$$

Substitute $\varphi = Y'$ in (18) and $\psi = p$ in (13) to obtain

$$\int_\Omega Y'(y - Y_d) \, dx = \int_\Gamma \{\nabla y \cdot \nabla p + yp - fp\} \, V(0) \cdot n \, d\Gamma \tag{19}$$

The substitution of the last identity in (16) finally yields

$$dJ(\Omega; V(0)) = \int_\Gamma \{1/2|y - Y_d|^2 + \nabla y \cdot \nabla p + yp - fp\} \, V(0) \cdot n \, d\Gamma \tag{20}$$

Notice that the above computations use the existence of a smooth extension Y of y and the hypothesis that Y' belongs to $H^1(\Omega)$. For instance, this approach would break down for the homogeneous Dirichlet problem. Yet the final expression (20) is only dependent on V and the solutions y and p to the variational equations (14) and (18). It is completely independent of Y' and the associated intermediary steps.

2.4. The Min Max formulation

A natural way to look at problem (1_t) - (3_t) would be to construct the Lagrangian functional

$$L(\Omega_t, \varphi, \psi) = F(\Omega_t, \varphi) + dE(\Omega_t, \varphi; \psi) \tag{21}$$

where φ and ψ belong to $H^1(\Omega_t)$ and

$$F(\Omega_t, \varphi) = 1/2 \int_{\Omega_t} |\varphi - Y_d|^2 \, dx \tag{22}$$

and notice that

$$J(\Omega_t) = \min_{\varphi \in H^1(\Omega_t)} \ \max_{\psi \in H^1(\Omega_t)} L(\Omega_t, \varphi, \psi) \tag{23}$$

since the Max is different from $+\infty$ only when φ is the solution of the variational equation (11). However this would now require a theorem on the differentiability of a Min Max with respect to a parameter $t \geq 0$ when the underlying spaces depend on t. This is probably one of the reasons why precise mathematical justifications to the Lagrangian approach have not appeared so far. A general theory which is applicable to the specific type of problems encountered in Shape Sensitivity Analysis would be extremely useful, but is difficult.

Fortunately by carefully rethinking the parametrization of the problem, it is possible to reformulate the Min Max problem over spaces which are independent of t. The key idea is the a priori parametrization of the spaces. To illustrate this point first consider the energy functional (10) defined on the space $H^1(\Omega_t)$. For appropriately smooth vector fields V, the transformation T_t, $t \geq 0$, is a diffeomorphism. As a result the space $H^1(\Omega_t)$ can be "parametrized" in the following way

$$H^1(\Omega_t) = \{\varphi \circ T_t^{-1} : \varphi \in H^1(\Omega_t)\} . \tag{24}$$

This naturally suggests to define a new energy functional

$$\underline{E}(t, \varphi) = E(\Omega_t, \varphi \circ T_t^{-1}) , \quad \varphi \in H^1(\Omega) , \tag{25}$$

on the fixed space $H^1(\Omega)$. It is not too difficult to check that the minimization problem

$$\text{Inf } \{\underline{E}(t, \varphi) : \varphi \in H^1(\Omega)\} \tag{26}$$

has a unique solution in $H^1(\Omega)$ which coincides with the transported solution from Ω_t to Ω,

$$y^t = y_t \circ T_t , \tag{27}$$

as defined in (8). Moreover

$$E(\Omega_t, y_t) = E(\Omega_t, (y_t \circ T_t) \circ T_t^{-1}) = \underline{E}(t, y^t) . \tag{28}$$

This new parametrization of the energy functional offers some interesting simplification after the change of variable from x_t to $x = T_t^{-1} x_t$

$$\begin{aligned} \underline{E}(t, \varphi) &= 1/2 \int_{\Omega_t} [|\nabla (\varphi \circ T_t^{-1})|^2 + |\varphi \circ T_t^{-1}|^2 - 2 f (\varphi \circ T_t^{-1})] \, dx \\ &= 1/2 \int_\Omega \{(A(t) \nabla\varphi) \bullet \nabla\varphi + [|\varphi|^2 - 2 (f \circ T_t) \varphi] J(t)\} \, dx , \end{aligned} \tag{29}$$

where DT_t is the Jacobian matrix associated with the transformation T_t,

$$J(t) = \det (DT_t) , \quad A(t) = J(t) \, ((DT_t)^{-1})^* \, (DT_t)^{-1} \tag{30}$$

and * denotes the transposed matrix.

If we want to work with $\underline{E}(t, \Omega)$ and y^t, we must also transform the functional F into a new functional on $H^1(\Omega)$

$$\varphi \to \underline{F}(t, \varphi) = F(\Omega_t, \varphi \circ T_t^{-1}) : H^1(\Omega) \to \mathbb{R} \tag{31}$$

As a result the cost function

$$J(\Omega_t) = F(\Omega_t, y_t) = F(\Omega_t, (y_t \circ T_t) \circ T_t^{-1}) = F(\Omega_t, y^t \circ T_t^{-1}) = \underline{F}(t, y^t) \tag{32}$$

Moreover the same change of variable as in (29) yields

$$\underline{F}(t, \varphi) = 1/2 \int_{\Omega_t} |\varphi \circ T_t^{-1} - Y_d|^2 \, dx = 1/2 \int_\Omega |\varphi - Y_d \circ T_t|^2 J(t) \, dx . \tag{33}$$

It is important to notice that we have not changed in any way the cost function $J(\Omega_t)$ in the above constructions. Moreover the new expressions for $\underline{E}$ and $\underline{F}$ do not involve integrals which depend on the parameter $t \geq 0$.

In this new formulation

$$J(\Omega_t) = \underline{F}(t, y^t) \tag{34}$$

where y^t is the unique minimizing element in $H^1(\Omega)$ of the energy functional $\underline{E}(t, \varphi)$. It is completely characterized by the variational equation

$$y^t \in H^1(\Omega), \; d\underline{E}(t, y^t; \varphi) = 0, \; \forall \varphi \in H^1(\Omega) . \tag{35}$$

We now introduce the new Lagrangian functional

$$\underline{L}(t, \varphi, \psi) = \underline{F}(t, \varphi) + d\underline{E}(t, \varphi; \psi), \; \varphi \in H^1(\Omega), \; \psi \in H^1(\Omega) \tag{36}$$

and it is readily seen that

$$J(\Omega_t) = \underset{\varphi \in H^1(\Omega)}{\text{Min}} \; \underset{\psi \in H^1(\Omega)}{\text{Max}} \; L(t, \varphi, \psi) \tag{37}$$

So we have succeeded in constructing a Min Max formulation where the spaces involved are independent of t. In addition the new functionals $\underline{F}$ and $\underline{E}$ given by (33) and (29) only contain integrals on the fixed domain Ω and the parameter t simply appears inside the integrals.

We shall see in the next section that the above construction combined with appropriate theorems on the differentiability of a Min Max with respect to a parameter provides a very convenient and powerful tool for Shape Sensitivity Analysis.

3. THEOREMS ON THE DIFFERENTIABILITY OF A MIN MAX WITH RESPECT TO A PARAMETER

This is not a new topic. It seems that the pioneering work in this area was done by V.F. DEM'JANOV [1] in 1968 in the finite dimensional case. Since then many versions of such theorems have appeared in the literature under various hypotheses motivated by specific applications. In this paper, our objective is not to review the literature, but to single out theorems which are well suited and applicable to our problem.

For our purpose we give two theorems: one which does not assume the existence of a saddle point to the Min Max problem (cf. DELFOUR and ZOLESIO [1,2,3]) and one which does (cf. CORREA and SEEGER [1]).

3.1. No saddle point hypothesis

Let $A \subset X$ and $B \subset Y$ be subsets of two topological spaces X and Y and let $\tau > 0$ be a real number. Given a map

$$G : [0, \tau] \times X \times Y \to R,$$

we consider the following functions:

$$H(t, x) = \operatorname{Sup}\{G(t, x, y) : y \in B\},\ t \in [0, \tau],\ x \in A \tag{1}$$

$$g(t) = \operatorname{Inf}\{H(t, x) : x \in A\},\ t \in [0, \tau]\ . \tag{2}$$

As a result

$$g(t) = \operatorname{Inf}\{\operatorname{Sup}[G(t, x, y) : y \in B] : x \in A\}\ . \tag{3}$$

We wish to show that under appropriate hypotheses the function g is differentiable at $t = 0$ from the right

$$\lim_{t \to 0^+} (g(t) - g(0))/t \text{ exists}\ . \tag{4}$$

We shall need for $t \geq 0$ the set

$$A(t) = \{ x \in A : g(t) = H(t, x)\} \tag{5}$$

and for x in A the set

$$B(t, x) = \{y \in B : H(t, x) = G(t, x, y)\} \tag{6}$$

There are four sets of hypotheses. The first one ensures that the Sup and the Inf problems have solutions.

H1 $\exists \tau > 0,\ \forall t,\ 0 \leq t \leq \tau,$

(i) $A(0) \neq \emptyset,\ \forall x_0 \in A(0),\ B(t, x_0) \neq \emptyset$

(ii) $A(t) \neq \emptyset,\ \forall x_t \in A(t),\ B(0, x_t) \neq \emptyset$.

The second one yields upper and lower bounds on the differential quotient.

H2 (i) $\forall x_0 \in A(0)$, the function

$s \to G(s, x_0, y)$

is differentiable in a neighborhood of $t = 0$ for all y in

$\cup\{B(t, x_0) : 0 \leq t \leq \tau\}$

(ii) $\forall t,\ 0 \leq t \leq \tau,\ \forall x_t \in A(t)$, the function

$s \to G(s, x_t, y)$

is differentiable in a neighborhood of $t = 0$ for all y in

$\cup\{B(0,x_t) : 0 \le t \le \tau\}$.

The third and fourth sets of hypotheses are essentially continuity hypotheses.

H 3 $\exists$ a topology τ_Y on Y such that for all $x_0 \in A(0)$,

(i) for all sequences $t_n \to 0$, $t_n > 0$, $\exists$ $y_0 \in B(0, x_0)$ and a subsequence of $\{t_n\}$, still denoted $\{t_n\}$, such that $\forall n$, $\exists$ $y_n \in B(t_n, x_0)$, and

$y_n \to y_0$ in the τ_Y-topology

(ii) the map

$(t, y) \to \partial_t G(t, x_0, y)$

is upper semi continuous in $\{0\} \times \cup\{B(t, x_0) : 0 \le t \le \tau\}$ in the τ_Y-topology.

Remark 3.1. H3(i) is the Kuratovsky hypothesis at $t = 0$ for the set-valued function $t \Rightarrow B(t, x_0)$. ♦

H 4 There exist topologies τ_X on X and τ_Y on Y such that

(i) for all sequences $t_n \to 0$, $t_n > 0$, $\exists$ $x_0 \in A(0)$, $\forall y_0 \in B(0, x_0)$, $\exists$ a subsequence of $\{t_n\}$, still denoted $\{t_n\}$, such that $\forall n$, $\exists$ $x_n \in A(t_n)$, $\exists$ $z_n \in B(0, x_n)$, such that

$x_n \to x_0$ in the τ_X-topology and $z_n \to y_0$ in the τ_Y-topology

(ii) the map $(t,x,y) \to \partial_t G(t,x,y)$ is lower semi continuous in $\{0\} \times \{ (x,y) : x \in A(0), y \in B(0, x)\}$ in the $\tau_X \times \tau_Y$-topology.

Remark 3.2. Hypothesis H4(i) is verified when the following two hypotheses are verified:

H4(i_1) $\exists$ a topology τ_X on X such that for all sequences $t_n \to 0$, $t_n > 0$, $\exists$ $x_0 \in A(0)$, $\exists$ a subsequence of $\{t_n\}$, still denoted $\{t_n\}$, and $\forall n$, $\exists$ $x_n \in A(t_n)$ such that

$x_n \to x_0$ in the τ_X-topology

H4(i_2) $\exists$ a topology τ_Y of Y for which the set-valued function $x \Rightarrow B(0,x)$ is lower semi continuous on A(0) in the sense of J.P. AUBIN [1, Déf. 9.4, p. 121] :

for all convergent sequences $x_n \to x_0$ in X and all z^* in $B(0, x_0)$, there exists a sequence $z_n^* \in B(0, x_n)$ such that $z_n^* \to z^*$ in the τ_Y-topology.

Hypothesis H4(i_1) is the Kuratovsky condition at $t = 0$ for the set-valued map $t \Rightarrow A(t)$. ♦

We now state our main result

THEOREM 3. Under hypotheses (H1) to (H4), we have

$$dg(0) = \lim_{t \to 0^+} (g(t) - g(0))/t = \operatorname*{Inf}_{x \in A(0)} \ \operatorname*{Sup}_{y \in B(0, x)} \partial_t G(0, x, y). \ ♦ \tag{7}$$

The set of hypotheses H1 to H4 generalizes the one used in DELFOUR and ZOLESIO [1,2]. For an extended discussion and more details the reader is referred to DELFOUR and ZOLESIO [3].

3.2. Saddle point hypothesis

First introduce the function

$$h(t) = \operatorname{Sup}\{ \operatorname{Inf}[G(t, x, y) : x \in A] : y \in B\} \tag{8}$$

the associated sets

$$B(t) = \{y \in B : h(t) = \operatorname{Inf}[G(t, x, y) : x \in A]\} \tag{9}$$

$$A(t, y) = \{\underline{x} \in A : \text{Inf}\,[G(t, x, y) \; : \; x \in A] = G(t, \underline{x}, y)\}, \; y \in Y \tag{10}$$

and the saddle points

$$S(t) = \{(x_t, y_t) \in A \times B : g(t) = G(t, x_t, y_t) = h(t)\}\,. \tag{11}$$

Then the following lemma is immediate.

LEMMA 3. If $S(t) \neq \emptyset$ for some $t \geq 0$, then

$$S(t) = A(t) \times B(t), \; A(t) \neq \emptyset, B(t) \neq \emptyset, \tag{12}$$

$$\forall\, x_t \in A(t), B(t, x_t) = B(t) \text{ and } \forall\, y_t \in B(t), A(t, y_t) = A(t)\,. \; \blacklozenge \tag{13}$$

THEOREM 5. (CORREA and SEEGER [1]) Assume that there exists $\tau > 0$ such that the following hypotheses are verified:

HH1 $S(t) \neq \emptyset, 0 \leq t \leq \tau$.

HH2 For all (x, y) in $\cup\{A(t) \; : \; 0 \leq t \leq \tau\} \times \cup\{B(t) \; : \; 0 \leq t \leq \tau\}$, the map $t \rightarrow G(t, x, y)$ is differentiable everywhere in $[0, \tau]$.

HH3 There exists a topology τ_X on X such that

(i) $\forall\, t_n \rightarrow 0, 0 \leq t_n \leq \tau, \exists\, x_0 \in A(0), \exists$ a subsequence of $\{t_n\}$, still denoted $\{t_n\}$, and $\forall\, n, \exists\, x_n \in A(t_n)$, such that $x_n \rightarrow x_0$ in the τ_X-topology

(ii) $\forall\, y \in \cup\{B(t) \; : \; 0 \leq t \leq \tau\}$,

$(t, x) \rightarrow \partial_t G(t, x, y)$

is lower semicontinuous at $\{0\} \times A(0)$ for the τ_X-topology.

HH4 There exists a topology τ_Y on Y such that

(i) $\forall\, t_n \rightarrow 0, \; 0 \leq t_n \leq \tau, \; \exists\, y_0 \in B(0)\, \exists$ a subsequence of $\{t_n\}$, still denoted $\{t_n\}$, and $\forall\, n, \exists\, y_n \in B(t_n)$, such that $y_n \rightarrow y_0$ in the τ_Y-topology

(ii) $\forall\, x \in \cup\{A(t) \; : \; 0 \leq t \leq \tau\}$,

$(t, y) \rightarrow \partial_t G(t, x, y)$

is upper semicontinuous at $\{0\} \times B(0)$ for the τ_Y-topology.

Then

$$dg(0) = \lim_{t \rightarrow 0^+} (g(t) - g(0))/t = \operatorname*{Inf}_{x \in A(0)} \; \operatorname*{Sup}_{y \in B(0)} \partial_t G(0, x, y) = \operatorname*{Sup}_{x \in B(0)} \; \operatorname*{Inf}_{y \in A(0)} \partial_t G(0, x, y). \; \blacklozenge \tag{14}$$

The hypotheses are essentially the ones given in CORREA and SEEGER [1] except for the fact that we have used Kuratovsky's conditions HH3(i) and HH4(i) for the set-valued maps instead of the notion of **sequential semicontinuity** for a set-valued map $t \Rightarrow M(t)$ from $[0, \tau]$ to X:

$$\exists \text{ a topology } \tau_X \text{ on } X \text{ such that for all sequences } t_n \rightarrow 0, t_n > 0, \exists\, x_0 \in M(0)$$
$$\text{and } \forall n, \exists\, x_n \in M(t_n) \text{ such that } x_n \rightarrow x_0 \text{ in the } \tau_X\text{-topology}. \tag{15}$$

3.3. Application to the example of section 2.4.

We go back to the Min Max formulation (2.37) where the functionals $\underline{E}$ and $\underline{\underline{E}}$ are given by (2.33) and (2.29). We use the Lagrangian (2.36) and notice that for $t \geq 0$ in a neighborhood of 0 it has a unique saddle point $(y^t, p^t) \in H^1(\Omega) \times H^1(\Omega)$ solution of the saddle point equations:

$$\int_\Omega \{A(t)\, \nabla y^t \cdot \nabla \varphi + J(t)\, [y^t \varphi - (f \circ T_t)\, \varphi]\}\, dx = 0, \forall \varphi \in H^1(\Omega)\,, \tag{16}$$

$$\int_\Omega (y^t - Y_d \circ T_t)\, \psi\, dx + \int_\Omega \{(A(t)\, \nabla \psi \cdot \nabla p^t + J(t)\, \psi p^t\}\, dx = 0, \forall \psi \in H^1(\Omega)\,. \tag{17}$$

So we are in the situation of Theorem 2 with

$$A(t) = \{y^t\} \quad , \quad B(t) = \{p^t\}\,. \tag{18}$$

Hypothesis HH1 is clearly verified. Hypotheses HH3(i) and HH4(i) follow by continuity of y^t and p^t with respect to t at $t = 0$. In fact all we need to show is that they are bounded for $t \geq 0$ in a neighborhood of 0 and that there exist subsequences which converge weakly to y and p. As for Hypotheses HH2, HH3(ii) and HH4(ii) they are verified for sufficiently smooth vector fields V. More precisely

$$\partial_t L(t,\varphi,\psi) = \int_\Omega \{A'(0)\,\nabla\varphi \cdot \nabla\psi + \operatorname{div} V(0)\,\varphi\psi - \psi\,[\nabla f \cdot V(0) + f \operatorname{div} V(0)]\}\,dx$$
$$+ 1/2 \int_\Omega [-(\varphi - Y_d)\,\nabla Y_d \cdot V(0) + (\varphi - Y_d)^2 \operatorname{div} V(0)]\,dx \tag{19}$$

where

$$A'(0) = \operatorname{div} V(0)\; I_d - [DV(0) + DV(0)^*], \tag{20}$$

I_d is the identity matrix and $DV(0)^*$ is the transposed matrix of $DV(0)$. By direct application of Theorem 2 we get

$$dJ(\Omega; V(0)) = \int_\Omega \{A'(0)\,\nabla p \cdot \nabla y + \operatorname{div} V(0)\,py - p\,[\nabla f \cdot V(0) + f \operatorname{div} V(0)]\}\,dx$$
$$+ 1/2 \int_\Omega [-(y - Y_d)\,\nabla Y_d \cdot V(0) + (y - Y_d)^2 \operatorname{div}(V(0)]\,dx, \tag{21}$$

where y and p are the solutions of (14) and (18) or equivalently

$$-\Delta y + y = 0 \text{ in } \Omega, \quad \partial y/\partial n = 0 \text{ on } \Gamma \tag{22}$$

$$-\Delta p + p + y - Y_d = 0 \text{ in } \Omega, \quad \partial p/\partial n = 0 \text{ on } \Gamma. \tag{23}$$

When y and p belong to $H^2(\Omega)$ expression (21) reduces to the boundary integral (2.20) obtained by the classical approach. To see that set $\varphi = \nabla y \circ V$ in (2.18) and $\psi = \nabla p \circ V$ in (2.14), add the two equations and rearrange and integrate by parts the appropriate terms. The result is an identity between the right-hand-sides of (2.20) and (21).

Remark 3.3. The above results can also be obtained from Theorem 1 by using a special construction (cf. DELFOUR and ZOLESIO [1,2]).

3.4. Application to the homogeneous Dirichlet problem

Recall that in the simple example the classical approach was difficult to apply since it was necessary to define a smooth extension of the solution on Ω. This becomes even more difficult when the boundary condition is changed from a Neumann to a homogeneous Dirichlet condition

$$-\Delta y + y = f \text{ in } \Omega \qquad y = 0 \text{ on } \Gamma \tag{24}$$

with the same objective of minimizing the cost function (2.3). With the Min Max approach it suffices to change the space $H^1(\Omega)$ to $H_0^{\,1}(\Omega)$ and the final expression for the "Shape Gradient" is precisely (21) where y and p are now the solutions of

$$-\Delta y + y = f \text{ in } \Omega, \quad y = 0 \text{ on } \Gamma, \tag{25}$$

$$-\Delta p + p + y - Y_d = 0 \text{ in } \Omega, \quad p = 0 \text{ on } \Gamma. \tag{26}$$

3.5. Application to non-differentiable cost functions

When the function $F(\Omega,\varphi)$ associated with the cost function

$$J(\Omega) = F(\Omega, y_\Omega) \tag{27}$$

is non-differentiable with respect to φ but can be expressed as the Sup of a differentiable functional $F^*(\Omega,\varphi,\mu)$

$$F(\Omega, \varphi) = \operatorname{Sup}\{F^*(\Omega, \varphi, \mu) : \mu \in M\} \tag{28}$$

our techniques can readily be applied.

To illustrate this point, go back to the Neumann problem

$$-\Delta y + y = f \text{ in } \Omega, \quad \partial y/\partial n = 0 \text{ on } \Gamma \tag{29}$$

and choose the cost function

$$J(\Omega) = F(\Omega, y) \quad , \quad F(\Omega, \varphi) = \int_\Omega |y - Y_d|\, dx . \tag{30}$$

Here the non-differentiable functional can be expressed as a Sup

$$F(\Omega, \varphi) = \text{Sup}\, \{\int_\Omega \mu(y - Y_d)\, dx \ : \ \mu \in M\} \tag{31}$$

over the subset

$$M = \{\alpha \in L^2(\Omega) \ : \ |\, \alpha(x)\, | \leq 1 \quad , \quad \text{a.e. in } \Omega\} . \tag{32}$$

As a result

$$J(\Omega) = \text{Inf}\, \{\text{Sup}\, [L(\Omega, \varphi, (\mu, \psi)) \ : \ (\mu, \psi) \in M \times H^1(\Omega)] \ : \ \varphi \in H^1(\Omega)\} \tag{33}$$

where the Lagrangian functional is given by

$$L(\Omega, \varphi, (\mu, \psi)) = \int_\Omega [\mu(\varphi - Y_d) + \nabla\varphi \cdot \nabla\psi + \varphi\psi - f\psi]\, dx . \tag{34}$$

It is readily seen that the saddle point of L is achieved at the points $(y, (\mu,p))$ which are completely characterized by the following set of equations

$$-\Delta y + y = f \text{ in } \Omega \quad , \quad \partial y/\partial n = 0 \text{ on } \Gamma \tag{35}$$

$$\mu_\alpha = \text{sgn}\, y + \alpha\chi_{\Omega_0} , \quad \forall \alpha \in M \quad , \quad \Omega_0 = \{x \in \Omega \ : \ y(x) = 0\} \tag{36}$$

$$-\Delta p_\alpha + p_\alpha + \mu_\alpha = 0 \text{ in } \Omega \quad , \quad \partial p_\alpha/\partial n = 0 \text{ on } \Gamma . \tag{37}$$

In this example the saddle point is not unique, but the previous constructions and theory apply. The Shape Gradient is

$$dJ(\Omega; V(0)) = \underset{\alpha \in M}{\text{Sup}} \int_\Omega \{\mu_\alpha\, y \,\text{div}\, V(0) + A'(0)\, \nabla y \cdot \nabla p_\alpha + \text{div}\, V(0)\, y p_\alpha - [\text{div}\, V(0)\, f + \nabla f \cdot V(0)]\, p_\alpha\}\, dx \tag{38}$$

3.6. Application to problems with state constraint

Another class of problems which can be handled by the techniques presented in this paper are problems with state constraints. For instance the recent thesis by A. SOUSSI [1] contains many examples of Shape Optimal Design Problems for hydraulic turbines and diffuser where he systematically uses the Velocity Field Method to obtain Shape Gradient when the state equations are Stokes or Navier-Stokes equations.

For instance in the velocity-pressure expression of Stokes equations, the divergent of the velocity is zero. Using this as a constraint it is possible to show that the velocity u and the pressure p are solutions of the mixed formulation

$$\begin{aligned} & a(u, v) + b(v, p) = 0 \quad , \quad \forall v \in V_0 \\ & b(u, q) = 0 \quad , \quad \forall q \in Q \end{aligned} \tag{39}$$

+ boundary conditions on u

for appropriate functions space V_0 and Q and bilinear functionals a and b. Define the augmented energy functional

$$E(\Omega, v, q) = 1/2\, a(v, v) + b(v, q) \tag{40}$$

and the pair (u,p) is the solution of

$$E(\Omega, u, p) = \text{Inf}\, \{\text{Sup}\, [E(\Omega, v, q) \ : \ q \in Q] \ : \ v \in V\} \tag{41}$$

for some appropriate function space V. With the above formulation, A. SOUISSI [1] did the Shape Sensitivity Analysis of the efficiency of a hydraulic diffuser.

The technique of augmenting the state to handle state constraint equalities is quite general and natural in our framework. Another simple example is the non-homogeneous Dirichlet problem:

$$-\Delta y + y = f \text{ in } \Omega \quad , \quad y = g \text{ on } \Gamma \quad , \quad g \in H^{1/2}(\Gamma) , \tag{42}$$

which is equivalent to the Min Max problem

$$\mathrm{Min}\,\{\mathrm{Max}\,[E(\Omega, \varphi, \mu) \;:\; \mu \in H^{-1/2}(\Gamma)] \;:\; \varphi \in H^1(\Omega)\} = E(\Omega, y, \lambda) \tag{43}$$

where

$$E(\Omega, \varphi, \mu) = \int_\Omega 1/2\, [\, |\nabla\varphi|^2 + \varphi^2 - 2f\varphi]\, dx + \langle \mu, g - \varphi|_\Gamma \rangle_{H^{1/2}(\Gamma)}\,. \tag{44}$$

It is easy to show that y is the solution of (42) and that the multiplier λ is the normal derivative of y.

4. CONCLUSIONS

The techniques described in this paper combined with the use of theorems on the differentiability of a Min Max with respect to a parameter provide a powerful tool for Shape Sensitivity Analysis without the study and characterization of the Shape or Material derivative. This also provides a precise mathematical justification to many existing results in the literature. In addition classes of non-differentiable cost functions and some types of state constraint equalities can be handled.

As stated in the introduction the Min Max formulation is not the only one to justify a Lagrangian construction. Recently DELFOUR and ZOLESIO [4,5] used a penalization technique to handle non-linear state equations and some situations where the state is the solution of a variational inequality. Finally all this is not limited to state problems with an associated energy functional. The same techniques are applicable to evolution problems where the state is the solution of a parabolic or an hyperbolic equation.

REFERENCES

J.P. AUBIN [1], L'analyse non linéaire et ses motivations économiques, Masson, Paris, New-York, 1984.

N.V. BANICHUK [1], Optimization of the Shapes of Elastic Bodies, Nauka, Moscow, 1980 (Engl. Transl. 1984).

J. CEA [1], Conception optimale ou identification de formes. Calcul rapide de la dérivée directionnelle de la fonction coût, R.A.I.R.O. 20 (1986), 371-402.

[2], Problems of Shape Optimal Design, in "Optimization of Distributed Parameter Structures", Vol. II, pp. 1005-1048, Sijthoff and Noordhoff, Alphen aan den Rijn, Netherlands 1980.

[3], Numerical Methods of Shape Optimal Design, in "Optimization of distributed parameter structures", E.J. Haug and J. Céa, eds.,pp. 1049-1087, Sijthoff and Noordhoff, Alphen aan den Rijn, Netherlands 1980.

R. CORREA and A. SEEGER [1], Directional derivatives of a minimax function, Nonlinear Analysis, Theory, Methods and Applications 9 (1985), 13-22.

M.C. DELFOUR and J.P. ZOLESIO[1], Dérivation d'un Min Max et application à la dérivationpar rapport au contrôle d'une observation non différentiable de l'état, C.R. Acad. Sc. Paris 302, Sér. I (1986), 571-574.

[2], Differentiability of a Min Max and Application to Optimal Control and Design Problems. Parts I and II., in Proc. IFIP WG 7.2 on "Control Problems for Systems described by Partial Differential Equations", I. Lasiecka, ed., Gainesville, Fla, Feb. 1986, Springer-Verlag, New York, to appear.

[3], Shape Sensitivity Analysis via Min Max Differentiability, SIAM J. on Control and Optimization, to appear.

[4], Further developments in shape sensitivity analysis via a penalization method, in "Boundary Control and Bondary Variations", J.P. Zolésio, ed., Springer, Verlag, New York, 1987.

[5], Shape sensitivity analysis via a penalization method, Annali di Matematica Pura i Applicata, to appear.

K. DEMS and Z. MROZ [1], Variational approach by means of adjoint systems to structural optimization and sensitivity analysis, Part 2. Structure shape variations, Int. J. of Solids and Structures 20 (1984), 527-552.

V.F. DEM'YANOV [1], Differentiability of a Maximin function. I, USSR Comp. Math. and Math.Phys. 8 (1968), 1-15 (transl from Z. Vychisl. Mat. i Mat. Fiz. 8 (1968), 1186-1195).

I. EKELAND and R. TEMAM [1], Analyse convexe et problèmes variationnels, Dunod,Gauthiers-Villars, Paris, Bruxelles, Montréal, 1974.

R.H. GALLAGHER and O.C. ZIENKIEWICZ [1], Optimum structural design theory and applications, John Wiley and Sons, 1972; Springer-Verlag, Berlin 1973.

E.J. HAUG [1], A review of Distributed Parameter Structural Optimization Literature, in"Optimization of Distributed Parameter Structures", Vol. I, pp. 3-74, (cf. Haug and Céa [1]).

E.J. HAUG and J.S. ARORA [1], Applied Optimal Design, Wiley-Interscience, New York,1979.

E.J. HAUG and J. CEA [1], Optimization of Distributed Parameter Structures, Vol. I and II,Sijthoff and Noordhoff, Alphen aan den Rijn, The Netherlands, Rockville, Maryland, USA, 1981.

E.J. HAUG , K.K. CHOI and V. KOMKOV[1], Design Sensitivity Analysis of Structural Systems, Academic Press, New York, 1986.

A. MYSLINSKI and J. SOKOLOWSKI [1], Nondifferentiable optimization problems for elliptic problems, SIAM J. Control and Optimization 23 (1984), 632-648.

O. PIRONNEAU [1], Optimal Design for Elliptic Systems, Springer-Verlag, New York, 1984.

W. PRAGER [1], Introduction to Structural Optimization, Courses and Lectures: International Centre for Mechanical Sciences, Udine, No 212, Springer-Verlag, Vienna, 1974.

G.I.N. ROZVANY [1], Optimal Design of Flexural Systems, Pergamon Press, New York,1976.

A. SAWCZUK and Z. MROZ [1], Optimization in Structural Design, Springer-Verlag, New York, 1975.

J. SOKOLOWSKI [1], Optimal control in coefficients of boundary value problems with unilateral constraints, Bull. Pol. Acad. Sc. Tech. Sc. 31 (1983), 71-81.

J. SOKOLOWSKI and J.P. ZOLESIO [1], Dérivée par rapport au domaine de la solution d'un problème unilatéral, C.R. Acad. Sc. Paris 301 (1985), Sér. I, pp. 103-106.

[2]. Shape sensitivity analysis of an elasto-plastic torsion problem, Bull. Pol. Acad. Sc. Tech. Sc. 33 (1985), 579-586.

A. SOUISSI [1], Quelques problèmes d'optimisation de formes en hydrodynamique, Ph.D. Thesis, Université Laval, Québec (Québec), Canada, Décembre 1986.

J.P. ZOLESIO [1], Identification de domaine, Thèse de doctorat d'état, Nice, 1979.

[2], Semi-Derivatives of Repeated Eigenvalues, in "Optimization of Distributed Parameter Structures", Vol. II, pp. 1457-1473 (cf. HAUG and Cea [1]).

[3], An optimal design procedure for optimal control support, in "Convex Analysis and its Applications", A. Auslender, ed., pp.207-219, Springer-Verlag, Berlin, Heidelberg ,New-York, 1977.

OPTIMAL DESIGN IN FLUID MECHANICS.
THE FINITE ELEMENT APPROACH.

E. Fernández Cara
University of Seville,
C/ Tarfia s/n, 41012 SEVILLA (SPAIN).

1. INTRODUCTION.

Optimum design problems are natural in fluid mechanics: for instance, it may be important to know whether a given profile minimizes drag or simply yields "good" velocities, etc ... Unfortunately, the numerical computation of optimum profiles is very difficult.

The reasons are threefold. First, one often deals with complicate nondifferentiable criteria - such as the position of the forward transition point. Secondly, for "real" problems, the state equations are strongly nonlinear, with coefficients depending upon small parameters, such as the kinematic viscosity. This frequently gives rise to a complex composite structure for the solution. Finally, a large number of variables is usually involved; consequently, unless the descent algorithm is chosen carefully, the practical solution will need much work.

In the sequel, I shall present the rigorous formulation of some particular optimal design problems arising in fluid mechanics. More precisely, we will be concerned with the search of an "optimal profile" around which a given "real" or viscous fluid must flow (see Fig. 1). The fluid will be assumed to fill a given large domain $\mathcal{O}$ (an approximation to $\mathbb{R}^N$ with N = 2 or 3) and possess uniform velocity $\vec{u}_\infty$ near infinity. It will be my aim:

a) To establish - at least in some particular cases - an existence result and
b) To show that the FEM approach - which reduces the task to the solution of an optimal control problem in $\mathbb{R}^n$ - can be helpful for the design.

2. THE FORMULATION OF AN OPTIMAL DESIGN PROBLEM RELATED TO THE SEARCH OF AN OPTIMAL PROFILE.

For the formulation of a problem of the kind above, it will be necessary to settle without ambiguity:

i) The family $\mathcal{S}$ of "admissible" (and physically relevant) profiles,
ii) The regime of the flow (incompressible or not, stationary or not,...). This will be accomplished by a mapping

$$S \to \vec{u}(S), p(S), \ldots \qquad (2.1)$$

This determines, for each profile S in $\mathcal{S}$, the corresponding velocity field $\vec{u}(S)$, pressure distribution p(S) and, eventually, other thermodynamical variables, such as density or temperature, necessary for the description of the flow. From the viewpoint of control theory, $(\vec{u}(S),p(S),\ldots)$ is the state associated to the control S ; of course, it will be obtained by solving the so-called state equation.

iii) The meaning of optimality. As usual, a profile S will be considered optimal if it minimizes a given cost function $J = J(S;\vec{u}(S),p(S),\ldots)$.

Concerning the choice of the family $\mathcal{S}$, from now on it will be assumed that it has already been fixed as a subset of $\mathcal{S}_0$, with

$$\mathcal{S}_0 = \{ S \mid S=\bar{B} \text{ for an open set B satisfying } \emptyset \neq D_0 \subset B \subset D_1 \subset\subset \mathcal{O}, \; \partial S \text{ is locally Lipschitz-continuous} \tag{2.2}$$

Here, D_0 and D_1 are two open sets (see Fig. 2). Such an assumption is expected to be sufficient to ensure well-posedness for the state equation and make $J(S;\ldots)$ meaningful.

A key point in the formulation concerns the choice of the mapping (2.1) or, equivalently, the choice of the state equation. For simplicity, it will be assumed that the mechanical and thermodynamical characteristics of the flow only depend on two adimensional parameters: the Reynolds number Re and the Mach number at infinity M_∞. Many flows considered for real problems have this feature (see |1|) If the components u,p,... of the state variable have been adimensionalized, Re is the reciprocal of the kinematic viscosity ν and M_∞ is the reciprocal of the speed of propagation of sound at infinity. Once S is fixed, $\vec{u}$, p, the density ρ and the temperature θ are given by a solution of the compressible Navier-Stokes pde's:

$$\rho_{,t} + \nabla\cdot(\rho\,\vec{u}) = 0 \tag{2.3}$$

$$\rho\,(\vec{u}_{,t} + (\vec{u}\cdot\nabla)\vec{u}) + \nabla p = \nabla\cdot\underset{\sim}{\tau}\,, \quad \underset{\sim}{\tau} = \frac{1}{Re}\left(\underset{\sim}{\nabla\vec{u}} + \underset{\sim}{\nabla\vec{u}^T} - \frac{2}{3}(\nabla\cdot\vec{u})\,\underset{\sim}{Id.} \right) \tag{2.4}$$

$$\rho\,(\theta_{,t} + u\cdot\nabla\theta) = \ldots \quad \text{(the energy equation in terms of } \theta\text{)} \tag{2.5}$$

$$p = R\rho\theta \qquad (\text{ R is a positive constant }) \tag{2.6}$$

Clearly, (2.3)-(2.6) must be completed with a set of initial and boundary conditions. For simplicity, the following will be imposed:

$$\begin{bmatrix} \rho \\ \vec{u} \\ \theta \end{bmatrix}_{t=0} = \begin{bmatrix} \rho_0 \\ \vec{u}_0 \\ \theta_0 \end{bmatrix} \tag{2.7}$$

$$\rho|_{\Gamma_\infty^-} = \rho_\infty \quad \text{with } \Gamma_\infty^- = \{ x \mid x \in \partial\mathcal{O},\ \vec{u}\cdot\vec{n} < 0 \} \tag{2.8}$$

$$\vec{u}|_{\Gamma_\infty^-} = \vec{u}_\infty\,, \quad \vec{u}|_{\partial S} = 0 \tag{2.9}$$

$$\theta|_{\Gamma_\infty^-} = \theta_\infty\,, \quad \theta|_{\partial S} = \theta_* \tag{2.10}$$

In (2.7)-(2.10), the functions ρ_o , u_o and θ_o , the positive constants $\rho_\infty, \theta_\infty, \theta_*$ and the vector $\vec{u}_\infty$ are given.

When M_∞ is near zero (see |2|), the flow can be considered incompressible. This essentially means that

$$\nabla \cdot \vec{u} = 0 \quad \text{in } \mathcal{O} \setminus S \tag{2.11}$$

As a consequence, if the fluid is homogeneous, the density must be a constant whose value is known. This leads to the important fact that the variables $\vec{u}$ and p can be obtained from (2.4) and (2.11) separately from θ. Thus, they are given by a solution of the incompressible Navier-Stokes equations ($\rho_o = 1$ for simplicity):

$$\vec{u}_{,t} + (\vec{u} \cdot \nabla)\vec{u} + \nabla p = \nabla \cdot \underset{\sim}{\tau} = \frac{1}{Re} \Delta \vec{u} \tag{2.12}$$

$$\nabla \cdot \vec{u} = 0 \tag{2.13}$$

This gives a good approximation to (2.3)-(2.4) when M_∞ is near zero. This is no longer true for larger values of M_∞.

On the other hand, for low or moderate Re, the flow can be considered laminar, while high values of Re give rise to a turbulent (in particular, extremely irregular) motion of he fluid.

When Re is low ($Re < Re_* \simeq 10^3$-10^5) and $M_\infty \simeq 0$, it is natural to assume that the flow is time-independent or stationary, because a quasi-stationary state is reached after a small time interval. Hence, $\vec{u}$ and p satisfy:

$$-\frac{1}{Re} \Delta \vec{u} + (\vec{u} \cdot \nabla)\vec{u} + \nabla p = 0 \tag{2.14}$$

$$\nabla \cdot \vec{u} = 0 \tag{2.15}$$

Furthermore, for very low Re and $M_\infty \simeq 0$, the inertia terms $(\vec{u} \cdot \nabla) u_i$ are much smaller than the other terms in (2.14) and can be completely neglected. This leads to the Stokes equations

$$-\frac{1}{Re} \Delta \vec{u} + \nabla p = 0 \tag{2.16}$$

$$\nabla \cdot \vec{u} = 0 \tag{2.17}$$

On the contrary, for the highest values of Re occurring in real problems, it is not expectable to solve numerically either the compressible or incompressible Navier-Stokes equations accurately (see | 3 |). Due to turbulence effects, one is forced to work with averaged quantities, such as for example

$$\bar{\vec{u}} = \frac{1}{T} \int_0^T \vec{u}(x,t)\, dt \ , \quad p = \frac{1}{T} \int_0^T p(x,t)\, dt \ , \quad \text{etc...}$$

(with $[0,T]$ being the time interval of observation).

For instance, when M_∞ is small, time-averaging of (2.14)-(2.15) leads to the incompressible Reynolds equations:

$$(\bar{u}\cdot\nabla)\bar{u} + \nabla p = \nabla\cdot(\underset{\sim}{\tau}_\ell + \underset{\sim}{\tau}_t) \tag{2.18}$$

$$\nabla\cdot\bar{u} = 0 \tag{2.19}$$

Here, $\underset{\sim}{\tau}_\ell$ and $\underset{\sim}{\tau}_t$ are the time-averaged laminar and Reynolds stress tensors, resp. given by

$$\underset{\sim}{\tau}_\ell = \frac{1}{Re}\,(\underset{\sim}{\nabla\bar{u}} + \underset{\sim}{\nabla\bar{u}^T}) \tag{2.20}$$

$$\underset{\sim}{\tau}_t = -\overline{\underset{\sim}{u'\otimes u'}}\ , \qquad u' = \vec{u} - \bar{u} \tag{2.21}$$

To close the system (2.18)-(2.21), one needs additional hypotheses on the flow, i.e. a relationship between $\underset{\sim}{\tau}_t$ and $\bar{u}$. Even in the simplest models, this relationship is rather complicate. The most usual is given by Boussinesq's hypothesis:

$$\underset{\sim}{\tau}_t = \nu_t(\bar{u}, |\underset{\sim}{\nabla\bar{u}} + \underset{\sim}{\nabla\bar{u}^T}|)\,(\underset{\sim}{\nabla\bar{u}} + \underset{\sim}{\nabla\bar{u}^T}) \tag{2.22}$$

In (2.22), ν_t is a strongly nonlinear function of $\bar{u}$ and $|\underset{\sim}{\nabla\bar{u}} + \underset{\sim}{\nabla\bar{u}^T}|$ (the eddy viscosity of the fluid; here, nonlinearities are not local) whose behavior is essentially linear with $|\underset{\sim}{\nabla\bar{u}} + \underset{\sim}{\nabla\bar{u}^T}|$ near ∂S. There are many possibilities for the choice of ν_t in (2.22) (in many models,the dependance above is made through new variables, such as the turbulent kinematic energy $1/2\ \overline{u'\cdot u'}$; recent reviews of turbulent models are |3-6|).

In a similar way, time-averaging can be used to deduce the compressible Reynolds equations, which govern the motion of the fluid when Re and M_∞ are large. Frequently, the unknowns are $\bar{p}$, $\bar{\rho}$ and the mass-averaged quantities

$$\tilde{u} = \overline{\rho u}/\bar{\rho}\ , \quad \check{\theta} = \overline{\rho\theta}/\bar{\rho}\ .$$

Along this discussion, use has been made of the critical Reynolds numbers Re_* and Re_{**}. Many experiments confirm the fact that below Re_* the flow is never turbulent, while laminar flow never emerges above Re_{**}. Otherwise, when $Re_* < Re < Re_{**}$, there is the transition regime. For transitional Reynolds number, even the choice of an appropriate state equation is a very complicate task.

Unfortunately, for the kind of problems stated above, we do not possess, generally speaking, satisfactory existence/uniqueness results. For instance, it is not clear whether (2.12)-(2.13), together with an initial condition and the boundary conditions (2.9), possesses more than one solution when N = 3 (cf. |7|).

The third and last thing we have to do is to fix the cost function J. Again, there are many possibilities, each one corresponding to a specific meaning of the word optimality. A first example is given by the distance to a desired state:

$$J = \frac{a}{2}\int_{\mathcal{O}\setminus S}|\vec{u}-\vec{u}_d|^2\,dx + \frac{b}{2}\int_{\mathcal{O}\setminus S}|\underset{\sim}{\nabla\vec{u}} - \underset{\sim}{\nabla\vec{u}_d}|^2\,dx + \frac{c}{2}\int_{\mathcal{O}\setminus S}|\rho-\rho_d|^2\,dx + \frac{d}{2}\int_{\mathcal{O}\setminus S}|\theta-\theta_d|^2\,dx + \ldots \tag{2.23}$$

where the nonnegative parameters a,b,c, ... and the functions u_d, ρ_d, θ_d, ... are prescribed. Clearly, when dealing with instationary state equations, it is appropriate to integrate also with respect to time in (2.23).

A more interesting J is provided by the dissipation energy. With (2.14)-(2.15), this is given (up to a coefficient) by

$$J = Re \int_{O\setminus S} \tau_{ij}\tau_{ij}\, dx = \frac{1}{Re} \int_{O\setminus S} \left(\frac{\partial u_i}{\partial x_j} + \frac{\partial u_j}{\partial x_i}\right)\left(\frac{\partial u_i}{\partial x_j} + \frac{\partial u_j}{\partial x_i}\right) dx \qquad (2.24)$$

In fact, (2.24) can be used as an approximation of a very realistic cost function: The drag, i.e. the $\vec{u}_\infty$-component of the force exerted on the body (see | 8 |). Actually, the drag is given by (recall that $\vec{u}_\infty$ is a unit vector):

$$-\left(\int_{\partial S} (-p\, \underset{\sim}{Id}. + \underset{\sim}{\tau})\cdot \vec{n}\, d\Gamma \right)\cdot \vec{u}_\infty \qquad (2.25)$$

Let us finally indicate a particularly complicate example of cost function. When Re is large, it has been said that the flow becomes turbulent. Frequently, turbulence is generated near S, due to the roughness of ∂S. As a consequence, the closest layer is initially laminar in the streamwise direction. For mechanical and thermodynamical reasons, it is often interesting to reduce the <u>turbulent layer</u> (from the transition point P_{tr} to the trailing edge in Fig. 2). So we set

$$J = \text{dist}\,(S;P_{tr},P_{te}) \qquad (2.26)$$

Here, for given $P,Q \in \partial S$, dist(S;P,Q) stands for the "distance along ∂S" from P to Q. In (2.26), P_{te} is fixed (and may be chosen independently from S), while P_{tr} is usually determined by an empirical equation

$$G(S;u,p,\dots;P_{tr}) = G_* \qquad (2.27)$$

(the chice of G is itself a nontrivial problem; in general, (2.27) is complicate).

We have thus seen that the optimal design problems we are interested in may be difficult to solve due to:

1. Possible nonexistence/nonuniqueness and complexity of the state equation,
2. Possible complexity of the objective function (notice that e.g. (2.26) is nondifferentiable).

3. AN EXISTENCE RESULT.

A design problem of the kind above is determined by fixing $\mathcal{S} \subset \mathcal{S}_o$, the state equation and J. In order to obtain existence, one needs a distance d(• ,•) on $\mathcal{S}_o$ leading to, roughly speaking, <u>compactness</u> of $\mathcal{S}$ and <u>lower-semicontinuity</u> of J. To fis ideas, let us strict ourselves to the following problem:

$$\begin{aligned} &\underline{\text{Minimize}}\ J(S;\vec{u},p) \\ &\underline{\text{Subject to}}\ S \in \mathcal{S},\ (\vec{u},p) \in \mathcal{U}(S) \end{aligned} \qquad (3.1)$$

Here, $J(S;\vec{u},p)$ is "the drag experienced by S, $\vec{u}$ and p" (to be defined rigorously),

$\mathcal{S} \subset \mathcal{S}_o$ with $\mathcal{S}_o$ given by (2.2) and $\mathcal{U}(S)$ is, for fixed S, the set of all solutions $(\vec{u},p) \in H^1(\mathcal{O} \setminus S)^N \times L^2_o(\mathcal{O} \setminus S)$ of (2.12)-(2.13) in the usual weak sense:

$$\frac{Re}{4} \int_{\mathcal{O} \setminus S} \tau_{ij}(\vec{u})\, \tau_{ij}(\vec{w})\, dx + \int_{\mathcal{O} \setminus S} (\vec{u}\cdot\nabla)\vec{u}\cdot\vec{w}\, dx - \int_{\mathcal{O} \setminus S} p\, (\nabla\cdot\vec{w})\, dx = 0 \qquad \forall \vec{w} \in H^1_o$$

$$\int_{\mathcal{O} \setminus S} q\, (\nabla\cdot\vec{u})\, dx = 0 \qquad \forall q \in L^2_o$$

It is well known that $\mathcal{U}(S) \neq \emptyset$ $\forall S \in \mathcal{S}_o$. Also, any $(\vec{u},p) \in \mathcal{U}(S)$ solves the incompressible Navier-Stokes equations in the distribution sense (see |9|).

Our main tasks are to give a sense to J in (3.1) and to choose $d(\cdot,\cdot)$ and $\mathcal{S}$ satisfying:

$$\begin{aligned} &S^m \to S \text{ in } (\mathcal{S}_o, d)\,,\ (\vec{u}^m, p^m) \in \mathcal{U}(S^m) \text{ and } (\vec{u},p) \in \mathcal{U}(S) \\ &\underline{\text{implies}}\ \ \underline{\lim}\ J(S^m; \vec{u}^m, p^m) \geqslant J(S; \vec{u}, p), \end{aligned} \tag{3.3}$$

$\mathcal{S}$ <u>is compact (with respect to $d(\cdot,\cdot)$)</u> <u>and as large as possible.</u> (3.4)

In the sequel, let us take $\sigma = 4/3$ if N = 3 and let σ be any real number in (1,2) if N = 2. Then, it is straightforward to show that

$$-p\,\delta_{ij} + \tau_{ij}(\vec{u}) \in L^2(\mathcal{O} \setminus S) \quad \text{and} \quad \frac{\partial}{\partial x_\ell}(-p\,\delta_{ij} + \tau_{ij}(\vec{u})) \in L^\sigma(\mathcal{O} \setminus S)$$

for all $(\vec{u},p)$ (S). We state without proof the following

<u>LEMMA Let Z be the linear space</u>

$$Z = Z(\mathcal{O} \setminus S) = \left\{ \underset{\sim}{\sigma} \mid \sigma_{ij} \in L^2(\mathcal{O} \setminus S)\,,\ \frac{\partial}{\partial x_\ell}\sigma_{ij} \in L^\sigma(\mathcal{O} \setminus S) \right\}.$$

<u>Z is a reflexive Banach space for its natural norm. Furthermore, the normal trace mapping</u>

$$\underset{\sim}{\sigma} \in \mathcal{D}(\overline{\mathcal{O} \setminus S})^{N\times N} \to \underset{\sim}{\sigma}\cdot\vec{n}\,|_{\partial(\mathcal{O} \setminus S)} \in H^{-\frac{1}{2}}(\partial(\mathcal{O} \setminus S))^N$$

<u>is linear and continuous when $\mathcal{D}(\overline{\mathcal{O} \setminus S})^{N\times N}$ is endowed with the norm of Z; it can thus be extended continuously to the whole of Z.</u>

This Lemma allows us to define $(-p\,\underset{\sim}{Id}. + \underset{\sim}{\tau}(\vec{u}))\cdot\vec{n}$ as an element of $H^{-1/2}$. Now, an appropriate (generalized) expression for the drag experienced by S, $\vec{u}$ and p is:

$$\begin{aligned} &- \langle(-p\,\underset{\sim}{Id}. + \underset{\sim}{\tau}(\vec{u}))\cdot\vec{n}\,,\ \vec{u}_\infty\rangle_{\partial S} = \langle(-p\,\underset{\sim}{Id}. + \underset{\sim}{\tau}(\vec{u}))\cdot\vec{n}\,,\ \vec{u}_\infty\rangle_{\partial\mathcal{O}} \\ &= \frac{1}{2\nu}\int_{\mathcal{O} \setminus S} \tau_{ij}(u)\,\tau_{ij}(u)\, dx \end{aligned} \tag{3.5}$$

(see (2.25)). Of course, this gives a rigorous sense to (3.1).

Let us now take

$$d(S_1,S_2) = \int_{\mathcal{O}} |\chi_{S_1} - \chi_{S_2}|\, dx \equiv |S_1 \Delta S_2| \quad \forall\, S_1,S_2 \in \mathcal{S}_o \tag{3.6}$$

Here, for any set R, χ_R is the usual characteristic function.

Obviously, $d(\cdot,\cdot)$ is a distance on $\mathcal{S}_o$. For r,k > 0, let us introduce the family $\mathcal{S}(r,k)$ of all $S = \bar{B} \in \mathcal{S}_o$ satisfying the following property:

relatively simplified case, but it can certainly be applied to more complicate design problems (instationary Navier-Stokes equations, Reynolds equations, other cost functions, etc... ; see |13|).

b) Up to our knowledge, there exists no uniqueness result for design problems in this context.

c) Unfortunately, this Theorem is not completely satisfactory: we have had to introduce two "artificial" constants r and k to ensure existence. Of course, from the viewpoint of engineering sciences, r is un upper bound for the size of the different pieces, while k is a lower bound for all assembling angles.

4. FINITE ELEMENT APPROXIMATION.

In this Section, we briefly describe how problem (3.1) can be approximated by using a standard FEM. For simplicity, assume that $\mathcal{O}$ is a poligonal domain in $\mathbb{R}^2$ and, for each $h \in (0,1]$, set

$$\mathcal{S}_{oh} = \left\{ \{a^k\}_{k=1}^{K(h)} \mid a^k \text{ is the } k^{th} \text{ vertex point of a standard triangulation } \mathcal{T}_h \text{ of } \mathcal{O}\setminus S_h \text{ with } S_h \in \mathcal{S}_o \right\}$$

$$\mathcal{S}_h = \mathcal{S}_{oh} \cap \mathcal{S}$$

(in particular, all $S_h \in \mathcal{S}_h$ are poligonal).

Then, the discrete problem is:

$$\begin{array}{ll} \underline{\text{Minimize}} & -\left(\int_{\partial S_h} (-p_h \underset{\sim}{Id}. + \underset{\sim}{\tau}(\vec{u}_h)).\vec{n}\, d\Gamma \right).\vec{u}_\infty \\ \underline{\text{Subject to}} & \{a^k\} \in \mathcal{S}_h , \quad (\vec{u}_h, p_h) \in \mathcal{U}_h(S_h) \end{array} \qquad (4.1)_h$$

Here, $\mathcal{U}_h(S_h)$ is the set of all solutions of a discrete Navier-Stokes problem associated with $\mathcal{T}_h$. For instance, we may take $\mathcal{U}_h(S_h)$ as the set of all pairs $(\vec{u}_h, p_h)$ satisfying:

$$\frac{1}{2\nu} \int_{\mathcal{O}\setminus S_h} \tau_{ij}(\vec{u}_h)\, \tau_{ij}(\vec{w}_h)\, dx + \int_{\mathcal{O}\setminus S_h} u_{jh} \frac{\partial u_{ih}}{\partial x_j} w_{ih}\, dx - \int_{\mathcal{O}\setminus S_h} p_h(\nabla\cdot\vec{w}_h)\, dx = 0$$

$$\int_{\mathcal{O}\setminus S_h} q_h(\nabla\cdot\vec{u}_h)\, dx = 0 \qquad \forall\, \vec{w}_h \in X_h , \quad \forall\, q_h \in M_h ,$$

$$X_h = \{ \vec{v}_h \mid \vec{v}_h \in C^0,\ v_h|_{\tilde{T}} \in P_1 ,\forall\, \tilde{T} \in \tilde{\mathcal{T}}_h ,\ v_h|_{\partial\mathcal{O}} = v_h|_{\partial S_h} = 0 \}$$

$$M_h = \{ q_h \mid q_h \in L^2,\ q_h|_T\ \ P_0 \ \ \forall\, T \in \mathcal{T}_h \}$$

The vertex points of the triangles $\tilde{T} \in \tilde{\mathcal{T}}_h$ are $a^1, a^2, \ldots$ and the mid-points of the edges of the triangles $T \in \mathcal{T}_h$. A convergence result is the following:

For each $x \in \partial B$, there exists an open set $U_x \subset \mathbb{R}^N$ and
a bijective k-bilipschitzian mapping $T_x : B(x;r) \to U_x$ such that (3.7)
$T_x(B(x;r) \cap B) \subset U_x \cap (\mathbb{R}^{N-1} \times (0,+\infty))$.

Here, to say that T_x is k-bilipschitzian means:

$$\frac{1}{k} |y - z| \leq |T_x(y) - T_x(z)| \leq k\, |y - z| \qquad \forall\, y,z \in B(x;r) \tag{3.8}$$

According to a result from D. Chenais |10| (see also |11|), $\mathcal{S}(r,k)$ is compact in the matric space $(\mathcal{S}_o, d)$. We can now state and proof our existence result:

<u>THEOREM Assume</u> $\mathcal{S} = \mathcal{S}(r,k)$ <u>for some</u> $r,k > 0$. <u>Then there exists at least solution of</u> (3.1).

<u>SKETCH OF THE PROOF:</u> Let $\{(S^m, u^m, p^m)\}$ be a minimizing sequence. For each $m \geq 1$, let $\tilde{u}^m$ be the extension by zero of u^m. From Korn's inequality and (3.5), one deduces that

$\tilde{u}^m$ is uniformly bounded in $H^1(\mathcal{O})^N$.

Hence, it can be assumed that $\tilde{u}^m$ converges weakly in H^1 to a function $\tilde{u}$.

Clearly, it can also be assumed that

$$\lim d(S^m, S_*) = \lim \rho(S^m, S_*) = 0 \text{ for some set } S_* \in \mathcal{S}, \tag{3.9}$$

where $\rho(\cdot,\cdot)$ is the usual Haussdorf distance (see |12|).

This is sufficient to prove that, for $\vec{u}_* = \tilde{u}|_{\mathcal{O}\setminus S_*}$, one has:

$$\begin{aligned} &\nu \int_{\mathcal{O}\setminus S_*} \tau_{ij}(\vec{u}_*)\, \tau_{ij}(\vec{w})\, dx + \int_{\mathcal{O}\setminus S_*} (\vec{u}_* \cdot \nabla)\vec{u}_* \cdot \vec{w}\, dx = 0 \\ &\forall \vec{w} \in H^1_0(\mathcal{O}\setminus S_*)^N \text{ such that } \nabla \cdot \vec{w} = 0, \\ &\vec{u}_* \in H^1(\mathcal{O}\setminus S_*)^N, \ \nabla \cdot \vec{u}_* = 0, \ \vec{u}_*|_{\partial\mathcal{O}} = \vec{u}_\infty, \ \vec{u}_*|_{\partial S_*} = 0 \end{aligned} \tag{3.10}$$

Furthermore, the convergence of $\tilde{u}^m$ towards $\tilde{u}$ is in fact strong (for details, see |13|).

On the other hand, by putting

$$p^m = q^m + \alpha^m \text{ with } \alpha^m \in \mathbb{R}, \quad \int_{\mathcal{O}\setminus D_1} q \, dx = 0,$$

it is not difficult to see that q^m converges strongly in L^2 to a function q.
For $q_* = q|_{\mathcal{O}\setminus S_*}$, it is straightforward to show that $(\vec{u}_*, q_*) \in \mathcal{U}(\mathcal{O}\setminus S_*)$.

It is now clear that $(S_*, \vec{u}_*, q_*)$ provides a solution:

$$\begin{aligned} \lim J(S^m; \vec{u}^m, p^m) &= \lim J(S^m; \vec{u}^m, q^m) = \lim \langle(-\tilde{q}^m \underset{\sim}{Id}. + \underset{\sim}{\tau}(\tilde{u}^m))\cdot\vec{n}, \vec{u}_\infty\rangle_{\partial\mathcal{O}} \\ &= \langle(-q_* \underset{\sim}{Id}. + \underset{\sim}{\tau}(\vec{u}_*))\cdot\vec{n}, \vec{u}_\infty\rangle_{\partial\mathcal{O}}. \end{aligned}$$

<u>REMARKS:</u>

a) In a related (but different) context, the general argument leading to existence was introduced by J.L. Lions |14|. More recently, it was used by D. Chenais |11| and by O. Pironneau |15| among others. It has been presented here in a

THEOREM Assume $\{\mathcal{T}_h\}_{h\in(0,1]}$ is "regular" in the following sense:

$$\delta(T) \leq h \quad \text{and} \quad \delta(T)\,/\,\rho(T) \leq \text{Const.} \quad \forall\, T\in\mathcal{T}_h\,,\quad \forall\, h\in(0,1] \qquad (4.2)$$

($\delta(T) \equiv$ diameter of T; $\rho(T) \equiv$ diameter of the biggest ball $B \subset T$).
For $h \in (0,1]$, let (S_h^*, u_h^*, p_h^*) be a solution of $(4.1)_h$. Then, a subsequence $\mathcal{H}' \subset (0,1]$, $\mathcal{H}' \to 0$ exists such that

$$\begin{aligned} & S_{h'}^* \to S^* \quad \text{in } (\mathcal{S}_0, d) \text{ and for the Haussdorf distance,} \\ & u_{h'}^* \to u^* \quad \text{weakly in } H^1 \text{ and } p_{h'}^* \to p^* \quad \text{weakly in } L^2, \text{ as } h' \in \mathcal{H}',\ h' \to 0 \end{aligned} \qquad (4.3)$$

with (S^*, u^*, p^*) being a solution of (3.1).

For the proof, see |16|. The key point is obviously a uniform estimate for u_h in H^1.

5 SOME FINAL COMMENTS ON THE NUMERICAL TECHNIQUES.

Once the discrete problem (such as $(4.1)_h$) has been introduced, the task is reduced to the numerical solution of a constrained finite-dimensional extremal problem. It is thus natural to apply gradient methods.

The determination of the gradient requires some work. It was achieved by A. Marrocco and O. Pironneau |17| for the solution of a magneto-static problem and by F. Angrand |18| in the framework of fluid mechanics among others. In the case of problem $(4.1)_h$, after a tedious computation, one finds the following formula for every admissible displacement vector $\vec{\alpha}$:

$$\frac{\partial J_h}{\partial a^k} \cdot \vec{\alpha} = L_h \cdot \vec{\alpha} \quad , \text{ with } \quad L_h = L_h(\vec{u}_h, p_h, \vec{w}_h, q_h)$$

Here, L_h^k is a rather complicate expression which involves the state function $(\vec{u}_h, p_h)$ as well as an appropriate "adjoint state" $(\vec{w}_h, q_h)$ (see |16,19|).

In practice, only a few a^k are treated as control variables in (4.1) , the others being fixed. Indeed, it is expected that the displacement of a vertex point which is far from S does not lead to a significative change in the value of J_h . Furthermore, the free a^k 's are allowed to move only along prescribed curves (in order to avoid degeneracy ...).

Let us finally indicate that a more realistic, interesting and difficult problem in this framework can be found in |13| . It is concerned with the boundary layer approximation of the Navier-Stokes equations (see also |20|).

REFERENCES:

1. L. Landau et E. Lifchiz .- Mécanique de Fluides. Mir, Moscou 1967.
2. H. Schlichting .- Boundary-Layer Theory (7th edition). McGraw - Hill, New York 1979.
3. T. Cebeci and A.M. Smith .- Analysis of Turbulent Boundary Layers. Academic Press, London 1974.
4. W.S. Reynolds .- Physical and analytical foundations, concepts and new directions in turbulence modelling and simulation. Proc. Ecole d'Eté d'Analyse Numérique, Clamart 1982.
5. J.H. Herziger .- Simulation of incompressible turbulent flows. J. Comp. Phys. 69 (1987), 1 - 48.
6. M. Nallasamy .- Turbulence models and their applications to the prediction of internal flow. Comp. Fluid, 230 (1987), 218 - 236.
7. J.L. Lions .- Quelques méthodes de résolution des problèmes aux limites non linéaires. Dunod, Gauthiers - Villars, Paris 1969.
8. O. Pironneau .- Optimum profiles in Stokes flow. J. Fluid Mech. 59 (1973). See also: J. Fluid Mech. 64 (1974).
9. O.A. Ladyzhenskaya .- Mathematical Theory of Viscous Incompressible Flow. Gordon and Breach, New York 1963.
10. D. Chenais .- On the existence of a solution in a domain identification problem. J. Math. Anal. Appl. 52 (2), 1975,
11. D. Chenais .- Sur une famille de varietés à bord lipschitziennes. Application à un problème d'identification de domaines. Ann. Inst. Fourier Grénoble, 27, 4 (1977), 201 - 231.
12. C. Dellachérie .- Analytical Sets, Capacities and Haussdorf Measures. Lecture Notes in Appl. Math. 295. Springer - Verlag, Berlin 1972.
13. J.A. Bello - Tesis, Univ. de Sevilla, to appear.
14. J.L. Lions .- Some aspects of the optimal control of distributed parameter systems. RESAM No. 6 SIAM, 1972.
15. O. Pironneau .- Optimal shapes design for elliptic systems. Springer - Verlag, New York 1984.
16. J.A. Bello and E. Fernández Cara , to appear.
17. A. Marrocco and O. Pironneau .- Optimal design with Lagrangian finite elements: Design of an electromagnet. Comp. Math. Appl. Mech. Eng. 15 (1978), 277 - 308.
18. F. Angrand .- These 3e Cycle, Univ. P. et M. Curie (Paris VI), 1980.
19. E. Fernández Cara .- Thése 3e Cycle, Univ. P. et M. Curie (Paris VI), 1981. See also: Collect. Math., Vol. 3, XXXIII (1982), 225 - 247.
20. R. Glowinski and O. Pironneau .- Towards the computation of minimum drag profiles. Appl. Math. Model., Vol. 1, Sept. 1976, 58 - 66

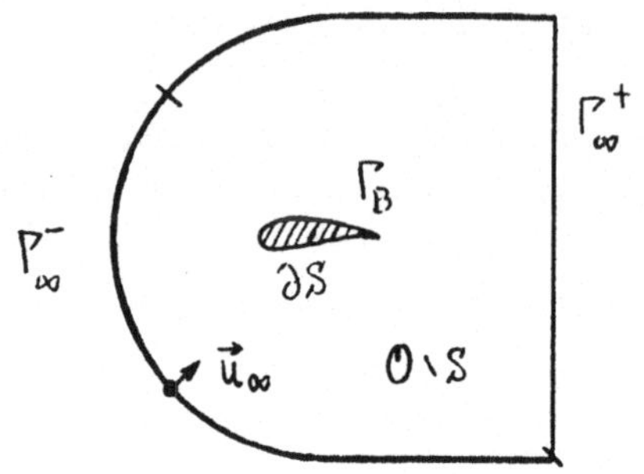

Fig. 1

Flow of a fluid around and past a rigid body. $\mathcal{O}$ is an approximation of $\mathbb{R}^2$. The fluid enters (resp. leaves) the domain through Γ_∞^- (resp. Γ_∞^+).

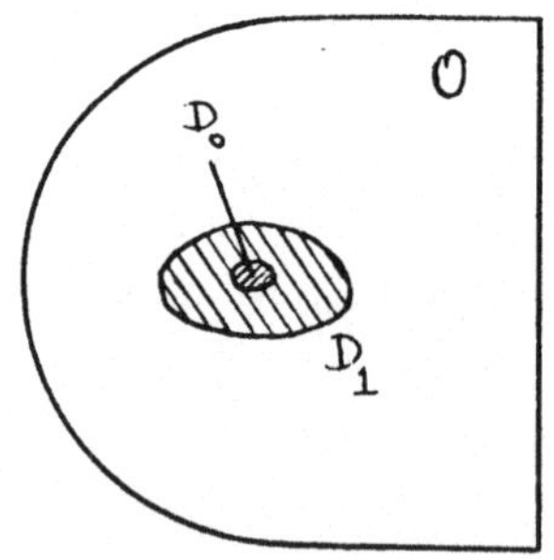

Fig. 2

The open sets D_o and D_1 arising in the definition of $\mathcal{S}_o$.

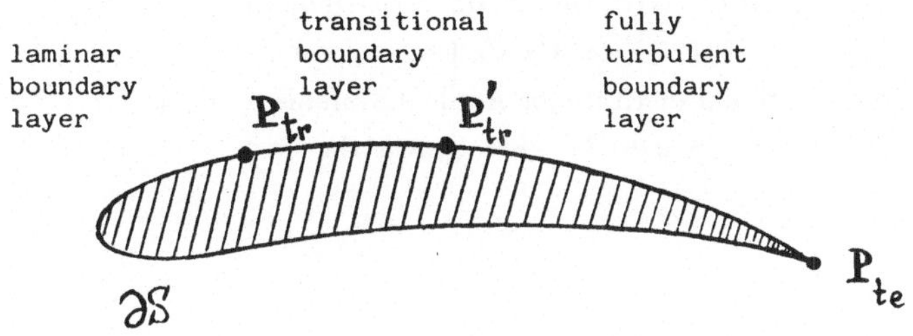

Fig. 3

The location of the "first" and the "second" transition points (from laminar to transitional and from transitional to turbulent regimes resp. in the boundary layer) and of the trailing edge point. Flow is from left to right.

On the Linearization of Nonlinear Control Systems and Exact Reachability

Halina Frankowska

CEREMADE, Université Paris-Dauphine
75775 Paris Cx 16, France, and
International Institute for Applied Systems Analysis
A-2361 Laxenburg, Austria

Abstract. We study the problem of exact local reachability of infinite dimensional nonlinear control systems. The main result shows that the exact local reachability of a linearized system implies that of the original system. Our main tool is an inverse mapping theorem for a map from a complete metric space to a reflexive Banach space.

1. Introduction.

Consider the following abstract control system

$$\begin{cases} x' = & Ax + f(x,u),\ u \in U \\ x(0) = & x_0 \end{cases} \tag{1.1}$$

where A is the infinitesimal generator of a strongly continuous semigroup of linear operators. Let $R(T)$ denote the reachable set of (1.1) at time T (by the mild solutions of (1.1)). Consider a mild solution $\bar{x}$ and let $\bar{u}$ be the corresponding control.

We address here the following two questions:

1. Does $\bar{x}(T) \in \text{Int}\, R(T)$?
2. Given a point y near $\bar{x}(T)$ how much we have to change the control $\bar{u}$ in order the trajectory x corresponding to this new control satisfies $x(T) = y$?

We study the above via a linearization technique and an inverse mapping theorem for a map whose domain of definition is a complete metric space.

Let us explain now what do we mean by linearization. In control theory linearization is usually understood as a substitution of the nonlinear system (1.1) by the linear system

$$\begin{cases} w'(t) = & Aw(t) + \frac{\partial f}{\partial x}(\bar{x}(t), \bar{u}(t))w(t) + \frac{\partial f}{\partial u}(\bar{x}(t), \bar{u}(t))u \\ w(0) = & 0 \end{cases} \tag{1.2}$$

where the controls u belong to the space spanned by U. To make use of such linearization it is usually required that $\bar{u}(t) \in \text{Int } U$ or even that U is a Banach space (see for example [19]).

Studying a particular trajectory $\bar{x}$ we can not guarantee such property (unless the set U is open, for example equal to a Banach space). Another linearization of (1.1) along $(\bar{x},\bar{u})$ is given by the following linear control system

$$\begin{cases} w'(t) = & Aw(t) + \frac{\partial f}{\partial x}(\bar{x}(t), \bar{u}(t))w(t) + v(t) \\ v(t) \in & \overline{co} f(\bar{x}(t), U) - f(\bar{x}(t), \bar{u}(t)) \\ w(0) = & 0 \end{cases} \tag{1.3}$$

Here we do not have any restrictions on the control $\bar{u}$ and thus we can apply it to any trajectory-control pair $(\bar{x},\bar{u})$.

The two linearizations (1.2) and (1.3) are related in the following way: If $\bar{u}(t) \in \text{Int } U$ then for every u, $\frac{\partial}{\partial u} f(\bar{x}(t), \bar{u}(t))u \in T_{\overline{co} f(\bar{x}(t),U)}(f(\bar{x}(t), \bar{u}(t)))$, where $T_K(x)$ denote the tangent cone (of convex analysis) to a convex set K at $x \in K$.

The second linearization was used in [22] to get local reachability of a nonlinear finite dimensional control system via the local reachability of the linearized system. It seems that the result of [22] does not have yet its analogue in infinite dimensional spaces and we shall prove it here in Section 3. Namely we show that if zero is an interior point of the reachable set $R^L(T)$ of the linear system (1.3) at time T, then $\bar{x}(T) \in \text{Int } R(T)$, i.e. we obtain a sufficient condition to answer positively the Question 1. We also show that $0 \in \text{Int } R^L(T)$ implies the existence of $L > 0$ such that for every y near $\bar{x}(T)$, there exists a trajectory-control pair (x,u) satisfying

$$x(T) = y, \ \mu(\{t : u(t) \neq \bar{u}(t)\}) \leq L\|x(T) - y\| \tag{1.4}$$

where μ denotes the Lebesgue measure. This second result seems to be unknown even in the finite dimensional case.

To prove the above, we need a very general inverse mapping theorem for maps whose domain of definition is a complete metric space. This will allow us to avoid difficult constructions of "fixed point argument" type proofs. Let us explain briefly how. We assume that f is so that to every admissible control u corresponds the unique mild solution x_u of (1.1). Consider the map $U_{ad} \ni u \to {}^G x_u(T)$.

In 1934 Ljusternik [13] proved that if a C^1-function $G : U \to X$ between two Banach spaces has a surjective derivative $G'(\bar{u})$ at a point $\bar{u} \in U$, then for all $h > 0$, $G(\bar{u}) \in \text{Int } G(B_h(\bar{u}))$ (i.e., the open mapping principle holds true) and the set-valued map G^{-1} is roughly speaking Lipschitzian at $\bar{u}$.

This is the THEOREM one would like to use to study the questions 1 and 2.

Unfortunately when the control set U is not a Banach space, the set of admissible controls neither does. Therefore the Ljusternik theorem can no longer be applied. We extend here this theorem to the case when U is a complete metric space, X is a reflexive Banach space and

$G : U \to X$ is a continuous function. Since we can no longer differentiate G, we shall deal with its variations instead of derivatives.

In the next section we introduce the variation $G^{(1)}(u)$ of G at u and prove that if $0 \in \text{Int}\, \overline{co}G^{(1)}(u)$ uniformly on a neighborhood of $\bar{u}$ then G^{-1} has a "regular" inverse at $(G(\bar{u}), \bar{u})$. In Section 3 we show that $\frac{1}{T}R^L(T) \subset \overline{co}G^{(1)}(\bar{u})$ and deduce from it an answer to questions 1 and 2. In Section 4 we apply the result to study the local exact reachability at zero of some nonlinear control systems.

2. An Inverse Function Theorem

Consider a complete metric space (U,d), a Banach space X and a continuous map $G : U \to X$. Throughout the whole paper we assume that the norm of X is Gâteaux differentiable away from zero. For all $u \in U$, $h > 0$ let $B_h(u)$ denote the closed ball in U of center u and radius h. Let $\bar{u} \in U$ be a given point. In this section we study a sufficient condition for:

$$\forall_{h>0}\; G(\bar{u}) \in \text{Int}\, G(B_h(\bar{u})) \quad (\textit{open mapping principle})$$

and the regularity of the inverse map $G^{-1} : X \to U$ given by

$$G^{-1}(x) = \{u \in U : G(u) = x\}$$

on a neighborhood of $(G(\bar{u}), \bar{u})$.

We first recall the notion of Kuratowski's limsup:

Let T be a metric space and $A_\tau \subset X, \tau \in T$ be a family of subsets of X. The Kuratowski limsup of A_τ at τ_0 is the closed set defined by

$$\limsup_{\tau \to \tau_0} A_\tau = \{v \in X : \liminf_{\tau \to \tau_0} \text{dist}\,(v, A_\tau) = 0\}$$

Definition 2.1. The contingent variation of G at $u \in U$ is the closed subset of X defined by

$$G^{(1)}(u) := \limsup_{h \to 0+} \frac{G(B_h(u)) - G(u)}{h}$$

In other words $v \in G^{(1)}(u)$ is and only if

$$\liminf_{h \to 0+} \text{dist}\,(v, \frac{G(B_h(u)) - G(u)}{h}) = 0$$

or equivalently if there exist sequences $h_i \to 0+$, $v_i \to V$ such that $G(u) + h_i v_i \in G(B_{h_i}(u))$. The word contingent is used because the definition reminds that of the contingent cone introduced by Bouligand (see, for example, [2]).

Clearly, $G^{(1)}(u)$ is starshaped at zero closed set, i.e. $0 \in G^{(1)}(u)$ and for all $v \in G^1(u)$, $\lambda \in [0,1]$, $\lambda v \in G^{(1)}(u)$. Let $\overline{co}$ denote the closed convex hull and B the closed unit ball in X.

Theorem 2.2 (Uniform Open Mapping Principle). Assume that for some $\epsilon > 0$, $\rho > 0$

$$\rho B \subset \bigcap_{d(u,\bar{u}) \leq \epsilon} \overline{co}\, G^{(1)}(u) \tag{2.1}$$

Then for every $u \in B_{\frac{\epsilon}{2}}(\bar{u})$ and $h \in [0, \frac{\epsilon}{2}]$ we have

$$G(u) + h\rho \overset{\circ}{B} \subset G(B_h(u))$$

where $\overset{\circ}{B}$ denotes the open unit ball in X.

Proof. Fix u,h as above and assume that there exists $y \in X$ satisfying

$$\| y - G(u) \| < h\rho, \; y \notin G(B_h(u)) \tag{2.2}$$

Set $\Theta^2 = \| y - G(u) \| / h\rho$. Then $0 < \Theta < 1$. Applying the Ekeland variational principle [5], [6] to the complete metric space $B_h(u)$ and the continuous function $x \to \| G(x) - y \|$ we prove the existence of $\bar{x} \in B_{\Theta h}(u)$ such that for all $x \in B_h(u)$

$$\| G(\bar{x}) - y \| \leq \| G(x) - y \| + \Theta \rho d(x,\bar{x}) \tag{2.3}$$

Observe that $\bar{x} \in \operatorname{Int} B_h(u)$ and, by (2.2), $y \neq G(\bar{x})$. Hence, by differentiability of the norm, there exists $p \in X^*$ of $\| p \| = 1$ such that for all $h_j \to 0+$, $v_j \to v$ we have

$$\| G(\bar{x}) + h_j v_j - y \| = \| G(\bar{x}) - y \| + <p, h_j v_j> + o(h_j) \tag{2.4}$$

where $\lim_{j \to \infty} o(h_j)/h_j = 0$. Fix $v \in G^{(1)}(\bar{x})$. Then from (2.3), (2.4) and Definition 2.1 we obtain

$$0 \leq <p, h_j v_j> + \Theta \rho h_j + o(h_j)$$

Dividing by h_j and taking the limit yields $<p,v> \geq -\Theta\rho$ for all $v \in G^{(1)}(\bar{x})$. Hence

$$<p,v> \geq -\Theta\rho \;\text{ for all }\; v \in \overline{co}\, G^{(1)}(\bar{x})$$

Since $d(\bar{x},\bar{u}) \leq d(\bar{x},u) + d(u,\bar{u}) \leq \Theta h + \frac{\epsilon}{2} < \epsilon$, by (2.1), $\rho B \subset \overline{co}\, G^{(1)}(\bar{x})$. This implies that

$$-\rho \geq \inf_{v \in \overline{co}\, G^{(1)}(\bar{x})} <p,v> \geq -\Theta\rho$$

Since $0 < \Theta < 1$ and $\rho > 0$ we obtained a contradiction. Hence (2.2) can not hold.

Theorem 2.3. Under all assumptions of Theorem 2.2 for all $u \in B_{\frac{\epsilon}{2}}(\bar{u})$, $x \in B_{\frac{\epsilon}{4}}(G(u))$ we have

$$\operatorname{dist}(u, G^{-1}(x)) \leq \frac{1}{\rho} \| G(u) - x \| \tag{2.5}$$

Theorem 2.3 follows from Theorem 2.2 and the general inverse function theorem which we prove below.

Theorem 2.4. Let G be a continuous map from a complete metric space (U,d) to a metric space (X,d_X) and let $\bar{u} \in U$. Assume that form some $\rho > 0, \epsilon > 0, 0 \leq \alpha < 1$ and all $u \in B_\epsilon(\bar{u})$, $h \in [0,\epsilon]$

$$\sup_{b \in B_{\rho h}(G(u))} \text{dist}\,(b, G(B_h(u))) \leq \alpha \rho h \tag{2.6}$$

Then for all $h > 0$ satisfying $h/(1-\alpha) + 2\rho h \leq \epsilon/2$ and all $u \in B_{\frac{\epsilon}{2}}(\bar{u})$, $x \in B_{\rho h}(G(u))$ we have

$$\text{dist}\,(u, G^{-1}(x)) \leq \frac{1}{1-\alpha} h$$

In particular, this implies that for all u near $\bar{u}$ and all x near $G(\bar{u})$

$$\text{dist}\,(u, G^{-1}(x)) \leq \frac{1}{\rho(1-\alpha)} d_X(G(u), x) \tag{2.7}$$

Proof. Fix h,u,x as above. We look for $y \in G^{-1}(u)$ as the limit of a sequence we shall built. Set $u_0 = u$. By (2.6) there exists u_1 such that

$$d(u_0,u_1) = d(u,u_1) \leq h,\ d_X(G(u_1),x) \leq \alpha \rho h$$

Assume that we already constructed $u_i, i = 1,\ldots,n$ such that

$$d(u_{i-1},u_i) \leq \alpha^{i-1} h,\ d_X(G(u_i),x) \leq \rho \alpha^i h \tag{2.8}$$

Then

$$d(u,u_i) \leq \sum_{j=1}^{i} d(u_{j-1},u_j) \leq h \sum_{j=0}^{i-1} \alpha^i < \frac{h}{1-\alpha} \tag{2.9}$$

and

$$d(\bar{u},u_i) \leq d(\bar{u},u) + d(u,u_i) \leq \frac{\epsilon}{2} + \frac{h}{1-\alpha} \leq \epsilon$$

By (2.6), there exists u_{n+1} such that $d(u_n, u_{n+1}) \leq \alpha^n h$ and $d_X(G(u_{n+1}),x) \leq \rho\alpha^{n+1} h$. Observe that (2.8) implies that $\{u_i\}$ is a Cauchy sequence and that $\lim_{i\to\infty} G(u_i) = x$. Let y be the limit of $\{u_i\}$. Since G is continuous, $G(y) = x$ and therefore $y \in G^{-1}(x)$. Moreover, by (2.9), $d(u,y) < h/(1-\alpha)$.

Remark. The method applied in the proof is due to Ljusternik [13] and Graves [10].

Corollary 2.5. Let $g: X \to Y$ be a function between two Banach spaces. Assume that g is continuously differentiable at some $x_0 \in X$ and

$$0 \in \text{Int}\, \overline{g'(x_0) B} \tag{2.10}$$

Then for all $h > 0$, $g(x_0) \in \text{Int}\, g(B_h(x_0))$ and there exists $L > 0$ such that for all (x,y) near $(x_0, g(x_0))$

$$\text{dist}\,(x, g^{-1}(y)) \leq L \|g(x) - y\|$$

Remark. We observe that the assumption (2.10) is verified whenever $g'(x_0)$ is surjective, i.e.

$$g'(x_0)X = Y$$

Indeed $g'(x_0)X = \bigcup_{n\geq 1} ng'(x_0)B$ and, by Baire's theorem, for some $n \geq 1$ the set $ng'(x_0)B$ has a nonempty interior. Hence $\operatorname{Int} g'(x_0)B \neq \emptyset$ and, using that $g'(x_0)B = -g'(x_0)B$ is a convex set we obtain

$$0 \in \operatorname{Int} g'(x_0)B \subset \operatorname{Int} \overline{g'(x_0)B}$$

Thus Corollary 2.5 extends Ljusternik's theorem [13].

3. Interior Points of Reachable Sets

Let U be a topological space, X be a reflexive Banach space with the norm Gâteaux differentiable away from zero and $f: X \times U \to X$ be a continuous, differentiable in the first variable function. We assume that

a) f is locally Lipschitz in the first variable uniformly on U, i.e. for all $x \in X$ there exist $L > 0$ and $\epsilon > 0$ such that for all $u \in U$, $f(\cdot, u)$ is L-Lipschitz on $B_\epsilon(x)$:

$$\|f(x', u) - f(x'', u)\| \leq L\|x' - x''\|, \text{ for all } x', x'' \in B_\epsilon(x)$$

b) For all $u \in U$ the derivative $\dfrac{\partial f}{\partial x}(\cdot, u)$ is continuous

c) For all $x \in X$ the set $f(x, U)$ is bounded

For all $T > 0$ a (Lebesgue) measurable function $u : [0,T] \to U$ is called an admissible control. Let Q_T denote the set of all admissible controls defined on the time interval $[0,T]$. Define a metric on Q_T by setting

$$d_T(u,v) = \mu(\{t \in [0,t] : u(t) \neq v(t)\})$$

where μ denotes the Lebesgue measure. Then the space (Q_T, d_T) is complete (see Ekeland [6]).

Let $\{S(t)\}_{t\geq 0}$ be a strongly continuous semigroup of linear operators from X to X and A be its infinitesimal generator, $x_0 \in X$. Consider the control system

$$\begin{cases} x'(t) = & Ax(t) + f(x(t), u(t)), \ u \in Q_T, \ T > 0 \\ x(0) = & x_0 \end{cases} \tag{3.1}$$

Recall that a continuous function $x: [0,T] \to X$ is called a mild trajectory of (3.1) if for some $u \in Q_T$ and all $0 \leq t \leq T$

$$x(t) = S(t)x_0 + \int_0^t S(t-s)f(x(s), u(s))\, ds \tag{3.2}$$

We denote by x_u the trajectory corresponding to the control u. Define the reachable set of (3.1) at time $T > 0$ by

$$R(T) = \{x(T) : x \in C(0,T;X) \text{ is a mild trajectory of } (3.1)\} .$$

Let z be a mild trajectory of (3.1) on $[0,T]$ and $\bar{u}$ be the corresponding control. In this section we provide a sufficient condition for

$$z(T) \in \operatorname{Int} R(T)$$

and for the regularity of the "inverse." Consider the linear control system

$$\begin{cases} w'(t) = & Aw(t) + \frac{\partial f}{\partial x}(z(t), \bar{u}(t))\, w(t) + v(t) \\ w(0) = & 0 \\ v(t) \in & \overline{co} f(z(t), U) - f(z(t), \bar{u}(t)) \end{cases} \tag{3.3}$$

and let $R^L(T)$ denote the corresponding reachable set of (3.3) at time T. Let $S_{\bar{u}}(t;s)$ denote the solution operator of the equation

$$Z'(t) = AZ(t) + \frac{\partial f}{\partial x}(z(t), \bar{u}(t)) Z(t)$$

where $S_{\bar{u}}(s;s) = Id$. Then

$$R^L(T) = \{\int_0^T S_{\bar{u}}(T;s) v(s)\, ds : v(s) \in \overline{co} f(z(s), U) - f(z(s), \bar{u}(s))\}$$

Theorem 3.1. Assume that $0 \in \operatorname{Int} R^L(T)$. Then $z(T) \in \operatorname{Int} R(T)$ and there exist $\epsilon > 0$, $L > 0$ such that for every control $u \in Q_T$ satisfying $d_T(u, \bar{u}) \leq \epsilon$ and all $b \in B_\epsilon(z(T))$ there exists a trajectory-control pair $(x_{\hat{u}}, \hat{u})$ which verifies

$$x_{\hat{u}}(T) = b;\ d_T(u, \hat{u}) \leqslant L \| b - x_u(T) \|$$

In particular for all $b \in B_\epsilon(z(T))$ there exists a control $u \in Q_T$ such that

$$\mu(\{t \in [0,T] : u(t) \neq \bar{u}(t)\}) \leq L \| b - z(T) \|$$

and the trajectory x_u corresponding to this control verifies $x_u(T) = b$.

Proof. Replacing t by t/T we may assume that $T = 1$. Set $Q = Q_1$, $d = d_1$. For all $u \in Q$, let x_u be the solution of (3.1) (when it exists on $[0,1]$) corresponding to the control u.

From the Gronwall inequality follows that for some $\delta > 0$ the map $\varphi(u) = x_u$ from $B_{2\delta}(\bar{u})$ to $C(0,1;E)$ is well defined and is Lipschitzian. For all $u \in B_\delta(\bar{u})$ and $s \in [0,1]$, let $S_u(\cdot;s)$ denote the solution operator of the equation

$$Z'(t) = AZ(t) + \frac{\partial f}{\partial x}(x_u(t)), u(t)) Z(t)$$

Fix $u \in B_\delta(\bar{u})$ and $v \in U$. For all $t_0 \in]0,1[$, $h > 0$ we consider the needle perturbations of controls

$$u_h(t) = \begin{cases} v & t_0 - h \le t \le t_0 \\ u(t) & \text{otherwise} \end{cases}$$

Let x_h denote the solution of (3.1) corresponding to the control u_h. It is well known that at every Lebesgue point t_0 of the function $f(x_u(\cdot), u(\cdot))$ we have

$$\lim_{h \to 0+} \frac{x_h(1) - x_u(1)}{h} = S_u(1;t_0)\,(f(x_u(t_0),v) - f(x(t_0), u(t_0))) \tag{3.4}$$

(see for example Fattorini [7]). Set $V_u(t) = f(x_u(t),U) - f(x_u(t),u(t))$ and define the continuous map $G: B_{2\delta}(\bar{u}) \to X$ by

$$G(u) = x_u(1)$$

Then, by (3.4), for all $u \in B_\delta(\bar{u})$, and for almost all $t_0 \in [0,1]$ and all $v \in V_u(t_0), S_u(1;t_0)v \in G^{(1)}(u)$. Therefore for all $v \in \overline{co}\, V_u(t_0), S_u(t;t_0)v \in \overline{co} G^{(1)}(u)$. Hence, by the mean value theorem, for all measurable selection $v(t) \in \overline{co}\, V_u(t)$

$$\int_0^1 S_u(1;t)v(t)\,dt \in \overline{co} G^{(1)}(u) \tag{3.5}$$

Let $\rho > 0$ be such that

$$\rho B \subset \{\int_0^1 S_{\bar{u}}(1;t)v(t)\,dt : v(t) \in \overline{co}\ V_{\bar{u}}(t)\} \tag{3.5}$$

The Gronwall inequality implies that $S_u(1;\cdot) \to S_{\bar{u}}(1;\cdot)$ uniformly when $u \to \bar{u}$ and

$$\lim_{u \to \bar{u}} \int_0^1 H(\overline{co}\ V_{\bar{u}}(t), \overline{co}\ V_u(t))\,dt = 0$$

where H stands for the Hausdorff distance. Since the right-hand side of (3.5) is convex and closed this yields that for some $\delta' > 0$ and all $u \in B_{\delta'}(\bar{u})$

$$\frac{\rho}{2} B \subset \{\int_0^1 S_u(1;t)y(t)\,dt : y(t) \in \overline{co}\ V_{\bar{u}}(t)\} \subset \overline{co}\ G^{(1)}(u)$$

Theorem 2.3 ends the proof.

Remark. Recall that in infinite dimensions the linear system

$$x' = Ax + Bu, \quad u \in U \tag{3.7}$$

where U is a Banach space and $B \in L(U,X)$ is not in general exactly controllable by $L^p(0,T;U)$ controls $p > 1$) (see R. Triggiani [20], [21], J.C. Louis and D. Wexler [14]). Therefore, when U is a bounded subset of a Banach space we can neither expect (in general) the reachable sets of (3.3) to have a nonempty interior. The results from [4], [17] give an idea of what has to be assumed about the semigroup S and the operator B to get the exact local reachability of (3.3) at zero. They also indicate how narrow the class of such systems is. In the next section we apply

Theorem 3.1 to a nonlinear problem of local exact reachability.

4. A Local Reachability Problem

Let X be a reflexive, E be a separable reflexive Banach space, A be the infinitesimal generator of a C_0-semigroup $S(t) \in L(X,X)$, $t \geq 0, B \in L(E,X)$. Consider a topological space U and a continuous function $f: X \times U \to E$. We assume that f satisfies all the assumptions from Section 3. We study here the control system

$$\begin{cases} x' = Ax + Bf(x,u), \; u \in U \\ x(0) = 0 \end{cases} \tag{4.1}$$

Theorem 4.1. Assume that for some $\bar{u} \in U$, $f(0,\bar{u}) = 0$, $\frac{\partial f}{\partial x}(0,\bar{u}) = 0$ and that

$$\inf_{p \in X^*, \|p\|=1} \int_0^T \sup_{u \in U} \langle B^* S(s)^* p, f(0,u) \rangle ds > 0 \tag{4.2}$$

where $T > 0$ is a given time. Then for some $L > 0$ and all x_0 near zero there exists a measurable control $u(s) \in U$ such that the corresponding trajectory x_u satisfies

$$x_u(T) = x_0 \text{ and } \mu(\{t : u(t) \neq 0\}) \leq L\|x_0\|$$

Remark. Observe that for all $p \in X^*$

$$\sup_{u \in U} \langle B^* S(s)^* p, f(0,u) \rangle = \sup_{v \in \overline{co} f(0,U)} \langle B^* S(s)^* p, v \rangle$$

Therefore, from [17, Proposition 2.2] we deduce that the function $s \to \sup_{u \in U} \langle B^* S(s)^* p, f(0,u) \rangle$ is integrable. Hence the integral in (4.2) is well defined.

Some corollaries are in order.

Corollary 4.2. Assume that for some $\bar{u} \in U$, $f(0,\bar{u}) = 0$, $\frac{\partial f}{\partial x}(0,\bar{u}) = 0$ and $0 \in \text{Int}\, \overline{co} f(0,U)$. If

$$\inf_{p \in X^*, \|p\|=1} \|B^* S(\cdot)^* p\|_{L^1(0,T;E^*)} > 0 \tag{4.3}$$

then the conclusion of Theorem 4.1 is valid.

Corollary 4.3. Under the assumptions of Corollary 4.2 assume that B is surjective and that for some $t_0 > 0, S(t_0)$ is surjective. Then the conclusion of Theorem 4.1 is valid.

Corollary 4.4. In Theorem 4.1 assume that U is a bounded subset of a separable reflexive Banach space E and $f(x,u) = g(x) + u$, where g is C^1 on a neighborhood of zero. If $g(0) = 0, g'(0) = 0, 0 \in U$ and

$$\inf_{p \in X^*, \|p\|=1} \int_0^T \sup_{u \in U} \langle B^* S(s)^* p, u \rangle ds > 0$$

then the conclusion of Theorem 4.1 is valid.

Proof of Theorem 4.1. By Theorem 3.1 we have to show that $0 \in \mathrm{Int} R^L(T)$, where $R^L(T)$ denotes the reachable set at time T of the linear system.

$$x' = Ax + v, \; v \in \overline{co}\; Bf(0,U) \tag{4.4}$$

The set $\overline{co}f(0,U)$ being weakly sequentially compact, we know that $B\;\overline{co}f(0,U)$ is a closed convex set. Hence $\overline{co}\;Bf(0,U) \subset B\;\overline{co}\;f(0,U)$. Moreover $B\;cof(0,U) \subset \overline{co}\;Bf(0,U)$. Therefore we proved that $\overline{co}\;Bf(0,U) = B\;\overline{co}f(0,U)$. Thus the system (4.4) may be replaced by the linear control system

$$\begin{cases} x' = Ax + Bv, & v \in \overline{co}\; f(0,U) \\ x(0) = 0 \end{cases} \tag{4.5}$$

The admissible controls $U_{ad}(T)$ are measurable selections of $\overline{co}f(0,U)$ defined on the time interval $[0,T]$. By [17] the reachable set $R^L(T)$ of (4.5) at tome T is weakly compact. Clearly $R^L(T)$ is convex. Thus, by the separation theorem, we shall end the proof when we show that

$$\inf_{p\in X^*,\|p\|=1} \sup\{<p,x> : x \in R^L(T)\} > 0 \tag{4.6}$$

By [17], for all $p \in X$

$$\sup\{<p,x> : x \in R^L(T)\} = \sup\{\int_0^T <p,S(T-s)Bu(s)> ds : u \in U_{ad}(T)\}$$

$$= \int_0^T \sup_{u\in \overline{co}f(0,U)} <B^*S(T-s)^*p,u> ds = \int_0^T \sup_{u\in U} <B^*S(T-s)^*p, f(0,u)> ds$$

and therefore (4.6) follows from the assumption (4.2).

Proof of Corollary 4.2. Let $\gamma > 0$ be such that $\{v \in E : \|v\|_E \le \gamma\} \subset \overline{co}f(0,U)$. Then $\sup_{u\in U} <B^*S(s)^*p, f(0,u)> \ \ge \gamma\|B^*S(s)^*p\|$ and therefore (4.3) implies (4.2). Theorem 4.1. ends the proof.

Proof of Corollary 4.3. Since $S(t_0)$ is surjective, by [14], $S(t)$ is surjective for all $t > 0$ and therefore $S(t)B$ is surjective. Let $\gamma > 0$ be as in the proof of Corollary 4.2. By a Banach theorem, for every $t \ge 0$ there exists $\rho(t) > 0$ such that

$$\{x \in X : \|x\| \le \rho(t)\} \subset S(t)B(\{v \in E : \|v\|_E \le \gamma\})$$

Hence for all $t \ge 0, p \in X^*$ of $\|p\| = 1$ we have

$$\|B^*S(t)^*p\| = \frac{1}{\gamma} \sup_{\|v\|_E\le\gamma} <B^*S(t)^*p,v> = \frac{1}{\gamma} \sup_{\|v\|_E\le\gamma} <p, S(t)Bv> \ \ge \frac{1}{\gamma} \sup_{\|x\|\le\rho(t)} <p,x> = \rho(t)/\gamma$$

Therefore

$$\inf_{p\in X^*,\|p\|=1} \|B^*S(t)^*p\| > 0$$

and the proof follows from Corollary 4.2.

The proof of Corollary 4.4 is obvious.

Example 4.5. Consider the one dimensional hyperbolic equation with distributed control

$$\begin{cases} v_{tt} = & v_{xx} + f(v,u), u \in U, (t,x) \in [0,\infty[\times [0,1] \\ v(0,\cdot) = & 0; v_t(0,\cdot) = 0 \\ v(t,0) = 0 = v(t,1) \text{ for } t \geq 0 \end{cases} \tag{4.7}$$

where $f \in L^2(0,1) \times U \to L^2(0,1)$ satisfies all the assumptions of Section 3. The system (4.7) can be rewritten in the abstract form considered in Section 3: Set $Az = -z_{xx}$, $X = D(A^{1/2}) \times L^2(0,1)$

$$\hat{A} = \begin{bmatrix} 0 & I \\ -A & 0 \end{bmatrix} \quad B = \begin{bmatrix} 0 \\ I \end{bmatrix}$$

Then $\hat{A}$ generates a strongly continuous group on the Hilbert space X with the inner product $<w,\bar{w}>_X = \int_0^1 w_{1x}(x)\bar{w}_{1x}(x)dx + \int_0^1 w_2(x)\bar{w}_2(x)dx$ (see [4, pp. 47, 57]) and (4.7) can be written as an abstract control system

$$\begin{cases} x' = & \hat{A}x + Bf(x,u) \quad u \in U \\ x(0) = & 0 \end{cases} \tag{4.8}$$

where $f(x,u) = f(x_1,u)$ for $x = (x_1,x_2) \in X$.

Assume that

$$0 \in \text{Int } \overline{co}f(0,U) \text{ and } \exists \bar{u} \in U \text{ such that } f(0,\bar{u}) = 0, \quad f_x'(0,\bar{u}) = 0 \tag{4.9}$$

Let $T > 0$ be an arbitrary but fixed number. We claim that for some $L > 0$ and all $\varphi \in D(A^{1/2})$, $\psi \in L^2(0,1)$ of sufficiently small norm, there exists a measurable selection $u(s) \in U$, $s \in [0,T]$ such that $\mu(\{t : u(t) \neq 0\}) \leq L\|(\varphi,\psi)\|_X$ and the corresponding (mild) solution v of (4.7) satisfies $v(T,\cdot) = \varphi$, $v_t(T,\cdot) = \psi$.

Indeed, by Corollary 4.2 and the assumption (4.9) we have to show that

$$\inf_{p \in X^*, \|p\| = 1} \int_0^T \|B^* S(t)^* p\|_{L^2(0,1)} dt > 0 \tag{4.10}$$

By [4, p. 58] there exists $\gamma > 0$ such that for all $p \in X^*$

$$\|B^* S(\cdot)^* p\|_{L^2(0,T;L^2(0,1))} \geq \gamma \|p\| \tag{4.11}$$

On the other hand for some $M > 0$ and all $p \in X^*$ of $\|p\| = 1$

$$\int_0^T \|B^* S(t)^* p\|_{L^2(0,1)} dt \geq M \int_0^T \|B^* S(t)^* p\|^2_{L^2(0,1)} dt = M \|B^* S(\cdot)^* p\|^2_{L^2(0,1;L^2(0,1))} \geq M\gamma^2$$

This implies (4.10) and ends the proof of our claim.

References

[1] Aubin J.P. [1982] Comportement Lipschitzien des solutions de problemes de minimisation convexes. CRAS 295, 235-238.

[2] Aubin J.P. and I. Ekeland [1984] *Applied Nonlinear Analysis*. Wiley Interscience, New York.

[3] Aubin J.P. and H. Frankowska [1987] On inverse function theorems for set-valued maps. J. Math Pure Appl. 66, pp. 71-89.

[4] Curtain R.F., A.J. Pritchard [1978] *Infinite Dimensional Linear Systems Theory*, Lecture Notes in Control and Information Sciences, Springer Verlag.

[5] Ekeland I. [1974] On the variational principle, J. Math. Anal. Appl. 47 pp. 324-358.

[6] Ekeland I. [1979] Nonconvex minimization problems, Bull. Am. Math. Soc. 1, pp. 443-474.

[7] Fattorini M. [1987] A unified theory of necessary conditions for nonlinear nonconvex control systems, Applied Mathematics and Optimization, Vol. 2, pp. 141-184.

[8] Frankowska H. An open mapping principle for set-valued maps, J. of Math. Analysis and Appl. Vol.127, No.1 (1987), pp.172-180.

[9] Frankowska H. Local controllability of control systems with feedback. J. of Optimization Theory and Applications (to appear).

[10] Graves L.M. [1950] Some mapping theorem, Duke Math. J. 17, pp. 111-114.

[11] Lions J.L. [1971] *Optimal Control of Systems Described by Partial Differential Equations*, Springer.

[12] Lions J.L. and E. Magenes [1968-1970] *Problèmes aux limites non homogènes*, 3 Vols, Dunod, Paris.

[13] Ljusternik L.A. [1934] Conditional extrema of functionals. Mat. Sb. 41, pp. 390-401.

[14] Louis J.L. and D. Wexler, On exact controllability in Hilbert spaces, in *Trends in Theory and Practice of Nonlinear Differential equations*, ed. Lakshmikantham, Lecture Notes in Pure and applied Mathematics, Vol. 90, Marcel Dekker INC, Bew York and Basel.

[15] Magnusson K., Pritchard A.J. and M.D. Quinn [1985] The application of fixed point theorems to global nonlinear controllability problems, in Mathematical control theory, Banach Center Publications, Vol. 14, pp. 319-343.

[16] Magnusson K. and A.J. Pritchard [1981] Local Exact Controllability of Nonlinear Evolution Equations, in Recent Advances in Differential Equations, Academic Press, pp. 271-280.

[17] Peichl G. and W. Schappacher [1986] Constrained Controllability in Banach Spaces, SIAM J. on Control and Optimization, (24), pp. 1261-1275.

[18] Quinn M.D. and N. Carmichael [1984-85] An approach to non-linear control problems using fixed point methods, degree theory and pseudo-inverses, Numer. Funct. Anal. and Optimiz. 7(283) pp. 197-219.

[19] Russel D. [1978] Controllability and stability theory for linear partial differential equations: recent progress and open questions, SIAM Review, pp. 639-739.

[20] Triggiani R. [1975] Controllability and observability in Banach space with bounded operators, SIAM J. on Control and Optimization, (13), pp. 462-491.

[21] Triggiani R. [1980] A note on the lack of exact controllability for mild solutions in Banach spaces, SIAM J. on Control and Optimization, (18) , pp. 98-99.

[22] Yorke J. [1972] The maximum principle and controllability of nonlinear differential equations, SIAM J. of Control and Optimization, (10) pp. 334-338.

Invariant Imbedding and the Reflection of Elastic Waves

William W. Hager
Department of Mathematics
Pennsylvania State University
University Park, PA 16802 USA

and

Rouben Rostamian
Department of Mathematics
University of Maryland Baltimore County
Catonsville, MD 21228 USA

Abstract

Formulas are derived for the reflection and transmission tensors associated with a plane elastic wave impinging obliquely upon a stratified slab interposed between two homogeneous half-spaces. Both solid-solid and solid-liquid interfaces are considered.

1. Introduction.

This paper summarizes results contained in [5] concerning the reflection and transmission tensors for a stratified slab sandwiched between two homogeneous half-spaces. By a stratified medium, we mean an isotropic, linearly elastic material whose mechanical properties such as Lamé moduli and density vary in only one direction. The way these properties vary in this direction can be quite general (for example, bounded and measurable, not merely piecewise constant as is often assumed in the literature). This general nature for the stratification is essential for applications to the design and optimization of absorbent coatings. In these applications, the mechanical properties of the stratified layer are unknown independent variables whose specific structure cannot be prescribed *a priori*. In [4] we study in one dimension the problem of choosing the mechanical properties of a coating in order to minimize the amplitude of a reflected wave. When the minimization is subject to upper and lower bound constraints on both the Lamé moduli and the density, we observe that the optimal mechanical properties have a bang-bang structure over a portion of the coating while they vary continuously over the remaining part of the coating.

*This work was supported by Grant DMS-8602006 from the National Science Foundation and by Grant N00014-86-K-0498 from the Office of Naval Research. Part of this research was carried out while the second author was visiting and supported by the Institute for Mathematics and Its Applications at the University of Minnesota.

Our analysis of reflection and transmission tensors is related to the method of *invariant imbedding*, introduced by Bellman and Kalaba in [2] where they consider the problem of time-harmonic wave propagation in a one dimensional medium. If the x axis denotes the propagation direction, then the regions $x < 0$ and $x > a$ are assumed homogeneous while an arbitrary inhomogeneous material occupies the slab $0 < x < a$. To determine the reflectivity of the interface at $x = a$, Bellman and Kalaba partition the slab into thin homogeneous layers and analyze the reflections and refractions as a wave reverberates in a thin layer. They note that high order reflections and refractions can be ignored when one passes to the limit and they show that the reflectivity can be obtained from the solution to a Riccati differential equation whose initial condition is easily evaluated.

In theory, this strategy can be applied to obliquely incident waves; however, the analysis gets rather complicated. The method presented in this paper is in the spirit of invariant imbedding as described in [1]. We analyze a thin homogeneous layer in order to determine the reflection tensor for an interface. However, we do not analyze the reverberations of waves in the homogeneous layer; instead, solutions are patched together, preserving the relevant continuity conditions, to express the reflectivity in terms of an impedance tensor. The impedance is differentiated to obtain a Riccati equation. In one dimension, we get the Bellman-Kalaba reflectivity formula. For oblique incidence, we obtain new formulas for both the reflection and the transmission tensors (see Theorems 5.1 and 6.1). Both solid-solid and solid-liquid interfaces are analyzed. Although the coefficients appearing in the Riccati equation for oblique incidence seem complicated, there are enormous simplifications when these coefficients are evaluated relative to a natural coordinate system. Moreover, even though the Riccati equation for the impedance tensor is not symmetric, a simple linear transformation of the impedance tensor satisfies a symmetric equation.

2. Analysis in one dimension.

To present the fundamental ideas involved in the determination of reflection and transmission coefficients, we first give the one dimensional analysis corresponding to a wave at normal incidence. The analysis which follows is patterned after the two dimensional analysis in [5] although the algebraic manipulations in one dimension are vastly easier than the algebraic manipulations in two dimensions. Let us consider an isotropic material in one dimension. In the regions $x < 0$ and $x > a$ the material is homogeneous while in the region $0 < x < a$, the mechanical properties depend on x. If we consider waves with a harmonic time dependence which propagate perpendicular to the interfaces $x = 0$ and $x = a$, the displacement $u(x,t)$ has the form $u(x,t) = u(x)e^{i\omega t}$. Letting $\kappa(x)$ and $\rho(x)$ denote the stiffness and density at position x, the equation of motion reduces to

$$[\kappa(x)u'(x)]' + \omega^2\rho(x)u(x) = 0 \tag{2.1}$$

for $-\infty < x < +\infty$ where the coefficients κ and ρ are constant for $x < 0$ and $x > a$:

$$\kappa(x) = \kappa_0 \quad \text{and} \quad \rho(x) = \rho_0 \quad \text{for} \quad x < 0,$$
$$\kappa(x) = \kappa_1 \quad \text{and} \quad \rho(x) = \rho_1 \quad \text{for} \quad x > a.$$

In the region $x > a$, the solutions of (2.1) are linear combinations of $e^{i\omega s_1 x}$ and $e^{-i\omega s_1 x}$ where

$$s_1 = \sqrt{\rho_1/\kappa_1}\,.$$

After normalizing by a constant, the solution to (2.1) in the region $x > a$ can be expressed

$$u(x) = e^{i\omega s_1(x-a)} + re^{-i\omega s_1(x-a)} \tag{2.2}$$

so that the time varying displacement is

$$u(x,t) = e^{i\omega[s_1(x-a)+t]} + re^{-i\omega[s_1(x-a)-t]}\,. \tag{2.3}$$

The first term on the right side of (2.3) corresponds to the incoming wave while the second term is its reflection. Let us compute r, the ratio between the amplitude of the incoming wave and the amplitude of the outgoing wave. The parameter r is the reflectivity of the material in the region $x \leq a$.

We compute r using an auxiliary problem. Consider a stratified half-space $x < 0$ attached to a homogeneous slab $0 \leq x \leq \tau$ attached to a homogeneous half-space $x > \tau$. Let κ and ρ denote the stiffness and density corresponding to the homogeneous slab and let κ_1 and ρ_1 denote the stiffness and density corresponding to the homogeneous half-space. Assuming that the reflectivity $r(0)$ of the stratified half-space $x \leq 0$ is known, we will compute the reflectivity $r(\tau)$ corresponding to the half-space $x \leq \tau$. As in (2.2), the spatial component of the incident wave in the region $x > \tau$ is

$$u(x) = e^{i\omega s_1(x-\tau)} + r(\tau)e^{-i\omega s_1(x-\tau)}\,.$$

In the homogeneous slab, the spatial component has the form

$$u(x) = t_+e^{i\omega s(x-\tau)} + t_-e^{-i\omega s(x-\tau)} \quad \text{where} \quad s = \sqrt{\rho/\kappa}\,. \tag{2.4}$$

The unknowns t_+, t_-, and $r(\tau)$ are computed from the continuity of displacement and stress.

The continuity of displacement at $x = \tau$ implies that

$$t_+ + t_- = 1 + r(\tau)\,.$$

Since stress is κ times the derivative of u, the continuity of stress at $x = \tau$ yields the relation

$$\kappa_1 s_1 (1 - r(\tau)) = \kappa s (t_+ - t_-) .$$

This gives us two equations for the three unknowns. A third relation is obtained at the interface $x = 0$.

The t_+ component in (2.4) corresponds to a wave moving to the left while the t_- component corresponds to a wave moving to the right. The amplitude of the left propagating wave at $x = 0$ is $t_+ e^{-i\omega s\tau}$. Therefore, the amplitude of the reflected (right propagating) wave at $x = 0$ is $r(0) t_+ e^{-i\omega s\tau}$, which must equal the t_- component of (2.4) evaluated at $x = 0$:

$$r(0) t_+ e^{-i\omega s\tau} = t_- e^{i\omega s\tau} .$$

Rearranging this relation, we have

$$t_- . = r(0) e^{-2i\omega s\tau} t_+ .$$

This expression for t_- combined with the equations for continuity of displacement and stress imply that

$$\frac{1 - r(\tau)}{1 + r(\tau)} = \frac{\kappa s (t_+ - t_-)}{\kappa_1 s_1 (t_+ + t_-)} = \frac{\kappa s}{\kappa_1 s_1} \frac{(1 - r(0) e^{-2i\omega s\tau})}{(1 + r(0) e^{-2i\omega s\tau})} = \frac{1}{\kappa_1 s_1 G(\tau)}$$

where

$$G(\tau) = \frac{1 + r(0) e^{-2i\omega s\tau}}{\kappa s (1 - r(0) e^{-2i\omega s\tau})} .$$

Expressing $r(\tau)$ in terms of $G(\tau)$, we have

$$r(\tau) = \frac{G(\tau) - \dfrac{1}{\kappa_1 s_1}}{G(\tau) + \dfrac{1}{\kappa_1 s_1}} .$$

The objective in the manipulations above is to write the reflection coefficient in terms of two expressions, the expression $G(\tau)$ which depends on properties of the material in the region $x \le \tau$ and the expression $\kappa_1 s_1$ which depends on properties of the material in the region $x > \tau$. Letting τ tend to zero, we observe that

$$G(0) = \frac{1 + r(0)}{\kappa s\,(1 - r(0))} . \tag{2.5}$$

Differentiating G and letting τ tend to zero yields

$$G'(0) = \frac{2r(0)}{\kappa s} \frac{(-2i\omega s)}{(1-r(0))^2} . \tag{2.6}$$

Using equation (2.5) to express $r(0)$ in terms of G, we obtain

$$r(0) = \frac{\kappa s G(0) - 1}{\kappa s G(0) + 1} \quad \text{and} \quad 1 - r(0) = \frac{2}{\kappa s G(0) + 1} .$$

With these substitutions, equation (2.6) simplifies to

$$G'(0) = i\omega \left[\frac{1}{\kappa} - \rho G(0)^2 \right] .$$

Now return to the original problem where a stratified slab of thickness a is interposed between two homogeneous half-spaces. We can "build" the slab of thickness a by starting with a slab of thickness zero and adding infinitesimally thin sheets of homogeneous veneer. It follows that the reflectivity of the inhomogeneous slab of thickness a is given by

$$r(a) = \frac{G(a) - \dfrac{1}{\kappa_1 s_1}}{G(a) + \dfrac{1}{\kappa_1 s_1}}$$

where G satisfies the differential equation

$$G'(x) = i\omega \left[\frac{1}{\kappa(x)} - \rho(x) G(x)^2 \right], \quad 0 \le x \le a ,$$

with the boundary condition

$$G(0) = \frac{1}{\kappa_0 s_0} \quad \text{where} \quad s_0 = \sqrt{\rho_0/\kappa_0} .$$

(The boundary condition is obtained by noting that for two sheets of identical homogeneous material welded together, there is no reflection at the interface. We put $r(0) = 0$ in (2.5) to obtain $G(0)$.)

3. Oblique incidence.

In the framework of linear elasticity, the equation of motion for an isotropic elastic material is (see Gurtin's treatise [3])

$$\rho\frac{\partial^2 \mathbf{u}}{\partial t^2} = \text{div}\left[\mu(\nabla\mathbf{u}+(\nabla\mathbf{u})^T) + \lambda(\text{div }\mathbf{u})\mathbf{I}\right] \tag{3.1}$$

where $\mathbf{u} = \mathbf{u}(\mathbf{x},t)$ is the displacement vector, the superscript T indicates transpose, ρ is the density, and the coefficients μ and λ are the Lamé moduli. If the mechanical properties ρ, μ, and λ vary with position, we say that the material is *inhomogeneous* while if ρ, μ, and λ are constants in some region, the material is homogeneous in that region. In this paper, we analyze the reflection and refraction of waves for a stratified slab. By a stratified slab, we mean an inhomogeneous isotropic elastic material in three dimensional Euclidean space that lies between two parallel planes and the mechanical properties λ, μ, and ρ of the material depend only on the distance from a face of the slab.

A homogeneous, isotropic linearly elastic material with a strongly elliptic elasticity tensor admits exactly two types of waves: *dilatational waves*, in which the directions of displacement and propagation coincide, and *shear waves*, in which the directions of displacement and propagation are orthogonal to each other. Let D and S denote the dilatational and shear *slowness* (terminology of [6]) given by

$$D = \sqrt{\frac{\rho}{2\mu+\lambda}} \quad \text{and} \quad S = \sqrt{\frac{\rho}{\mu}}.$$

Given a unit vector $\mathbf{d}$, the expression $\mathbf{u}(\mathbf{x},t) = \mathbf{d}f(t-D\mathbf{x}\cdot\mathbf{d})$ defines a dilatational wave which formally satisfies the equation of motion (3.1). Similarly, given two unit vectors $\mathbf{s}$ and $\mathbf{p}$ where $\mathbf{s}\cdot\mathbf{p} = 0$, the expression $\mathbf{v}(\mathbf{x},t) = \mathbf{p}g(t-S\mathbf{x}\cdot\mathbf{s})$ defines a shear wave which formally satisfies (3.1). The vectors $\mathbf{d}$ and $\mathbf{s}$ are the *propagation vectors* for these waves and the shear wave is said to be *polarized* in the direction $\mathbf{p}$. Throughout this paper, we consider harmonic waves; in principle, waves of more general form can be synthesized by the superposition of harmonic waves. The motion of harmonic waves is described by the real or the imaginary parts of the expressions

$$\mathbf{v}(\mathbf{x},t) = \delta\mathbf{d}e^{i\omega(t-D\mathbf{d}\cdot\mathbf{x})} \quad \text{and} \quad \mathbf{v}(\mathbf{x},t) = \sigma\mathbf{p}e^{i\omega(t-S\mathbf{s}\cdot\mathbf{x})}.$$

Consider a plane interface I separating two distinct half-spaces of homogeneous, isotropic elastic materials. A dilatational wave striking the interface typically generates a reflected dilatational wave, a reflected shear wave, a refracted (that is, transmitted) dilatational wave, and a refracted shear wave. Similarly, a shear wave striking the interface typically generates waves of all four types. Therefore, when a combination of dilatational and shear waves impinges upon the interface, eight different waves are generated altogether. The plane formed by the propagation vector of an incident wave and the normal to the interface I is called the *plane of incidence* for the wave. The propagation vectors of the outgoing waves are determined by a set of equations known as *Snell's Laws* which we state as follows:

The propagation vectors $\mathbf{d}_r$ and $\mathbf{s}_r$ for a reflected wave and the propagation vectors $\mathbf{d}_t$ and $\mathbf{s}_t$ for the transmitted wave lie in the plane of incidence for the incoming wave. Moreover, if $\mathbf{m}$ is a unit vector in the intersection of the interface and the plane of incidence, then for an incident dilatational wave with propagation vector $\mathbf{d}$, we have

$$D\mathbf{d}\cdot\mathbf{m} = D\mathbf{d}_r\cdot\mathbf{m} = S\mathbf{s}_r\cdot\mathbf{m} = D_t\mathbf{d}_t\cdot\mathbf{m} = S_t\mathbf{s}_t\cdot\mathbf{m}\,, \tag{3.2}$$

and for an incident shear wave with propagation vector $\mathbf{s}$, we have

$$S\mathbf{s}\cdot\mathbf{m} = D\mathbf{d}_r\cdot\mathbf{m} = S\mathbf{s}_r\cdot\mathbf{m} = D_t\mathbf{d}_t\cdot\mathbf{m} = S_t\mathbf{s}_t\cdot\mathbf{m}\,. \tag{3.3}$$

Throughout this paper, the subscript t is attached to parameters associated with the transmitted wave while the subscript r is associated with the reflected wave. Given the unit propagation vectors $\mathbf{d}$ or $\mathbf{s}$ of the incident waves, equations (3.2) and (3.3) determine the propagation vectors of the corresponding scattered waves. Note that if a pair of incident dilatational and shear waves share a common plane of incidence and if they satisfy the relation $D\mathbf{d}\cdot\mathbf{m} = S\mathbf{s}\cdot\mathbf{m}$, then the two reflected waves have the same direction and the two refracted waves have the same direction. In other words, there are four rather than eight outgoing waves. A pair $(\mathbf{d},\mathbf{s})$ of incident waves which lie in the same plane of incidence and which satisfy the relation $D\mathbf{d}\cdot\mathbf{m} = S\mathbf{s}\cdot\mathbf{m}$ will be called a *conjugate* pair of waves. Observe that both the reflected pair and the refracted pair corresponding to a conjugate incident pair will be conjugate. Moreover, any incident pair of waves can be expressed as the sum of conjugate pairs of waves.

4. Reflection and transmission tensors.

Let us consider a stratified half-space attached at the interface I to a homogeneous half-space. We assume that the displacement field in the homogeneous half-space is the superposition of a conjugate pair of waves:

$$\mathbf{v}(\mathbf{x},t) = \delta\mathbf{d}e^{i\omega(t-D\mathbf{d}\cdot\mathbf{x})} + (\sigma\mathbf{p}+\psi\mathbf{q})e^{i\omega(t-S\mathbf{s}\cdot\mathbf{x})} \tag{4.1}$$

where $\mathbf{p}$ lies in the plane of incidence and $\mathbf{q}$ is perpendicular to the plane of incidence. The interaction of these waves with the interface I generates a reflected wave of the form

$$\mathbf{v}_r(\mathbf{x},t) = \delta_r \mathbf{d}_r e^{i\omega(t-D\mathbf{d}_r\cdot\mathbf{x})} + (\sigma_r \mathbf{p}_r + \psi_r \mathbf{q}) e^{i\omega(t-S\mathbf{s}_r\cdot\mathbf{x})} \tag{4.2}$$

where the vectors $\mathbf{d}_r$ and $\mathbf{s}_r$ are obtained from Snell's Laws and the vector $\mathbf{p}_r$, which lies in the plane of incidence, is determined by the orthogonality condition $\mathbf{s}_r \cdot \mathbf{p}_r = 0$.

The "reflectivity" of the interface is essentially the linear transformation that relates the incident amplitudes δ, σ, and ψ to the reflected amplitudes δ_r, σ_r, and ψ_r. However, the analysis of wave reflection is simplified if an equivalent notion of reflectivity is employed. Observe that *if the interface I contains the origin*, then for $\mathbf{x}$ in I, the conjugacy assumption and Snell's Laws imply that

$$\mathbf{v}(\mathbf{x},t) = (\delta\mathbf{d}+\sigma\mathbf{p}+\psi\mathbf{q})e^{i\omega(t-D\mathbf{d}\cdot\mathbf{x})} \quad \text{and} \quad \mathbf{v}_r(\mathbf{x},t) = (\delta_r\mathbf{d}_r+\sigma_r\mathbf{p}_r+\psi_r\mathbf{q})e^{i\omega(t-D\mathbf{d}\cdot\mathbf{x})} .$$

We call the vectors $\mathbf{a} = \delta\mathbf{d} + \sigma\mathbf{p} + \psi\mathbf{q}$ and $\mathbf{a}_r = \delta_r\mathbf{d}_r + \sigma_r\mathbf{p}_r + \psi_r\mathbf{q}$ the *vector amplitudes* of the incident and reflected waves. The linear transformation $\mathbf{R} : \mathbf{a} \rightarrow \mathbf{a}_r$ is the *reflection tensor* or the *reflectivity* corresponding to the stratified half-space and the propagation vectors $\mathbf{d}$ and $\mathbf{s}$. Since there is an invertible linear mapping between $\mathbf{a}$ and the coefficients δ, σ, and ψ and between $\mathbf{a}_r$ and the coefficients δ_r, σ_r, and ψ_r, a formula for the reflectivity relative to the vector amplitudes is equivalent to a formula for the reflectivity relative to the coefficients. Although the reflection tensor for the stratified half-space depends on the material properties for the homogeneous half-space, our notation suppresses this dependence for brevity.

As a special case of a stratified half-space attached to a homogeneous half-space, let us consider a stratified slab which is interposed between two homogeneous half-spaces with boundaries I and I_0. The incident wave (4.1), propagating through one half-space and impinging on its boundary I, generates a refracted wave

$$\mathbf{v}_t(\mathbf{x}',t) = \delta_t \mathbf{d}_t e^{i\omega(t-D_t\mathbf{d}_t\cdot\mathbf{x}')} + (\sigma_t \mathbf{p}_t + \psi_t \mathbf{q}) e^{i\omega(t-S_t\mathbf{s}_t\cdot\mathbf{x}')} \tag{4.3}$$

where the primed coordinate system is a translation of the unprimed coordinate system chosen so that the origin in the primed system lies in I_0 and the vector connecting the unprimed origin to the primed origin is orthogonal to I. The vector amplitude of the transmitted wave is

$$\mathbf{a}_t = \delta_t\mathbf{d}_t + \sigma_t\mathbf{p}_t + \psi_t\mathbf{q} .$$

The linear transformation $\mathbf{T} : \mathbf{a} \rightarrow \mathbf{a}_t$ is the *transmission tensor* for the stratified slab and the direction vectors $\mathbf{d}$ and $\mathbf{s}$. Again, we suppress the dependence of $\mathbf{T}$ on the material properties for the homogeneous half-spaces.

Using the vectors **d**, **p**, and **q** in (4.1) and introducing the unit outward normal vector **n** for the homogeneous half-space, we form 3×3 matrices **A** and **B** in the following way:

$$\mathbf{A} = \left[\, \mathbf{d} \mid \mathbf{p} \mid \mathbf{q} \,\right] \quad \text{and} \quad \mathbf{B} = \left[\, D\{2\mu(\mathbf{d}\cdot\mathbf{n})\mathbf{d} + \lambda\mathbf{n}\} \mid S\mu\{(\mathbf{s}\cdot\mathbf{n})\mathbf{p} + (\mathbf{p}\cdot\mathbf{n})\mathbf{s}\} \mid S\mu(\mathbf{s}\cdot\mathbf{n})\mathbf{q} \,\right].$$

Although these matrices depend on the choice of the coordinate system, it is easily seen that the product $\mathbf{BA}^{-1}$ defines a linear transformation that is independent of the choice of coordinate system. We call the tensor $\mathbf{II} = \mathbf{BA}^{-1}$ the *local impedance tensor* for the wave (4.1). The stress corresponding to a displacement field **u** is given by

$$\mathbf{S} = \mu(\nabla\mathbf{u}+(\nabla\mathbf{u})^T) + \lambda(\mathrm{div}\,\mathbf{u})\mathbf{I}.$$

As the following proposition (established in [5]) indicates, the local impedance tensor transforms the vector amplitude of the incident wave into the traction vector at the boundary:

PROPOSITION 4.1. *The traction* **Sn** *at the interface I due to the wave (4.1) is given by*

$$\mathbf{Sn} = -i\omega\mathbf{IIa}e^{i\omega(t-D\mathbf{d}\cdot\mathbf{x})} \tag{4.4}$$

where $\mathbf{a} = \delta\mathbf{d} + \sigma\mathbf{p} + \psi\mathbf{q}$ *is the vector amplitude of the incident wave.*

By Snell's Laws, the propagation angle of an incident wave relative to the interface normal **n** is equal to the propagation angle of the reflected wave. The transformation $\mathbf{P} = \mathbf{I} - 2\mathbf{nn}^T$ reflects a vector across the plane perpendicular to **n**. Thus the propagation vectors $\mathbf{d}_r$ and $\mathbf{s}_r$ for the reflected wave are given by $\mathbf{d}_r = \mathbf{Pd}$ and $\mathbf{s}_r = \mathbf{Ps}$. Since the polarization direction is orthogonal to the propagation direction, it follows that $\mathbf{p}_r = \pm\mathbf{Pp}$; our convention is that $\mathbf{p}_r = \mathbf{Pp}$. Applying Proposition 4.1 to the reflected wave (4.2) and substituting $\mathbf{d}_r = \mathbf{Pd}$, $\mathbf{s}_r = \mathbf{Ps}$, and $\mathbf{p}_r = \mathbf{Pp}$, we have

PROPOSITION 4.2. *The traction* **Sn** *at the interface I due to the wave (4.2) is given by*

$$\mathbf{Sn} = i\omega\mathbf{PIIPa}_r e^{i\omega(t-D\mathbf{d}\cdot\mathbf{x})} \tag{4.5}$$

where $\mathbf{a}_r = \delta_r\mathbf{d}_r + \sigma_r\mathbf{p}_r + \psi_r\mathbf{q}$ *is the vector amplitude of the reflected wave.*

Comparing (4.4) to (4.5), we conclude that the relation between the local impedance tensor $\mathbf{II}_r$ for the reflected wave and the local impedance tensor **II** for the incident wave can be expressed $\mathbf{II}_r = -\mathbf{PIIP}$.

5. Solid-solid interfaces.

The reflection and transmission tensors for a stratified slab at oblique incidence can be computed in much the same way that we computed the reflectivity for normal incidence in Section 2. That is, we consider an auxiliary problem where the slab is homogeneous and we express the reflection tensor in terms of a function G depending on properties of the material in the region $x \le \tau$ and the local impedance tensor $\mathbf{H}_1$ corresponding to the region $x > \tau$. After differentiating G and letting the τ tend to zero, we obtain an expression for $\mathbf{G}'(0)$ in terms of $\mathbf{G}(0)$, which leads us to a formula for the reflection tensor. First, some notation: We consider a stratified slab of total thickness a interposed between a stratified half-space and a homogeneous half-space. Let $\mathbf{n}$ denote the outward normal to the homogeneous half-space and let τ measure distance to the stratified half-space. For an incident wave which strikes the stratified slab, the "local propagation directions" $\mathbf{d}(\tau)$ and $\mathbf{s}(\tau)$ at any point in the stratified slab are determined by Snell's Laws. Accordingly, at each cross-section of the stratified slab, we can define a local impedance tensor $\mathbf{H}(\tau) = \mathbf{B}(\tau)\mathbf{A}(\tau)^{-1}$ where $\mathbf{A}$ and $\mathbf{B}$ are defined in Section 4. Similarly, we define the local tensor

$$\mathbf{F}(\tau) = \mathbf{A}(\tau)\begin{bmatrix} D(\tau)\mathbf{d}(\tau)\cdot\mathbf{n} & 0 & 0 \\ 0 & S(\tau)\mathbf{s}(\tau)\cdot\mathbf{n} & 0 \\ 0 & 0 & S(\tau)\mathbf{s}(\tau)\cdot\mathbf{n} \end{bmatrix}\mathbf{A}(\tau)^{-1} .$$

With these definitions, the reflection and transmission tensors can be expressed in the following way:

THEOREM 5.1. $\mathbf{R}(a) = (\mathbf{P}\mathbf{H}_1\mathbf{P}+\mathbf{G}(a))^{-1}(\mathbf{H}_1-\mathbf{G}(a))$ and $\mathbf{T}(a) = \mathbf{Q}(a)(\mathbf{G}(a)+\mathbf{P}\mathbf{H}_1\mathbf{P})^{-1}(\mathbf{H}_1+\mathbf{P}\mathbf{H}_1\mathbf{P})$ *where* G and Q *satisfy the differential equations*

$$\begin{aligned}\mathbf{G}'(\tau) = i\omega\big[&(\mathbf{P}\mathbf{H}(\tau)+\mathbf{G}(\tau)\mathbf{P})\mathbf{F}(\tau)[\mathbf{H}(\tau)\mathbf{P}+\mathbf{P}\mathbf{H}(\tau)]^{-1}(\mathbf{H}(\tau)-\mathbf{G}(\tau)) \\ &+ (\mathbf{H}(\tau)-\mathbf{G}(\tau))\mathbf{F}(\tau)[\mathbf{H}(\tau)\mathbf{P}+\mathbf{P}\mathbf{H}(\tau)]^{-1}(\mathbf{H}(\tau)\mathbf{P}+\mathbf{P}\mathbf{G}(\tau))\big]\end{aligned}$$

and

$$\mathbf{Q}'(\tau) = i\omega\mathbf{Q}(\tau)\big[\mathbf{P}\mathbf{F}(\tau)[\mathbf{H}(\tau)\mathbf{P}+\mathbf{P}\mathbf{H}(\tau)]^{-1}(\mathbf{H}(\tau)-\mathbf{G}(\tau)) - \mathbf{F}(\tau)[\mathbf{H}(\tau)\mathbf{P}+\mathbf{P}\mathbf{H}(\tau)]^{-1}(\mathbf{H}(\tau)\mathbf{P}+\mathbf{P}\mathbf{G}(\tau))\big] .$$

To apply Theorem 5.1, we must know the value of G and Q at some interface in order to integrate the Riccati equation governing G and the linear equation governing Q. If the stratified half-space is homogeneous and if $\mathbf{H}_0$ denotes the corresponding local impedance matrix, then $\mathbf{G}(0) = \mathbf{H}_0$ and $\mathbf{Q}(0) = \mathbf{I}$. A fundamental assumption underlying Theorem 5.1 is that the incident waves form a conjugate pair. If the incident waves do not form a conjugate pair, then they must be expressed as the sum of two conjugate pairs and a separate G and Q must be computed for each conjugate pair.

6. Solid-liquid interfaces.

The fundamental difference between wave propagation in solids and wave propagation in inviscid fluids is that fluid media do not support shear waves. Thus the number of unknowns in a solid-liquid contact problem is reduced by one. On the other hand, at the solid-liquid interface, the continuity requirement is reduced since only the normal components of stress and displacement are continuous necessarily. These two reductions balance out so that the analysis of a solid-liquid interface is similar to the analysis of a solid-solid interface. Let us consider a stratified solid slab of thickness a interposed between a stratified solid half-space and a homogeneous fluid half-space. We characterize the fluid as an elastic material with Lamé moduli $\mu_1 = 0$ and $\lambda_1 > 0$ and with dilatational slowness D_1. Letting $\mathbf{n}$ denote the outward normal to the fluid half-space, consider a plane wave with propagation direction $\mathbf{d}_1$ traveling in the fluid medium and impinging upon the stratified solid medium. The following result is established in [5]:

THEOREM 6.1. *The reflection and transmission tensors for the solid stratified slab are*

$$\mathbf{R}(a) = (\mathbf{n}^T\mathbf{d}_1+\lambda_1 D_1\mathbf{n}^T\mathbf{G}(a)^{-1}\mathbf{n})^{-1}(\mathbf{n}^T\mathbf{d}_1-\lambda_1 D_1\mathbf{n}^T\mathbf{G}(a)^{-1}\mathbf{n})(\mathbf{I}-2\mathbf{n}\mathbf{n}^T)$$

and

$$\mathbf{T}(a) = 2\lambda_1 D_1(\mathbf{n}^T\mathbf{d}_1+\lambda_1 D_1\mathbf{n}^T\mathbf{G}(a)^{-1}\mathbf{n})^{-1}\mathbf{Q}(a)\mathbf{G}(a)^{-1}\mathbf{n}\mathbf{n}^T$$

where $\mathbf{G}$ *and* $\mathbf{Q}$ *satisfy the differential equations of Theorem 5.1.*

7. Computational simplifications.

Although the Riccati equation in Theorem 5.1 seems complicated, there are enormous simplifications when tensors are evaluated in a natural coordinate system. Let us consider the plane of incidence and orient one axis of the coordinate system along the interface normal. Let α denote the angle between the interface normal and the propagation vector for the incident dilatational wave and let β denote the angle between the interface normal and the propagation vector for the incident shear wave. It can be shown that

$$\mathbf{H} = \frac{1}{\cos(\alpha-\beta)}\begin{bmatrix} (2\mu+\lambda)D\cos\beta & -\mu S\sin(\alpha-2\beta) \\ \mu S\sin(\alpha-2\beta) & \mu S\cos\alpha \end{bmatrix},$$

$$\mathbf{PH} = \frac{1}{\cos(\alpha-\beta)}\begin{bmatrix} -(2\mu+\lambda)D\cos\beta & \mu S\sin(\alpha-2\beta) \\ \mu S\sin(\alpha-2\beta) & \mu S\cos\alpha \end{bmatrix},$$

$$\mathbf{HP} = \frac{-1}{\cos(\alpha-\beta)} \begin{bmatrix} (2\mu+\lambda)D\cos\beta & \mu S\sin(\alpha-2\beta) \\ \mu S\sin(\alpha-2\beta) & -\mu S\cos\alpha \end{bmatrix},$$

$$\mathbf{HP}+\mathbf{PH} = \frac{2}{\cos(\alpha-\beta)} \begin{bmatrix} -(2\mu+\lambda)D\cos\beta & 0 \\ 0 & \mu S\cos\alpha \end{bmatrix},$$

$$\mathbf{F} = \frac{1}{\cos(\alpha-\beta)} \begin{bmatrix} D\cos\beta & D\cos\alpha\sin(\alpha-\beta) \\ S\cos\beta\sin(\alpha-\beta) & S\cos\alpha \end{bmatrix}, \text{ and}$$

$$\mathbf{F}(\mathbf{HP}+\mathbf{PH})^{-1}\mathbf{P} = \frac{1}{\rho} \begin{bmatrix} D^2 & DS\sin(\alpha-\beta) \\ DS\sin(\alpha-\beta) & S^2 \end{bmatrix}.$$

Also note that even though the matrix **G** not symmetric, the matrix **PG** is symmetric when the starting condition **PG**(0) is symmetric. Thus when integrating the 2×2 Riccati equation for **PG**, we only need to compute three of the four elements of **PG**.

REFERENCES

[1] R. Bellman, *Methods of Nonlinear Analysis, Vol II*, Academic Press, New York, 1973.

[2] R. Bellman and R. Kalaba, "Functional equations, wave propagation and invariant imbedding," *J. Math. Mech.*, 8(1959), pp. 683–704.

[3] M. E. Gurtin, "Linear Theory of Elasticity," *Handbuch der Physik* VIa/2 (1972), pp. 1–295.

[4] W. W. Hager and R. Rostamian, "Optimal coatings, bang-bang controls, and gradient techniques," *Optimal Control: Applications and Methods*, 8(1987), pp. 1–20.

[5] W. W. Hager and R. Rostamian, "Reflection and refraction of elastic waves for stratified materials," to appear.

[6] B. L. N. Kennett, *Seismic Wave Propagation in Stratified Media*, Cambridge University Press, Cambridge, 1983.

IDENTIFICATION OF PARAMETERS IN DIFFUSION CONVECTION MODELS : APPLICATION TO THE ANALYSIS OF BLOOD FLOW IN DIGITAL SUBTRACTION ANGIOGRAPHY

J. Henry* - Y. Sadikou# - J.P. Yvon*#

* INRIA, Domaine de Voluceau, BP 105, Rocquencourt, 78153 Le Chesnay Cedex, France.
# Université de Technologie de Compiègne, BP 233, 60206 Compiègne Cedex, France.

1 - *PRINCIPLES OF DIGITAL SUBTRACTION ANGIOGRAPHY*

The purpose of angiography is to obtain X-rays images of the evolution of blood-flow in vessels. The standard procedure consists in injecting a contrast medium in the vessel and to get images at a given frequency of the part of the body which contains the vessel of interest.

Presently the images are purely bi-dimensionnal (in the near future stereoscopic views are envisaged). The image received by a video camera is digitalized (256 grey-levels classically) and stored. Due to the fact that images are very quickly memorized it is possible to get a temporal sequence of images (frequency range : 2-24 images/second). Thus the first interest lies in the fact that it is possible to obtain a <u>dynamical</u> information on the evolution of the blood flow in the vessel.

The second aspect of the DSA is the <u>subtraction</u>. The principle is the following. A first image is taken before injection which plays the role of a mask. After the injection of contrast medium in the vessel a "subtraction" of the reference image is made from the current image (the subtracting technique will be detailed below). It is hoped that this procedure will "wipe off" all the objects which are not affected by the experimentation such as : bones, organs, etc...

Actually this is not always the case because of eventual movements of the patient or displacements of the vessel due to the pulse.

The absorption of X-rays obeys the law

$$(1.1) \qquad A = \exp \int_0^{\ell} \mu(r)dr$$

where $\mu(r)$ is a coefficient proportional to the concentration of contrast medium and ℓ is the thickness of the vessel (see Fig.1). For this reason the subtraction is usually logarithmic :

(1.2) $$z_{i,j} = \ell n\, I_o(p_{ij}) - \ell n\, I(p_{ij}) = \int_o^{\ell} \mu(p_{ij},r)dr$$

where I_o and I are respectively the intensity at the pixel p_{ij} of the mask and of the current image. For each image the family of values $\{z_{ij}\}$ furnishes the observation.

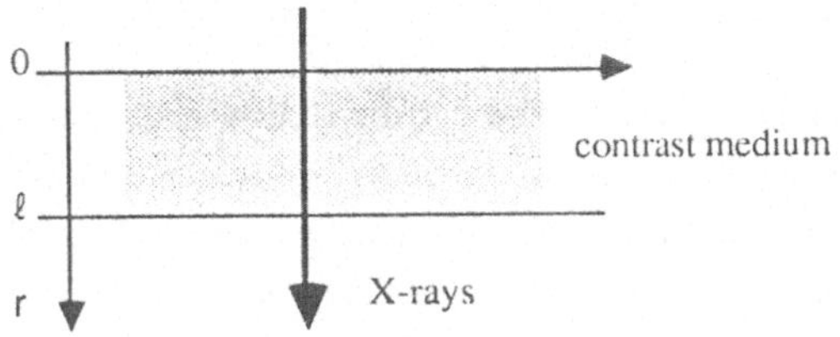

Figure 1

2 - MEDICAL PURPOSES AND DIFFICULTIES

2.1. Medical purpose

At the present time two main goal are envisaged :

- to examine the shape of vessels (essentially in order to detect stenosis),
- to measure global parameters like blood flow velocity.

2.2 Difficulties

Basically the main difficulties are related to the geometry of arteries :

- the cross section of a vessel is not circular,
- the axis of the vessel is not parallel to the plane of the image,
- some branching may exist.

The first problem is then to identify the cross section along the vessel.

The second problem will consist in calculating the of the flow-rate in order to get the average velocity.

Furthermore it is interesting to evaluate the mixing of the contrast medium in the blood which is related to the presence or absence of turbulence in the flow.

2.3 State of the art

The main studies of these problems consist in general in taking a single image (corresponding to the maximum value of concentration of contrast medium in some region of interest) and to calculate the cross section by pixel-densogram.

A first dynamical approach has been made by some authors [1], [4], [5] which uses a representation of the concentration as a "Γ-curve"

$$(2.1) \qquad \Gamma_x(t) = A(t - t_0)^\alpha \exp[- \frac{t - t_0}{\beta}]$$

where $\Gamma_x(t)$ denotes the mean concentration in a cross-section at a point x of the axis of the vessel. The essential drawback of this formalism is that it is very sensitive to the noise because there is no spatial coupling of the function Γ_x at different points x. Nevertheless this method can be used to calculate the time corresponding to the maximum value of concentration at a point x and then to compute the velocity (see n°4 infra).

3 - MODELS AND IDENTIFIABILITY STUDY

In this section, we present some simple phenomenological models of the advection and mixing of the contrast medium in blood vessels. The observation of the phenomenon is quite rich as we can acquire a two dimension projection image (512 × 512) at a relatively high frequency. So we will assume for the identifiability investigation that we have a continuous spatial 1D observation (along the axis of the vessel) or 2D observation and continuous in time. This allows us to derive some simple results of identifiability indicating which parameters are likely to be determined by the observation of the various models.

Let us first recall the definition of local identifiability. Let $\theta \in \Theta$ denote the parameters to be identified, and y the state of the system. The state equation is :

$$A(\theta \; ; y) = f$$

where f is the input. The observation z is defined by :

$$z = \Phi(y).$$

The system is said to be locally identifiable at θ_0 for the input f, if the mapping

$$F : \theta \rightarrow z$$

is locally one to one at θ_0, i.e. : for a neigbourhood O of θ_0

$$\theta, \theta' \in O \qquad z(\theta) = z(\theta') \Rightarrow \theta = \theta'.$$

Furthermore, the identification is said to be locally stable if F^{-1} from $F(O)$ to O is continuous.

3.1 The pure advection equation

The simplest model of a flow is given by the pure advection equation in 1D :

$$(3.1) \qquad \frac{\partial y}{\partial t} + C(x)\frac{\partial y}{\partial x} = 0 \qquad x \in [0,L] \qquad t \in [0,T]$$

with the boundary and initial conditions :

$$(3.2) \qquad y_{|x=0} = e(t) \in C^1([0,T]) \qquad y_{|t=0} = 0$$

The velocity C(x) is the unknown parameter :

$$C(x) = \frac{Q}{S(x)} \geqslant \alpha > 0$$

where Q is the flow rate and S(x) the cross-section area. C(x) is assumed to belong to a bounded set of $H^1(0,L)$. One has the explicit solution :

$$(3.3) \qquad y(x,t) = e\left(t - \int_0^x \frac{d\xi}{C(\xi)}\right), \qquad \text{if } t > t_0 = \int_0^L \frac{dx}{C(x)}.$$

If y and y* are solutions corresponding to C and C* :

$$(3.4) \qquad y(x,t) - y^*(x,t) = \int_0^x \left(\frac{1}{C(\xi)} - \frac{1}{C^*(\xi)}\right)d\xi \, \frac{de}{dt}(t - \tau)$$

where :

$$\min\left(\int_0^x \frac{d\xi}{C(\xi)}, \int_0^x \frac{d\xi}{C^*(\xi)}\right) \leqslant \tau \leqslant \max\left(\int_0^x \frac{d\xi}{C(\xi)}, \int_0^x \frac{d\xi}{C^*(\xi)}\right).$$

Then :

$$(3.5) \qquad |y(t) - y^*(t)|_{L^\infty(0,L)} \geqslant \inf_{0 \leqslant \tau \leqslant t} \left|\frac{de}{dt}(\tau)\right| \left|\int_0^x \frac{C^*(\xi) - C(\xi)}{C(\xi)C^*(\xi)} d\xi\right|$$

If $y_n \to y^*$ in L^∞, there is a subsequence $C_{\mu_n} \to \tilde{C}$ uniformly. If

$$(3.6) \qquad \inf_t \left|\frac{de}{dt}\right| > 0$$

by (3.5) one gets : $C^* = \tilde{C}$.

That is under the condition (3.6) and $t > t_o$, C(x) is locally identifiable from the observation of y at time t.

In practice it is sufficient to have (3.6) on some subinterval of]0,T[.

3.2 1-D convection diffusion model

The observation of DSA images shows that there is not only an advection of the contrast medium, but also a dispersion. So, we propose the following 1D convection diffusion model :

$$S(x)\frac{\partial x}{\partial t} + Q\frac{\partial y}{\partial x} - \frac{\partial}{\partial x}\left[D(x)S(x)\frac{\partial y}{\partial x}\right] = 0 \tag{3.7}$$

y(x,t) is the average value of the concentration of contrast medium in the section of area S(x) at position x along the axis of the vessel. Let e(t) be the concentration at the inlet, the boundary and initial conditions are :

$$y_{|x=0} = e(t) \qquad \frac{\partial y}{\partial x}_{|x=L} = 0 \qquad y_{|t=0} = 0. \tag{3.8}$$

The observation is given by the formula (1.2) but as the concentration is averaged over a section in this model, we define the observation variable z as :

$$z(t,x) = S(x)\, y(t,x). \tag{3.9}$$

a) Identifiability of D(x)

Let D and D* be two values of the diffusion coefficient and y and y* the corresponding values of y ;

$$\bar{D} = D - D^*, \quad \bar{y} = y - y^*, \quad \bar{z} = z - z^*.$$

One gets :

$$S\frac{\partial \bar{y}}{\partial t} + Q\frac{\partial \bar{y}}{\partial x} - \frac{\partial}{\partial x}\left(SD\frac{\partial \bar{y}}{\partial x}\right) = \frac{\partial}{\partial x}\left(S\bar{D}\frac{\partial y^*}{\partial x}\right). \tag{3.10}$$

Estimates on $\bar{D}$ can be obtained in two ways. First we apply the method proposed by Kunish [3] for elliptic equations. Let χ be the sign of $\bar{D}$. By multiplying (3.10) by $\chi\,(y^* - e(t))$ and integrating over $D =]0,L[\times]0,T[$:

$$\int_0^L S|\bar{D}| \int_0^T \left|\frac{\partial y^*}{\partial x}\right|^2 dxdt = -\int_D \left(S\frac{\partial y}{\partial t} + Q\frac{\partial \bar{y}}{\partial x} - \frac{\partial}{\partial x}\left(SD\frac{\partial \bar{y}}{\partial x}\right)\right)(y^* - e(t))\chi dxdt$$

which leads to the bound

(3.11) $$\int_0^L S|\bar{D}| \int_0^T |\frac{\partial y^*}{\partial x}|^2 \, dxdt \leqslant K \| y^* - e \|_{L^2(D)} \| \frac{\bar{z}}{S} \|_{H^{2,1}(D)}$$

if S remains in a bounded set of $C^o(0,L)$, $S \geqslant \beta > 0$, where

$$H^{2,1}(D) = \left\{ y \in L^2(0,T \; ; H^2(0,L)), \; \frac{dy}{dt} \in L^2(D) \right\}.$$

If we assume that :

$$\int_0^T |\frac{\partial y^*}{\partial x}(x,t)|^2 dt \geqslant \alpha > 0 \qquad \forall x \in [0,L] \; ,$$

then (3.11) gives the local identifiability of D, and the local stability of the identification of D in $L^1(0,L)$ with the weight $S \int_0^T | \frac{\partial y^*}{\partial x}(x,t)|^2 dt$ and the observation of z in $H^{2,1}(D)$.

This means in practice that a good accuracy of the identification can be expected at points x where $\int_0^T | \frac{\partial y^*}{\partial x}(x,t)|^2 \, dt$ is large, that is, in our problem, near the injection point.

Similar results can be obtained in another way : let V and W be defined by :

$$V = \{ y \in H^1(0,L) \; ; y(0) = 0 \; \}.$$
$$W(0,T) = \left\{ y \in L^2(0,T \; ; V) \; ; \frac{dy}{dt} \in L^2(0,T \; ; V) \right\}$$

From (3.10) we derive :

(3.12) $$\| S\bar{D} \frac{\partial y^*}{\partial x} \|_{L^2(D)} \leqslant K \| \frac{\partial}{\partial x} (S\bar{D} \frac{\partial y^*}{\partial x}) \|_{L^2(0,T \; ; V')} \leqslant K' \| \frac{\bar{z}}{S} \|_{W(0,T)}.$$

In both cases an analyticity regularity of y may be used to warrant that the number of points x where $\int_0^T | \frac{\partial y^*}{\partial x}(x,t)|^2 \, dt$ vanishes is at most finite, if e(t) is not constant.

b) Identifiability of S(x)

The equation satisfied by the observation is :

(3.13) $$\frac{\partial z}{\partial t} + \frac{\partial}{\partial x} (Rz) - \frac{\partial}{\partial x} (D \frac{\partial z}{\partial x}) = 0$$

with the boundary and initial conditions :

(3.14) $\quad z|_{t=0} = 0 \ , \quad z|_{x=0} = S(0)e(t) \ , \quad \frac{\partial z}{\partial x}|_{x=L} = 0,$

assuming that S'(L) = 0 and S(0) = S_0 is known. Here R denotes the new unknown parameter :

$$R = \frac{Q}{S} + D\frac{S'}{S}.$$

Calculating the difference $\bar{z}$ corresponding to two parameters R and R* we derive from (3.13) :

(3.15) $\quad \frac{\partial \bar{z}}{\partial t} + \frac{\partial}{\partial x}(R\bar{z}) - \frac{\partial}{\partial x}(D\frac{\partial \bar{z}}{\partial x}) = -\frac{\partial}{\partial x}(\bar{R}\ z^*).$

Assuming furthermore that S'(0) = 0 to have $\bar{R}(0) = 0$, we obtain by the previous method :

(3.16) $\quad \|\bar{R}\ z^*\|_{L^2(D)} \leqslant K\ \|\bar{z}\|_{W(0,T)}.$

This ensures the local identifiability of R and hence of S provided that

$$\int_0^T |z^*(t,x)|^2 dt \geqslant \alpha > 0.$$

Analyticity results may be used as previously.

c) Identifiability of both S(x) and D(x)

The equation for the difference $\bar{z}$ becomes now :

(3.17) $\quad \frac{\partial \bar{z}}{\partial t} + \frac{\partial}{\partial x}(R\bar{z}) - \frac{\partial}{\partial x}(D\frac{\partial \bar{z}}{\partial x}) = \frac{\partial}{\partial x}(\bar{D}\frac{\partial z^*}{\partial x} - \bar{R}\ z^*).$

hence :

(3.18) $\quad \|\bar{R}\ z^* + \bar{D}\frac{\partial z^*}{\partial x}\|_{L^2(D)} \leqslant K\ \|\bar{z}\|_{W(0,T)}.$

The following lemma is necessary to obtain the identifiability.

Lemma : Let $f = (f_i)_{i=1,\ldots,n}$ where f_i belongs to a Hilbert space H. Then :

$$\| \sum_{i=1}^{n} \lambda_i f_i \|_H^2 \geq \alpha(f) \left(\sum_{i=1}^{n} \lambda_i^2 \right)$$

where α is the smallest eigenvalue of the matrix :

$$\begin{pmatrix} \|f_1\|_H^2 & (f_1,f_2) & \dots & \dots \\ (f_1,f_2) & \|f_2\|_H^2 & \dots & \dots \\ \dots & \dots & \dots & \dots \\ \dots & \dots & \dots & \|f_n\|_H^2 \end{pmatrix}$$

Then we obtain, taking $H = L^2(0,T)$:

(3.19) $$\left[\|\bar{R}\|_{L^2(0,L)}^2 + \|\bar{D}\|_{L^2(0,L)}^2 \right]^{\frac{1}{2}} \inf_x \left[\alpha\left(z^*(x), \frac{\partial z^*}{\partial x}(x)\right) \right]^{\frac{1}{2}} \leq K \, |\bar{z}|_{W(0,T)}.$$

3.3 2-D diffusion convection model

For large vessels in the projection of the section of which there is a sufficiently large number of pixels, a 2D model is useful to analyze the variation of concentration along the radial coordinate denoted by r. The vessel is supposed to be axisymetric and the streamlines of the flow are assumed homothetical to the wall.

(3.20) $$\frac{\partial y}{\partial t} + V \frac{\partial y}{\partial x} + V_r \frac{\partial y}{\partial r} - D\left(\frac{\partial^2 y}{\partial r^2} + \frac{1}{r}\frac{\partial y}{\partial r}\right) = 0 \quad , \; x \in]0,L[\quad , \quad r \in]0,r_o(x)[$$

where $r_o(x)$ is the radius of the vessel. Initial and boundary conditions are :

(3.21) $$y|_{x=0} = e(t) \;, \quad \frac{\partial y}{\partial r}\Big|_{r=r_o} = 0 \;, \quad y_{t=0} = 0.$$

The assumptions on the streamlines give :

$$V_r = V \frac{dr_o}{dx} \frac{r}{r_o}.$$

The divergence free condition is not satisfied in general but it is true in mean on a section due to the condition :

(3.22) $$\int_0^{r_o} 2\pi \, r \, V(r,x) dr = Q$$

It is more convenient to transform the domain into a cylinder by taking as new radial coordinate $\frac{r}{r_o(x)}$:

$$\frac{\partial y}{\partial t} + V\frac{\partial y}{\partial x} - \frac{D}{r_o^2}\left(\frac{\partial^2 y}{\partial r^2} + \frac{1}{r}\frac{\partial y}{\partial r}\right) = 0 \quad , \quad r \in]0,1[\ , \tag{3.23}$$

$$\int_0^1 2\pi\, r\, V(r,x)dr = \frac{Q}{r_o^2(x)} \tag{3.24}$$

The observation is related to the absorption of X-ray which is proportional to the distance covered in an homogeneous medium and the concentration of contrast medium. If ρ denotes the coordinate orthogonal to the axis of the vessel in the image plane, the observation z is given by :

$$z(x,\rho,t) = 2\, r_o(x) \int_\rho^1 \frac{y(x,r,t)rdr}{(r^2-\rho^2)^{\frac{1}{2}}} \ . \tag{3.25}$$

This relation can be inverted in :

$$y(x,r,t) = \frac{-1}{\pi\, r_o(x)} \int_r^1 \frac{\partial z}{\partial \rho}\frac{1}{(\rho^2-r^2)^{\frac{1}{2}}}\, d\rho. \tag{3.26}$$

The parameters to be identified are the flow rate Q, the local radial diffusion coefficient D(x), the velocity profile V(r,x). The identifiability can be studied by the methods presented on the 1-D model. The parameters D(x) and V(r,x) are interesting from the medical viewpoint as they represent the local effect of mixing of the contrast medium in the blood which may be altered by the presence of a turbulent flow.

4 - NUMERICAL RESULTS AND COMMENTS

The following numerical results are presented in order to emphasize two points:

- the interest of 1-D distributed model, compared to the classical approach by " Γ-curves ",
- the possible interest of considering a 2-D model.

4.1 Identification of a 1-D model.

Figure 2 shows a comparison between the mean velocity calculated by two methods. The numerical experimentation was the following : the advection-diffusion of the contrast medium is *simulated* by a 1-D model. The results are pertubed by a random noise and the two identification processes are used to estimate the velocity. The first one is based upon the approximation at any point x of the time evolution of the concentration by a function $\Gamma_x(t)$ of the form (2.1). For two points x_1

and x_2 the corresponding functions are $\Gamma_{x_1}(t)$ and $\Gamma_{x_2}(t)$. Let us define t_1 and t_2 as the times where the concentration attains its maximum value, then it is possible to calculate the mean velocity between x_1 and x_2 by the formula

(4.1) $$V(x_2) \simeq \frac{x_2 - x_1}{t_2 - t_1} .$$

The second method empoyed here consists in solving the identification problem on the 1-D model itself. The presented results show that the first method is very sensitive to noise by comparison to an identification method.

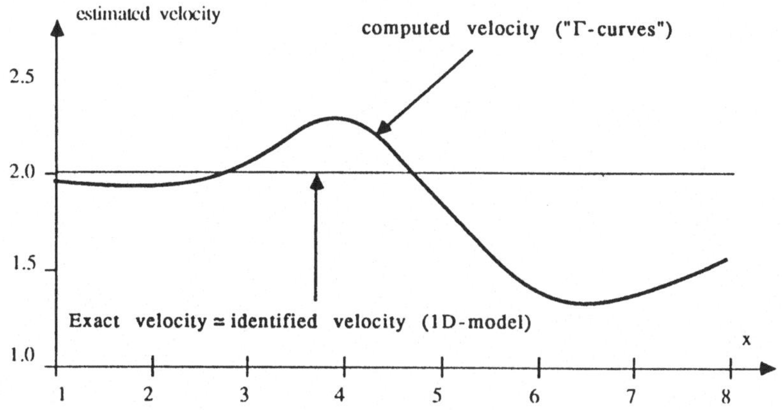

Figure 2

4.2 Interest of 2.D models.

The first question is the following : is there any interest to consider a 2-D model, or in other words, is there phenomena which can be interpreted only by introducing a radial coordinate? The Figure 3 is an attempt to prove radial inhomogeneity. On this figure we have plotted the values of *time* corresponding to the *maximal contrast* at different pixels obtained by a scanning of the vessel in a direction orthogonal to the axis (on this figure three cross sections are shown). It is clear, on this figure that the maximum value of contrast is not attained at the same time for various pixels in the same cross-section. This remark would be reinforced if it was possible to make the same observation along a radius of the vessel itself.

The second result illustrated here is the role of radial diffusivity. Figure 4 shows the value of radial diffusivity along an hypothetical vessel. We have simulated a vessel where at a given point the contrast medium is homogeneized. After identification of the diffusion coefficient it turns out that this coefficient presents great variation precisely at this point.

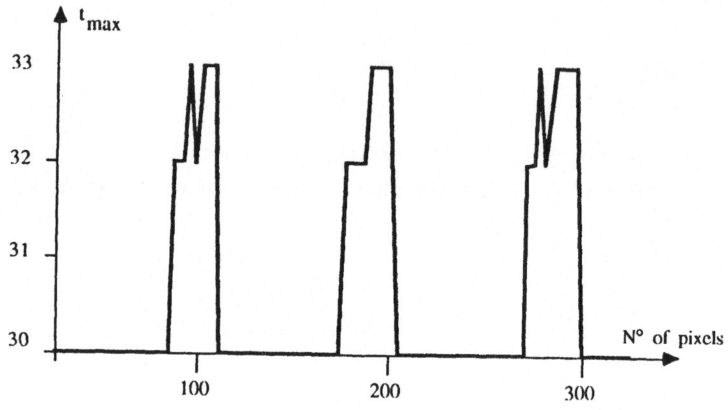

Figure 3

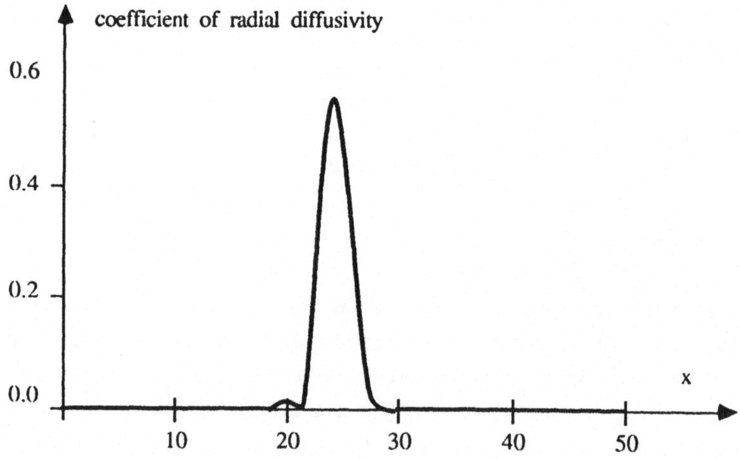

Figure 4

4.3 Some conclusions.

Is clear that distributed models (at least 1-D) may offer a good approach for study of DSA. The main problem in the lack of experimentations both on "phantoms" (physical models) and real blood vessels. It is less clear that the 2-D models can be used "on line" in angiography, but it is possible that they can offer a useful tool to investigate diseases which are less visible on images. Finally the introduction of stereoscopic angiography can offer a better framework to implement identification of parameters in distributed systems.

REFERENCES

[1] Brennecke R., Bursch J.H. (1984) Functional analysis of angiograms by digital image processing techniques. Lecture notes in medical informatics 23, 183-216, Springer 1984.
[2] Kitamura S., Nakagiri S. (1977) Identifiability of spatially-varying and constant parameters in distributed systems of parbolic type. SIAM J. of Control Optim. 15 785-802.
[3] Kunisch K. (1986) A survey of some recent results on the output least squares formulation oof parameter estimation problems. IFAC Control of distributed parmeter systems (Los Angeles USA) 17-23.
[4] Lemetais Ch. (1983) Etude de la dilution d'un produit de contraste dans le sang par le modèle continu. Rapport Thomson-Csf.
[5] Sadikou Y. (1986) Thèse de l'Univeristé de Technologie de Compiègne.

HEAT TRANSFER OPTIMIZATION IN INDUSTRIAL SYSTEMS WITH MULTIPLE STEADY STATES

G. Joly, J.P. Kernévez,

Université de Compiègne, B.P. 233, 60206 Compiègne, France

1. Introduction

The heat flux density q_i transferred from a heating surface to boiling water is a highly non-linear term as a function of the temperature difference θ between the heating surface temperature and the two-phase flow temperature. Due to this flux density shape, previous studies [1,2] revealed the existence of multiple states and of hysteresis effects for heat exchangers such as those heated by liquid sodium. In this paper we consider also an electrically heated steam generator which is used to do experimental measurements about the function q_i and the results are enlarged to liquid heated steam generators.

So, on the one hand, we want to compare both exchanger types and, on the other hand, we will study the electrically heated systems more precisely, in order to know, under which conditions, it is possible to have the best information about the function q_i.

In the figure 1, we have plotted the term q_i as a function of the temperature θ for various values of the mass quality χ of the heated fluid. These curves, called boiling curves, present two critical points corresponding to a sign change in the derivative. It is very difficult to obtain experimentally these points because the temperature grows very rapidly in their vicinity and the equipment may be damaged.

The first section of this paper is devoted to the physical problem presentation and the equations of the model, the second one is about solution existence and we find the first numerical results in the third one. Then we take up optimizing the behaviour of the system in the forth section, give the corresponding numerical results in the following one and we will end by a conclusion.

2. Problem formulation.

The first exchanger (fig.2) we consider is a sodium heated water evaporator. It is made up of two concentric tubes. Water circulates inside the inner tube and sodium flows in the opposite direction in the annular space between the two tubes. In this tube containing a boiling liquid and heated by another fluid, the exchanged heat flux is not the same throughout the tube and there is a thermal coupling between the heating liquid and the two-phase (water-steam) mixture in the inner tube.

The system of equations expresses energy conservation law in the fluids and in the wall and we are only interested by the steady-states of the system. So we have two first order differential equations: one equation for the sodium temperature T_s and one equation for the water mass quality χ.

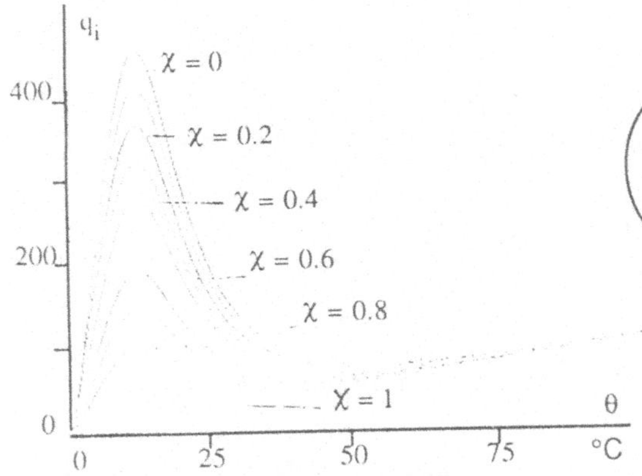

figure 1. Boiling curves

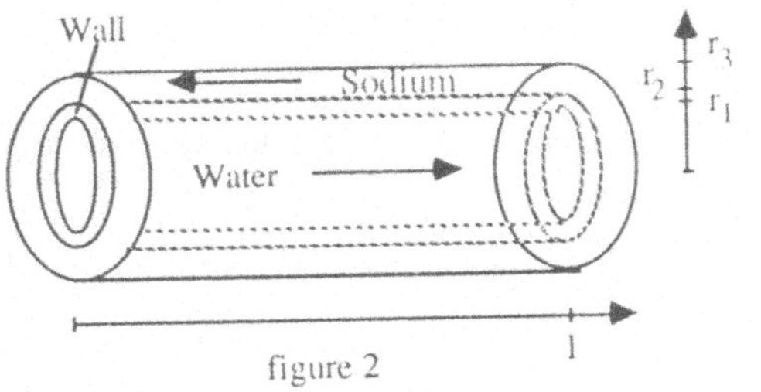

figure 2

$$\begin{cases} -G_sC_pT_s'(z) + \dfrac{2r_1}{r_3^2 - r_2^2} h_s\,(T_s(z) - T_w(z)) = 0\,, & 0 < z < 1 \\ T_s(1) = T_{se} \end{cases} \tag{2.1}$$

$$\begin{cases} GL\chi'(z) - \dfrac{2}{r_1} q_i(T_w(z) - T_{sat}, \chi(z)) = 0, & 0 < z < 1 \\ \chi(0) = 0\,. \end{cases} \tag{2.2}$$

Due to the axisymmetry of the problem, the wall temperature T_w depends on the radial and axial coordinates, and due to the small thickness of the tube wall, we can consider that it depends only on the axial coordonate z. In this case the heat transfer coefficient h_s is global in the sense that it takes into account the wall thickness.

$$\begin{cases} -\lambda_m T_w''(z) + \dfrac{2r_1}{r_2^2 - r_1^2}\left(q_i(T_w(z) - T_{sat}, \chi(z)) - hs\,(T_s(z) - T_w(z))\right) = 0, \; 0 < z < 1 \\ T_w'(0) = T_w'(1) = 0 \end{cases} \tag{2.3}$$

and

$$q_i(\theta,\chi) = h_l\theta + (1-\chi)\frac{k\theta}{1 + k'\theta^4}\,.$$

In the sequel we are always in the case where the boiling liquid is at the saturated state, so the mass quality is less or equal to one.

The solution of the equations (2.1) (2.2) (2.3) is obtained and followed when the sodium inlet temperature T_{se} varies.

The second device is an electric heating evaporator. It is made up of one tube and boiling liquid flows inside the tube. Due to the electric heating the external heat flux is uniform along the tube. Therefore the water-steam flow can become vapour flow. So we have two equations for the heated fluid :

$$\begin{cases} GL\chi'(z) - \dfrac{2}{r_1} q_i(T_w(z) - T_{sat}, \chi(z)) = 0, & 0 < z < z_f \\ \chi(0) = 0\,,\ \chi(z_f) = 1.\ \text{or}\ \ z_f = 1\ \text{if}\ \chi(z) < 1\ \forall z \in [0,1] \end{cases} \tag{2.4}$$

$$\begin{cases} G_sC_p'T_v'(z) - \frac{2}{r_1}h_f\,(T_w(z) - T_v(z)) = 0\,, & z_f < z < 1 \\ T_v(z_f) = T_{sat}\,. \end{cases} \quad (2.5)$$

The equation which gives the wall temperature is the same as (2.3) but is rewritten as

$$\begin{cases} -\lambda_m T_w''(z) + \frac{2r_1}{r_2^2 - r_1^2}(q - q_c) = 0,\ 0 < z < 1 \\ T_w'(0) = T_w'(1) = 0 \end{cases} \quad (2.6)$$

with

$$q = q_i(T_w(z) - T_{sat}, \chi(z)),\ 0 < z < z_f\,,\ \text{and}\ \ q = q_i(T_w(z) - T_p(z), 1)\,,\ \ z_f < z < 1\,,$$

and q_c is the electric flux density.

In these equations, the unknown T_v is the vapour temperature and the new variable f is the abscissa of the free boundary.

The solution of the equations (2.4) (2.5) (2.6) is obtained and we are seeking the curve which represents these solutions when the electric flux density q_c varies.

In the previous equations, many thermal constants are introduced : mass velocity (G_s, G), specific heat at constant pressure (C_p, C_p'), latent heat of vaporization (L), thermal conductivity (λ_m), heat transfer coefficient (h_s, h). The mathematical models that we have just described are based upon a number of simplifying assumptions such as : the physical properties of the sodium and of the wall are assumed to be constant, the coefficient h_s is also assumed to be constant, the water-steam flow is assumed to be governed by the homogeneous model and the pressure losses are disregarded.

3. Some theoretical results about the solution existence

We recall only the results that can be found in [2] about the first system (2.1) (2.2) (2.3).

Under the assumptions :

- $q_i(\theta,\chi) = 0$ if $\theta \le 0$, $\forall \chi \in [0,1]$,
- $\exists\, M > 0$ such that $b\, q_i(\theta,\chi) \le M\,\theta \quad \forall \chi \in [0,1]$
- If $\chi > 0$, $q_i(\theta,\chi) \longrightarrow +\infty$ as $\theta \longrightarrow +\infty$.

there exists at least a solution (T_s, T_w, χ) in $[C^{0,\alpha}(0,1)]^3$ such that

$$0 \le T_w(z) \le T_s(z) \le T_{se} \quad 0 \le z \le 1$$

and if T_{se} is greater than T_{sat} the temperature T_s and the mass quality χ are strictly increasing with respect to z.

To show similar results for the electric system, let us begin to reformulate the equations (2.4) (2.5) (2.6) as

$$\begin{cases} \chi'(x) - c\, q_i(T_p(x) - T_v(x), \chi(x))\, H(f-x) = 0, & 0 < x < 1 \\ \chi(0) = 0,\ \ \chi(f) = 1 \text{ ou } f = 1 \text{ si } \chi(x) < 1\ \forall x \in [0,1], \end{cases} \quad (3.1)$$

$$\begin{cases} T_v'(x) - d\, q_i(T_p(x) - T_v(x), \chi(x))\, H(x-f) = 0\,, & 0 < x < 1 \\ T_v(0) = T_{sat}\,, \end{cases} \quad (3.2)$$

$$\begin{cases} -T_p''(x) + b\,(q_i(T_p(x)-T_v(x),\chi(x)) - q_e) = 0\,, & 0 < x < 1 \\ T_p'(0) = T_p'(1) = 0\,, \end{cases} \qquad (3.3)$$

In these equations, the function H is the Heavyside function, we have brought the interval (0,l) to the interval (0,1) and we have introduced new simpler constants.

We begin the demonstration in introducing a "regularized" problem, that is to say we replace (3.3) by

$$\begin{cases} -T_p''(x) + \varepsilon T_p + b\,(q_i(T_p(x)-T_v(x),\chi(x)) - q_e) = 0\,, & 0 < x < 1 \\ T_p'(0) = T_p'(1) = 0\,. \end{cases} \qquad (3.4)$$

Let be the operator F : (P,X,V) —> (T_p,χ,T_v) defined as :

$$\begin{cases} -T_p''(x) + \varepsilon T_p = b\,(q_c - q_i(P(x)-V(x),X(x))),\ \ T_p'(0) = T_p'(1) = 0 \\ \chi'(x) = c q_i(P(x)-V(x),X(x))H(f-x)\,,\ \ \chi'(0)=0 \\ T_v'(x) + dT_v(x)H(x-f) = dP(x)H(x-f),\ T_v(O)=T_{sat}\,, \end{cases} \qquad (3.5)$$

and let B be the set

$$B = \{\ (P,X,V)\in [L^2(0,1)]^3 \mid \alpha \le P(x) \le \beta\,, 0 \le X(x) \le 1\,, 0 \le V(x) \le \beta\ \}$$

where $\beta = \frac{bq_e}{\varepsilon}$ and , since $P(x)-V(x) \le \beta - T_{sat}$, there exists γ such that $q(P-V,X)\le\gamma$ and $\alpha = \frac{b}{\varepsilon}(q_e-\gamma)$.

The operator applies B into $B\cap[H^1(0,1)]^3$ and if we define the sequence $(P^{n+1},X^{n+1},V^{n+1})=F(P^n,X^n,V^n)$ we can prove that it is possible to take the limit in the variational formulation of the equations (3.5) and therefore there exists a solution of (3.5).

Then "a priori" estimations can be proved directly in (3.4) i.e.

$$0 \le T_p \le T_{max}\,,\ \ 0 \le \chi \le 1\,, T_{sat} \le T_v \le T_{max}\,,$$

where T_{max} is such that $q(T_p-T_{sat},\chi) - q_e \ge 0\ \ \forall\chi\in(0,1),\ \forall T_p \ge T_{max}$.

By making ε —> 0 we get the existence of a solution of (3.1) (3.2) (3.3).

4. Numerical results.

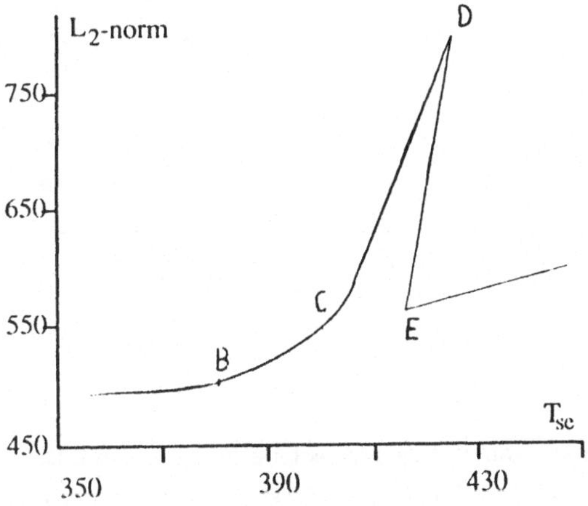

Figure 3.

The system of equations is rewritten as a system of first order differential equations and is solved by the program "AUTO" [3,4] which allows to follow the solution of such a system as a parameter varies. The adaptive mesh selection implemented in AUTO safeguards to some extent against the generation of spurious solutions. The differential equation is approximated by the method of collocation at m Gauss points with piecewise polynomials of C(0,1).

In the case of the sodium heated evaporator the figure 3 shows the presence of two limit points [5] in the curve giving the L_2-norm of the solution as a function of T_{sc}.

If we look at the wall temperature profiles (fig. 4,5,6,7) we remark that, at first the profiles are soft, then a front appears and we can observe its progression up to the inlet of the tube. Then, we find the first limit point corresponding to a wall temperature front situated in the tube inlet. Between this limit point and the second one the temperature front disappears.

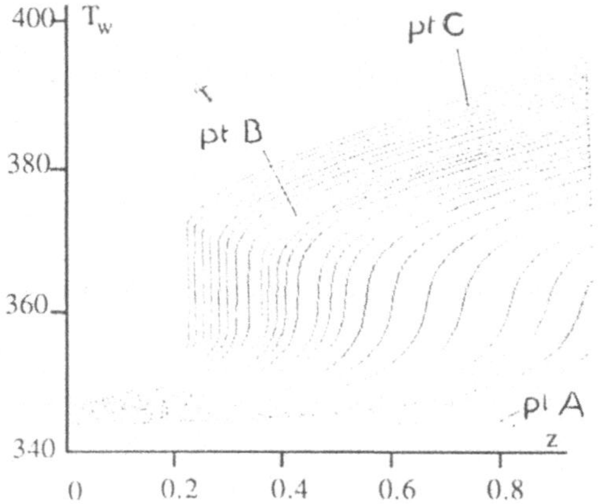

Figure 4.

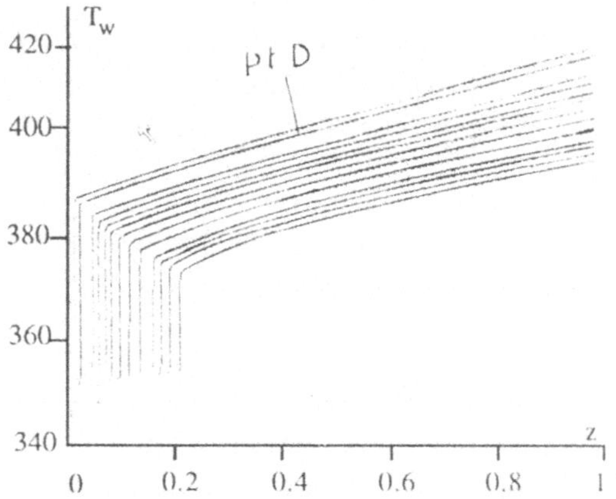

Figure 5.

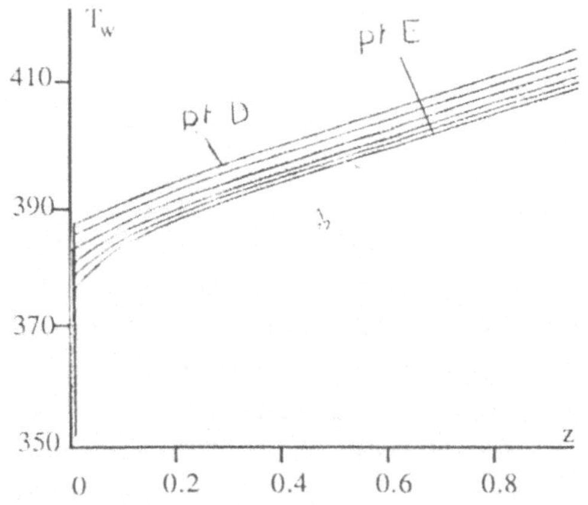

Figure 6.

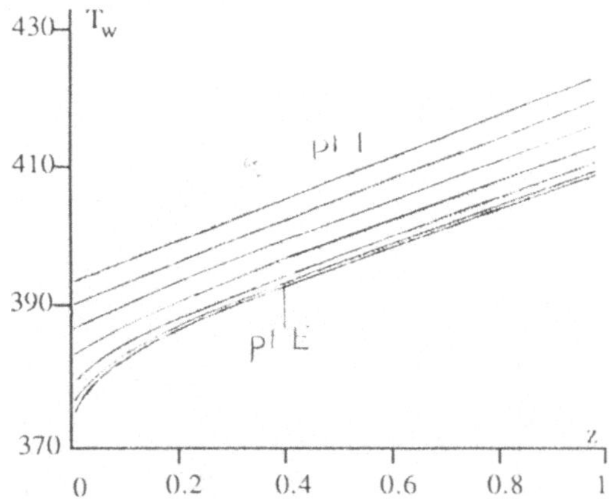

Figure 7.

In the case of an electric evaporator, the results are similar if the tube is long enough (11m.). But if the evaporator is three meters long for example, we see four limit points (fig. 8) due to the fact that the wall temperature front begins to be formed at the outlet of the tube (fig. 9).

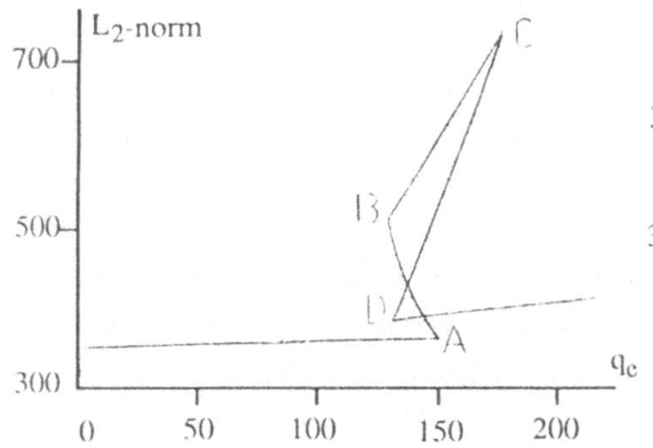

Figure 8.

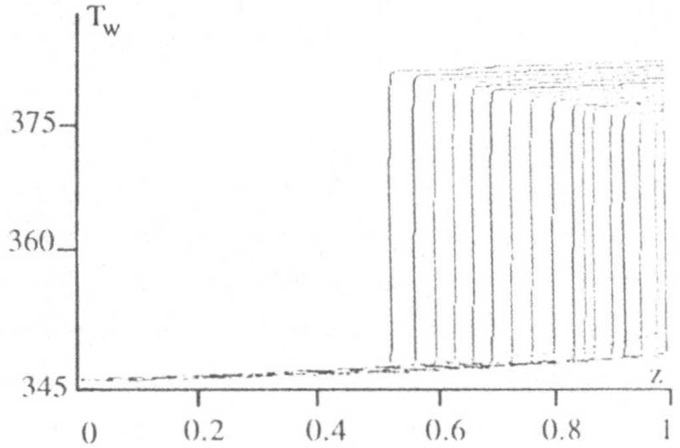

Figure 9.

Let us investigate the nature of the limit points :

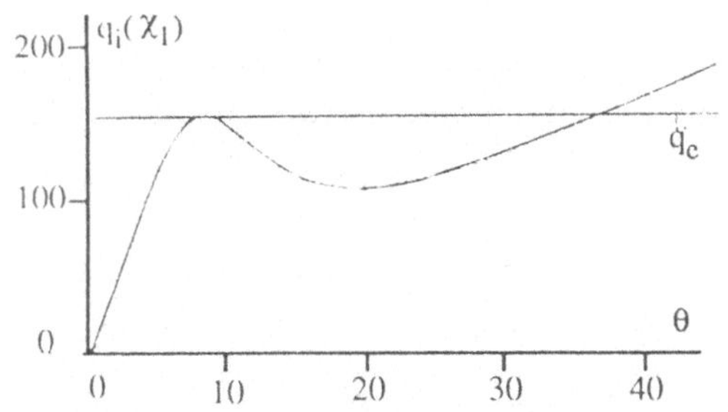

figure 10

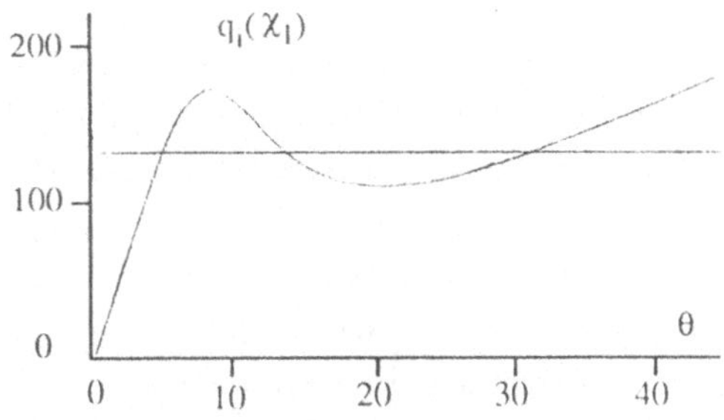

figure 11

- At the first limit point (fig. 10), the line $q=q_c$ is tangent to the boiling curve $q=q_i(\theta,\chi_l)$ at the burning point $\theta=\theta_b$ (which is the first point of zero derivative) and for the mass quality $\chi_l = \chi(l)$.
- At the second limit point (fig. 11), the line $q=q_c$ cuts $q=q_i(\theta,\chi_l)$ into two equal areas.

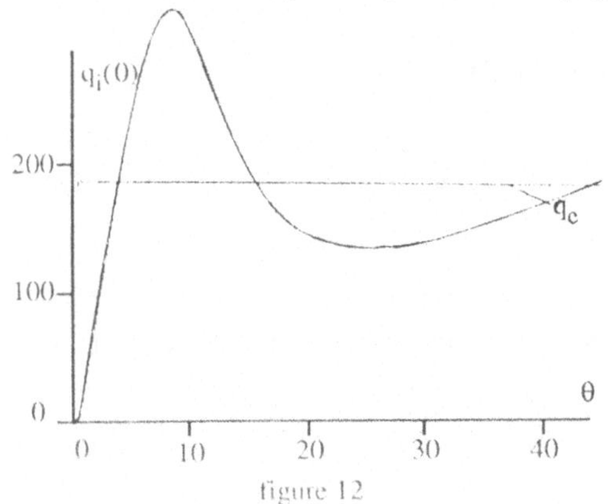

figure 12

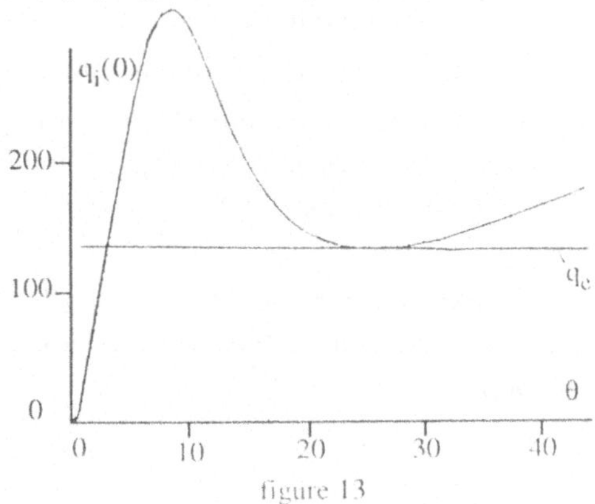

figure 13

- At the third limit point (fig. 12), the line $q=q_c$ cuts $q=q_i(\theta,0)$ into two equal areas.

• At the forth limit point (fig. 13), the line $q=q_e$ is tangent to the boiling curve $q=q_i(\theta,0)$ at the Leidenfrost point $\theta=\theta_r$ (which is the second point of zero derivative).

Now we recall that a limit point corresponds to a change stability of the dynamical system of which the equations (2.4) (2.5) (2.6) give the stationary solutions. But in the last case we do not know if the stationary solutions located between the second and the third limit points are stable or are "twice unstable". The only case for which we can be sure that these solutions are stable (and then reachable by the experimentation) is when the abscissa $q_{c,1}$ of the first limit point is smaller than the abscissa $q_{c,4}$ of the forth limit point. Indeed there in only a solution for the values of q_c such that $q_{c,1}<q_c<q_{c,4}$ and experimentally it can be then possible to get the states of the curve arc before the first limit point, the curve arc between the second and the third limit points, and the last arc after the forth limit point.

5. Optimization problem.

Before formulating the optimization, we observe that the two last limit points are given by the boundary condition on the mass quality $\chi(0)=0$, therefore they do not depend on the length of the tube. So the optimization problem could be expressed as :

" to find a tube length l_{opt} such that $q_{c,1}(l_{opt}) < q_{c,4}$ ".

The first thing to think of is the existence of l such that the continuation curve has four limit points. This has been done in an easier case [6]. Then each limit point is given by the solution of the equations (3.4) (3.5) (3.6) written in a general form as :

$$\begin{cases} u'(t) = f(u(t),q_c) \text{ , } u(t)\in R^4 \\ b(u(0),u(1),q_c) = 0 \text{ , } b(\ )\in R^4 \end{cases} \quad \Leftrightarrow \quad F(u,q_c) = 0 \text{ ,}$$

(5.1)

with p is the continuation parameter, and the equations giving $Ker(F_u)$

$$\begin{cases} v'(t) = f_u(u(t),q_c)v(t) \text{ , } v(t)\in R^4 \\ \int_0^1 v^*(t)v(t)dt - 1 = 0 \\ b_{u(0)}(u(0),u(1),q_c)v(0) + b_{u(1)}(u(0),u(1),q_c)v(1) = 0 \text{ .} \end{cases} \tag{5.2}$$

And the four limit points are ordered by the arc length.

A formulation of the optimization could be

$$J(l_{opt}) \le J(l) \quad \forall l\in [l_1,l_2]$$

where

$$J(l) = - [(q_{c,1}(l) - q_{c,4})^-]^2 \text{ and } (q_{c,1}(l) - q_{c,4})^- = \inf(0,q_{c,1}(l) - q_{c,4}). \tag{5.3}$$

If we could prove that the "admissible set" is not empty, we could demonstrate [7] the existence of an optimal control.

6. Limit point continuation

To solve numerically the optimization problem we compute the curve of the limit points as a function of the tube length. Then we seek for the solution of the system (5.1) (5.2) as a function of the tube length.

We have eight first order differential equations, eight boundary conditions and an integral equation. And we have eight unknown real functions and two real variables l and q_c. Then the system has one degree of freedom and the "AUTO" program allows to get the L_2-norm of the solution as a function of the parameter l.

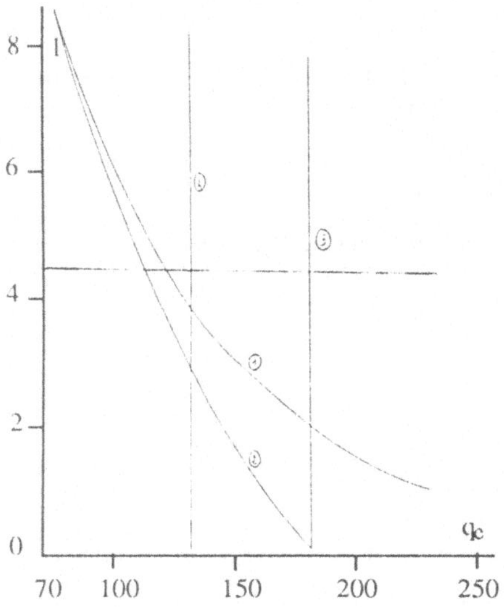

figure 14 .

In figure 14, we have not plotted the L_2-norm of the solution, but the tube length as a function of the q_c variable. And we see that if l is greater than 4 meters the first limit point is on the left of the forth limit point. But when the first limit point $q_{c,1}$ is at the farest distance on the left of $q_{c,4}$, it coalesces with the second limit point before disappearing. So the solution of the previous optimization problem should not give a "good" solution.

It seems that the length had to be chosen around 4.5 in order to give an admissible solution for the engineers.

7. Conclusion

With these results, it seems possible to get the critical points of the boiling curves, on the one hand, by modifying the mass quality of the fluid in the tube inlet, and on the other hand, by chosing the tube length.

In conclusion, the study highlights the prevailing impact of the heating system characteristics (heating fluid, geometrical dimensions of the tube) on the overall behavior of the system. To obtain such results it seems necessary to use not only mathematical modelling, but also numerical analysis of these models. And at last the path-following methods have proved a powerful tool for the analysis of non-linear effects met in thermal systems cooled by a boiling liquid and even to optimize the behaviour of these systems.

References.

[1] M. LLory, G. Joly, J.P. Kernévez, "Thoretical considerations on the boiling crisis in boiling systems under high heat flux conditions", The 8th International Heat Transfer Conference and Exhibition, Ed. C.L. Tien, V.P. Carey, J.K. Serell, H.P.C., Vol. 5, pp. 2191-2195, San Francisco 1986.

[2] E.J. Doedel, G. Joly, J.P. Kernévez, "Continuation of a steam temperature front in nonlinear heat transfer", in Semigroups, theory and applications, vol. 1, N°14, Eds Brezis, Crandall and Kapell, Pitman Research Notes in Math. Series, Longman Scientific and Technical, Karlow, UK, 1986.

[3] E.J. Doedel, J.P. Kernévez, "AUTO : Software for Continuation and Bifurcation Problems in Ordinary Differential Equations", Applied Mathematics Technical Report, CALTEC, 1986, 226 pages.

[4] H.B. Keller, "Numerical Solution of Bifurcation and Nonlinear Eigenvalues Problems" in Applications of Bifurcation Theory, pp. 359-384, P.H. Rabinowitz, Ed. Academic Press, New-York 1977.

[5] W.C. Rheinboldt, "Numerical Analysis of Parametrized Nonlinear Equations", University of Arkansas, Lecture Notes in the Mathematical Sciences, Wiley-Interscience, 1986.

[6] G. Joly, J.P. Kernévez, M. Llory, "Thermal Instability in pool boiling on wires at constant pressure", SIAM J. APPL. MATH., Vol. 43, N°6, 1294-1309, 1983.

[7] J.L. Lions, "Contrôle optimal de systèmes distribués singuliers", Dunod, Paris, 1983.

INFINITE HORIZON LINEAR–QUADRATIC REGULATOR PROBLEMS FOR BEAMS AND PLATES

John E. Lagnese
Department of Mathematics
Georgetown University
Washington, DC 20057 USA

1. Introduction and Setting of the Problem. We consider a thin, homogeneous, isotropic, elastic plate of uniform thickness h. Points within the plate will be represented by rectangular coordinates (x,y,z). It is assumed that the plate has a middle surface midway between its faces which, when the plate is in equilibrium, occupies a region Ω of the plane z=0. Let w(x,y,t) denote the z–component of the displacement vector of the point in the middle plane which has coordinates (x,y,0) when the plate is in equilibrium. The classical Kirchhoff model for the dynamics of the deflection w of the plate (in the absence of body forces) is

$$(1.1) \qquad \rho h w'' + D\Delta^2 w = 0 \qquad \text{in } \Omega\times(0,\infty)$$

where $'=d/dt$, Δ is the ordinary Laplacian in x,y variables, ρ is the mass density per unit volume and $D=Eh^3/12(1-\mu^3)$ is the *flexural rigidity* of the plate (assumed to be constant). E is Young's modulus and μ is Poisson's ratio. For elastic bodies one has $0<\mu<1/2$. Of course, (1.1) assumes that the deflections are "small". In addition, rotational inertia and, more importantly, transverse shear effects, are not represented in (1.1).

By the change of time scale $t\to t\sqrt{\rho h/D}$, (1.1) may be brought to the form

$$(1.2) \qquad w'' + \Delta^2 w = 0 \qquad \text{in } \Omega\times(0,\infty).$$

Henceforth we shall use (1.2) to describe the deflection w.

We consider the situation where the (middle plane of the) plate is clamped along a nonempty portion Γ_0 of its edge. (We assume throughout that $\partial\Omega$ is smooth.) On the remaining part Γ_1 of $\partial\Omega$ we suppose the plate to be free of bending moments but to be subject to a shear force (0,0,–v). Thus the *boundary conditions* on $\partial\Omega$ are (see G. DUVAUT and J.L. LIONS [2])

$$(1.3) \qquad w = \frac{\partial w}{\partial\nu} = 0 \qquad \text{on } \Gamma_0\times(0,\infty),$$

$$(1.4) \qquad \Delta w + (1-\mu)B_1 w = 0, \qquad \frac{\partial\Delta w}{\partial\nu} + (1-\mu)B_2 w = v \qquad \text{on } \Gamma_1\times(0,\infty)$$

where ν denotes the unit exterior normal to Γ and where

$$B_1 w = 2\nu_1\nu_2\frac{\partial^2 w}{\partial x\partial y} - (\nu_1)^2\frac{\partial^2 w}{\partial y^2} - (\nu_2)^2\frac{\partial^2 w}{\partial x^2},$$

$$B_2 w = \frac{\partial}{\partial\tau}\{[(\nu_1)^2 - (\nu_2)^2]\frac{\partial^2 w}{\partial x\partial y} + \nu_1\nu_2(\frac{\partial^2 w}{\partial y^2} - \frac{\partial^2 w}{\partial x^2})]\}, \qquad \tau = \{-\nu_2,\nu_1\}.$$

The initial conditions are

$$(1.5) \qquad w(0) = w^0, \qquad w'(0) = w^1.$$

In (1.4), v is the *control variable* through which the evolution of w is to be managed.

We shall assume that

(1.6) $\quad v \in L^2(\Gamma_1 \times (0,\infty)) = L^2(\Sigma_1),\ \Sigma_1 = \Gamma_1 \times (0,\infty).$

With regard to the initial data, we assume that

(1.7) $\quad w^0 \in V = \{\varphi: \varphi \in H^2(\Omega),\ \varphi = \frac{\partial \varphi}{\partial \nu} = 0 \text{ on } \Gamma_0\},\ \ w^1 \in H = L^2(\Omega).$

The *strain energy* P within the plate (due to bending) is expressed in terms of w by

$$P(v;t) = \frac{1}{2} \int_\Omega \left[\left[\frac{\partial^2 w}{\partial x^2}\right]^2 + \left[\frac{\partial^2 w}{\partial y^2}\right]^2 + 2\mu \left[\frac{\partial^2 w}{\partial x^2}\right]\left[\frac{\partial^2 w}{\partial y^2}\right] + 2(1-\mu)\left[\frac{\partial^2 w}{\partial x \partial y}\right]^2\right] dxdy$$

(see J.E. LAGNESE and J.L. LIONS [4], Chapter I) while the *kinetic energy* (in bending) is

$$K(v;t) = \frac{1}{2} \int_\Omega (w')^2 dxdy.$$

The *total energy* is given by

$$E(v;t) = P(v;t) + K(v;t).$$

One can show that (it is a version of Korn's Lemma; see [4], Chapter II)

$$\left\{ \int_\Omega \left[\left[\frac{\partial^2 w}{\partial x^2}\right]^2 + \left[\frac{\partial^2 w}{\partial y^2}\right]^2 + 2\mu \left[\frac{\partial^2 w}{\partial x^2}\right]\left[\frac{\partial^2 w}{\partial y^2}\right] + 2(1-\mu)\left[\frac{\partial^2 w}{\partial x \partial y}\right]^2\right] dxdy \right\}^{\frac{1}{2}}$$

defines a norm on V *equivalent* to the one induced by the standard topology of $H^2(\Omega)$. Therefore, the requirement (1.7) on $\{w^0,w^1\}$ is the same as

$$E(v;0) < \infty.$$

Let us introduce the *cost function*

(1.8) $$J(v) = \int_0^\infty E(v;t)dt + N\int_{\Sigma_1} v^2 d\Gamma dt$$

where $N>0$ is a constant. We consider the following *optimal control problem*:

(1.9) $\quad \inf\{J(v): v \in U(w^0,w^1)\}$

where the *admissible set* $U(w^0,w^1)$ is

(1.10) $\quad U(w^0,w^1) = \{v: v \in L^2(\Sigma_1),\ J(v) < \infty\}.$

Remark 1.1. For initial data $\{w^0,w^1\}$ and control v which satisfy (1.7) and (1.6), respectively, it is ***not true in general*** that $E(v;t) < \infty$ for all $t>0$, that is, solutions with finite initial energy may acquire infinite energy in finite time. Implicit in the definition of $U(w^0,w^1)$ is that we admit *only* those controls $v \in L^2(\Sigma_1)$ for which $E(v;t) < \infty$ for every $t>0$ (and, of course, $E(v;\cdot) \in L^1(0,\infty)$).

Remark 1.2. The "program" that we wish to achieve is this:

(I) Show that (1.2)–(1.10) admits a unique optimal control $u \in L^2(\Sigma_1)$.

(II) Realize the optimal control as a *feedback*

(1.11) $\quad u(t) = R\{y(t),y'(t)\}$

for an appropriate $R \in \mathscr{L}(V \times H; L^2(\Gamma_1))$.

If (I) and (II) are achieved, it follows *automatically* that the closed loop system (1.2)–(1.6) with v=u given by (1.11) has an *exponential rate of energy decrease*, i.e.,

(1.12) $$E(v;t) \le Ce^{-\omega t}E(0)$$

for some *positive* constant ω. In fact, a *feedback* control $v \in L^2(\Sigma_1)$ will be in $U(w^0,w^1)$ *if and only if* (1.12) holds. This is so because of a result of R. DATKO [1] which guarantees (1.12) once it is known that $E(v;\cdot) \in L^1(0,\infty)$.

The existence of a unique optimal control will be demonstrated in Section 2. However, we are unable to carry out part (II) of our program (because of a difficult technical issue which will be discussed in Section 3) except in the one–dimensional situation, i.e., an Euler beam clamped at one end and controlled at its free end. In that case, by utilizing a formalism developed by J.L. LIONS in [7] an optimal feedback operator R is obtained. Furthermore, R may be characterized as the solution of an algebraic operator equation with a quadratic nonlinearity. This analysis is carried out in Section 4.

2. Existence of a Unique Optimal Control. We first find a *feedback* control $v \in U(w^0,w^1)$. To this end, let us formally calculate the time derivative of the energy functional $E(v;t)$. One obtains

(2.1) $$\begin{aligned}\frac{d}{dt}E(v;t) &= \int_\Omega (w'' + \Delta^2 w)w'dxdy + (1-\mu)\int_\Omega [2\frac{\partial^2 w}{\partial x \partial y}\frac{\partial^2 w'}{\partial x \partial y} - \\ &\qquad \frac{\partial^2 w}{\partial x^2}\frac{\partial^2 w'}{\partial y^2} - \frac{\partial^2 w}{\partial y^2}\frac{\partial^2 w'}{\partial x^2}]dxdy + \int_{\Gamma_1} [\Delta w \frac{\partial w'}{\partial \nu} - \frac{\partial \Delta w}{\partial \nu} w']d\Gamma \\ &= \int_{\Gamma_1} \{[\Delta w + (1-\mu)B_1 w]\frac{\partial w'}{\partial \nu} - [\frac{\partial \Delta w}{\partial \nu} + (1-\mu)B_2 w]w'\}d\Gamma \\ &= -\int_{\Gamma_1} vw'd\Gamma.\end{aligned}$$

Since *any* admissible feedback control must lead to (1.11), (2.1) suggests as a candidate for an admissible control

(2.2) $$v = w'|_{\Sigma_1}.$$

If we define $E(t) = E(w'|_{\Sigma_1};t)$, it follows from (2.1), (2.2) that

$$E(t) + \int_0^t \int_{\Gamma_1} w'^2 d\Gamma dt = E(0),$$

thus, if $E(0) < \infty$, solutions of (1.2)–(1.4) with v given by (2.2) satisfy

(2.3) $$E \in L^\infty(0,\infty), \qquad w'|_{\Sigma_1} \in L^2(\Sigma_1).$$

One may prove by standard methods (as in J.L. LIONS and E. MAGENES [8], Chapter 3) that if $\dot{E}(0) < \infty$, then (1.2)–(1.5), with v given by (2.2), has a unique solution with $\{w,w'\} \in C([0,\infty);V\times H)$. That the choice (2.2) leads also to (1.11) was proved in J.E. LAGNESE [3] under some *geometric conditions* on Γ_0 and Γ_1. To describe the result, let us introduce the vector field

(2.4) $$m(x,y) = \{x,y\} - \{x_0,y_0\}$$

where $\{x_0,y_0\}$ is a fixed point of $\mathbb{R}^2$. We have the following.

THEOREM 2.1. *Assume that for some point* $\{x_0,y_0\}\in\mathbb{R}^2$ *one has*

(2.5) $\quad m\cdot\nu \leq 0 \quad$ *on* Γ_0, $\quad m\cdot\nu \geq 0 \quad$ *on* Γ_1.

and that $\Gamma_0\cap\Gamma_1=\emptyset$. *Assume also that the initial data satisfies* $E(0) < \infty$. *Let* w *be the solution of* (1.2)–(1.5) *with* v *given by* (2.2). *Then there are positive constants* C, ω, *independent of* w, *such that*

$$E(t) \leq Ce^{-\omega t}E(0).$$

Remark 2.1. In general, singularities will occur at points of $\Gamma_0\cap\Gamma_1$ regardless of the smoothness of the initial data. Therefore, if $\Gamma_0\cap\Gamma_1\neq\emptyset$, (1.2) will not in general admit classical (i.e., $H^4(\Omega)$) solutions. The hypothesis $\Gamma_0\cap\Gamma_1=\emptyset$ is made *only* to assure that solutions are sufficiently regular in the spatial variables whenever the initial data is regular.

THEOREM 2.2. *Under the hypotheses of Theorem 2.1, there exists a unique optimal solution* u, w(u), *to the problem* (1.2)–(1.10).

PROOF. From Theorem 2.1 we have that $v=w'|_{\Sigma_1}\in U(w^0,w^1)$. Set

$$U^*(w^0,w^1) = \{v\colon v\in L^2(\Sigma_1),\ J(v)\leq J(w'|_{\Sigma_1})\}.$$

Then $U^*(w^0,w^1)$ is a closed and convex subset of $L^2(\Sigma_1)$, hence $\inf\{J(v)\colon v\in U^*(w^0,w^1)\}$ admits a unique solution u.

3. The Reachable Set. In order to characterize the optimal control obtained in Section 2 as a linear feedback in $\{w,w'\}$, we shall first need to characterize the *reachable set* R_T defined, for each $T>0$, as

$$R_T = \{\{w(t),w'(T)\}\colon v\in L^2(\Gamma_1\times(0,T))\},$$

where w is the solution of (1.2)–(1.4) with initial data

$$w(0) = w'(0) = 0. \tag{3.1}$$

To do so, let us consider the problem

$$\varphi'' + \Delta^2\varphi = 0 \quad \text{in } \Omega\times(0,T), \tag{3.2}$$

$$\varphi = \frac{\partial\varphi}{\partial\nu} = 0 \quad \text{on } \Gamma_0\times(0,T), \tag{3.3}$$

$$\Delta\varphi + (1-\mu)B_1\varphi = \frac{\partial\Delta\varphi}{\partial\nu} + (1-\mu)B_2\varphi = 0 \quad \text{on } \Gamma_1\times(0,T), \tag{3.4}$$

$$\varphi(T) = \varphi^0\in H,\quad \varphi'(T) = \varphi^1\in V' \tag{3.5}$$

where V' is the dual of V with respect to H. Problem (3.2)–(3.5) has a unique solution $\{\varphi,\varphi'\}\in C([0,T];H\times V')$. Moreover, the map $\{\varphi^0,\varphi^1\}\to\{\varphi(t),\varphi'(t)\}$ is an *isomorphism* on $H\times V'$, $\forall t\in[0,T]$.

We define the set

$$F_T = \{\{\varphi^0,\varphi^1\}\colon \varphi|_{\Gamma_1\times(0,T)}\in L^2(\Gamma_1\times(0,T))\}.$$

where φ is the solution to (3.2)–(3.5). For $\{\varphi^0,\varphi^1\}\in F_T$ let us set

$$\|\{\varphi^0,\varphi^1\}\|_{F_T} = [\int_0^T\int_{\Gamma_1}\varphi^2\, d\Gamma dt]^{\frac{1}{2}}.$$

$\|\cdot\|_{F_T}$ is obviously a seminorm on F_T, and from trace theory

(3.6) $$\|\{\varphi^0,\varphi^1\}\|_{F_T} \le C_T\|\{\varphi^0,\varphi^1\}\|_{V\times H}.$$

In fact, under the geometric conditions (2.5), (2.6), $\|\cdot\|_{F_T}$ defines a *norm.*

Lemma 3.1. *Suppose that* $\{\varphi^0,\varphi^1\}\in F_T$ *and that* Γ_0, Γ_1 *satisfy* (2.5) and (2.6). *Then there are positive constants* T_0 *and* C *such that for* $T>T_0$ *one has*

(3.7) $$(T-T_0)\|\{\varphi^0,\varphi^1\}\|_{H\times V'} \le C\|\{\varphi^0,\varphi^1\}\|_{F_T}.$$

Remark 3.1. T_0 and C depend only on $\{x_0,y_0\}$, Ω and μ. Their values will be obtained in the course of the proof.

Remark 3.2. From (3.6) and (3.7) follows that

$V\times H \subset F_T \subset H\times V'$ *algebraically and topologically for* $T>T_0$.

Proof. We first consider the problem

(3.8) $$\psi'' + \Delta^2\psi = 0 \quad \text{in } \Omega\times(0,T),$$

(3.9) $$\psi = \frac{\partial\psi}{\partial\nu} = 0 \quad \text{on } \Gamma_0\times(0,T),$$

(3.10) $$\Delta\psi + (1-\mu)B_1\psi = \frac{\partial\Delta\psi}{\partial\nu} + (1-\mu)B_2\psi = 0 \quad \text{on } \Gamma_1\times(0,T),$$

(3.11) $$\psi(0) = \psi^0\in H^4(\Omega)\cap V, \quad \psi'(0) = \psi^1\in V.$$

We shall prove that

(3.12) $$(T-T_0)\|\{\varphi^0,\varphi^1\}\|^2_{V\times H} \le C\int_0^T\int_{\Gamma_1} \psi'^2 d\Gamma dt, \qquad T>T_0,$$

for some positive constants T_0, C.

We start from the equation

(3.13) $$\int_0^T\int_\Omega (\psi''+\Delta^2\psi)m\cdot\nabla\psi dXdt = 0$$

where dX=dxdy. We have

(3.14) $$\int_0^T\int_\Omega \psi''(m\cdot\nabla\psi)dXdt = \int_\Omega \psi'(m\cdot\nabla\psi)dX\Big|_0^T + \int_0^T\int_\Omega \psi'^2 dXdt - \frac{1}{2}\int_0^T\int_{\Gamma_1}(m\cdot\nu)\psi'^2 d\Gamma dt.$$

One may also verify with the aid of Green's Theorem that

(3.15) $$\int_0^T\int_\Omega \Delta^2\psi(m\cdot\nabla\psi)dXdt = \int_0^T\int_\Omega \left[\left[\frac{\partial^2\psi}{\partial x^2}\right]^2 + \left[\frac{\partial^2\psi}{\partial y^2}\right]^2 + 2\mu\left[\frac{\partial^2\psi}{\partial x^2}\right]\left[\frac{\partial^2\psi}{\partial y^2}\right] + \right.$$
$$\left. 2(1-\mu)\left[\frac{\partial^2\psi}{\partial x\partial y}\right]^2\right]dxdy + \frac{1}{2}\int_0^T\int_{\Gamma_1}(m\cdot\nu)\left[\left[\frac{\partial^2\psi}{\partial x^2}\right]^2 + \left[\frac{\partial^2\psi}{\partial y^2}\right]^2 + 2\mu\left[\frac{\partial^2\psi}{\partial x^2}\right]\left[\frac{\partial^2\psi}{\partial y^2}\right] + \right.$$
$$\left. 2(1-\mu)\left[\frac{\partial^2\psi}{\partial x\partial y}\right]^2\right]dxdy + \frac{1}{2}\int_0^T\int_{\Gamma_0}\left[(m\cdot\nu)(\Delta\psi)^2 - 2\Delta\psi\frac{\partial}{\partial\nu}(m\cdot\nabla\psi)\right]d\Gamma dt.$$

From (3.13)–(3.15) we obtain

$$(3.16)\qquad \int_\Omega \psi'(m\cdot\nabla\psi)dX\Big|_0^T + 2\int_0^T E(t)dt + \frac{1}{2}\int_0^T\int_{\Gamma_0}[(m\cdot\nu)(\Delta\psi)^2 - 2\Delta\psi\frac{\partial}{\partial\nu}(m\cdot\nabla\psi)]d\Gamma dt +$$
$$\frac{1}{2}\int_0^T\int_{\Gamma_1}(m\cdot\nu)\left[\left[\frac{\partial^2\psi}{\partial x^2}\right]^2 + \left[\frac{\partial^2\psi}{\partial y^2}\right]^2 + 2\mu\left[\frac{\partial^2\psi}{\partial x^2}\right]\left[\frac{\partial^2\psi}{\partial y^2}\right] + 2(1-\mu)\left[\frac{\partial^2\psi}{\partial x\partial y}\right]^2 - \psi'^2\right]dX = 0,$$

where $E(t)=P(t)+K(t)$ is the total energy defined in Section 1. For solutions of (3.8)–(3.11) one has $E(t)\equiv E(0)=E_0$.

Since $\psi=\partial\psi/\partial\nu=0$ on $\Gamma_0\times(0,T)$ one has

$$\frac{\partial}{\partial\nu}(m\cdot\nabla\psi) = (m\cdot\nu)\Delta\psi \quad \text{on } \Gamma_0\times(0,T).$$

Therefore, under the geometric conditions (2.5), (2.6) we obtain from (3.16)

$$(3.17)\qquad 2TE_0 \le \frac{R(X_0)}{2}\int_0^T\int_{\Gamma_1}\psi'^2 d\Gamma dt - \int_\Omega \psi'(m\cdot\nabla\psi)dX\Big|_0^T$$

where $R(X_0) = \max\{|X-X_0|: X\in\bar{\Omega}\}$. We have, for $\alpha>0$,

$$(3.18)\qquad \left|\int_\Omega \psi'(m\cdot\nabla\psi)dX\right| \le \frac{\alpha}{2}\int_\Omega \psi'^2 dX + \frac{R(X_0)}{2\alpha}\int_\Omega |\nabla\psi|^2 dX \le \frac{\alpha}{2}K(t) + \frac{\mu_0 R(X_0)}{2\alpha}P(t)$$

where μ_0 is the smallest constant for which

$$\int_\Omega |\nabla\chi|^2 dX \le \mu_0\int_\Omega\left[\left[\frac{\partial^2\chi}{\partial x^2}\right]^2 + \left[\frac{\partial^2\chi}{\partial y^2}\right]^2 + 2\mu\left[\frac{\partial^2\chi}{\partial x^2}\right]\left[\frac{\partial^2\chi}{\partial y^2}\right] + 2(1-\mu)\left[\frac{\partial^2\chi}{\partial x\partial y}\right]^2\right]dX$$

holds for all χ in V. We choose $\alpha=\sqrt{\mu_0 R(X_0)}$ in (3.18). We obtain

$$\left|\int_\Omega \psi'(m\cdot\nabla\psi)dX\right| \le \sqrt{\mu_0 R(X_0)}\,E(t).$$

Therefore, from (3.17) it follows that

$$(3.19)\qquad 2(T-T_0)E_0 \le \frac{R(X_0)}{2}\int_0^T\int_{\Gamma_1}\psi'^2 d\Gamma dt$$

where $T_0=\sqrt{\mu_0 R(X_0)}$. If we now introduce λ_0 as the smallest constant such that

$$\|\chi\|_V^2 \le \lambda_0\int_\Omega\left[\left[\frac{\partial^2\chi}{\partial x^2}\right]^2 + \left[\frac{\partial^2\chi}{\partial y^2}\right]^2 + 2\mu\left[\frac{\partial^2\chi}{\partial x^2}\right]\left[\frac{\partial^2\chi}{\partial y^2}\right] + 2(1-\mu)\left[\frac{\partial^2\chi}{\partial x\partial y}\right]^2\right]dX,\quad \forall\chi\in V,$$

we obtain from (3.19)

$$(3.20)\qquad \min(1,1/\lambda_0)(T-T_0)[\|\psi^0\|_V^2 + \|\psi^1\|_H^2] \le \frac{R(X_0)}{2}\int_0^T\int_{\Gamma_1}\psi'^2 d\Gamma dt$$

which proves (3.12), provided the initial data satisfies (3.11). (This was needed to assure sufficient regularity of the solution of (3.8)–(3.11) to justify the steps in the derivation of (3.20).)

We now introduce the norm

$$\|\{\psi^0,\psi^1\}\|_{G_T} = \left[\int_0^T\int_{\Gamma_1}\psi'^2 d\Gamma dt\right]^{\frac{1}{2}},$$

and define G_T to be the completion with respect to this norm of pairs $\{\psi^0,\psi^1\}$ which satisfy (3.11). From (3.20) we have $G_T\subset V\times H$ with continuous injection. Therefore, (3.20) *remains valid for pairs* $\{\psi^0,\psi^1\}\in G_T$.

Now let us consider the problem (3.2)–(3.5), and assume for the moment that

$\varphi^0 \in V$, $\varphi^1 \in H$. By making the change of variable $t \to T-t$, we may consider instead the problem (3.2)–(3.4) with *initial data*

$$\varphi(0) = \varphi^0, \quad \varphi'(0) = -\varphi^1.$$

We formally integrate (3.2) in t from 0 to t (all of the steps which follow can be justified). We obtain

$$\varphi'(t) + \varphi^1 + \Delta^2 \int_0^t \varphi(s)ds = 0.$$

Define

$$\psi(t) = \int_0^t \varphi(s)ds + \psi^0$$

where $\psi^0 \in H^4(\Omega) \cap V$ satisfies

$$\Delta^2 \psi^0 = \varphi^1 \quad \text{in } \Omega, \tag{3.21}$$

$$\psi^0 = \frac{\partial \psi^0}{\partial \nu} = 0 \quad \text{on } \Gamma_0, \tag{3.22}$$

$$\Delta \psi^0 + (1-\mu)B_1\psi^0 = \frac{\partial \Delta \psi^0}{\partial \nu} + (1-\mu)B_2\psi^0 = 0 \quad \text{on } \Gamma_1. \tag{3.23}$$

Then ψ is the solution of (3.8)–(3.10) with initial data

$$\psi(0) = \psi^0 \in H^4(\Omega) \cap V, \quad \psi'(0) = \varphi^0 \in V.$$

Therefore the estimate (3.20) is valid for ψ, that is

$$\min(1,1/\lambda_0)(T - T_0)[\|\psi^0\|_V^2 + \|\varphi^0\|_H^2] \le \frac{R(X_0)}{2} \int_0^T \int_{\Gamma_1} \varphi^2 \, d\Gamma dt. \tag{3.24}$$

But from elliptic variational theory, (3.21)–(3.23) defines an *isomorphism* $\psi^0 \to \varphi^1$ *of* V *onto* V′, and therefore from (3.24) we have

$$c(T - T_0)[\|\varphi^1\|_{V'}^2 + \|\varphi^0\|_H^2] \le \frac{R(X_0)}{2} \int_0^T \int_{\Gamma_1} \varphi^2 \, d\Gamma dt \tag{3.25}$$

for some $c>0$. (3.25) may obviously be extended from data $\{\varphi^0,\varphi^1\}$ in $V \times H$ to data in F_T. This completes the proof of Lemma 3.1.

We may now characterize the reachable set R_T.

THEOREM 3.1. *Assume* Γ_0, Γ_1 *satisfy* (2.5) *and* (2.6), *and that* $T > T_0$, *defined in Lemma 3.1. Then* $\{w(T),w'(T)\} \in R_T$ *if and only if* $\{w'(T),-w(T)\} \in F'_T$. *The map* $v \to \{w'(T),-w(T)\}$: $L^2(\Gamma_1 \times (0,T)) \to F'(T)$ *is continuous.*

PROOF. Let w be the solution of (1.2)–(1.4) with initial data

$$w(0) = w'(0) = 0,$$

and let φ be the solution of (3.2)–(3.4) with the final data

$$\{\varphi(T),\varphi'(T)\} = \{\varphi^0,\varphi^1\} \in F_T.$$

Multiply (1.2) by φ and integrate over $\Omega \times (0,T)$. After integration by parts in t and application of Green's formula we obtain

$$\int_\Omega (w'(T)\varphi^0 - w(T)\varphi^1)dX = -\int_0^T \int_{\Gamma_1} v\varphi d\Gamma dt, \tag{3.26}$$

and therefore

(3.27) $$|\langle\{w'(T),-w(T)\},\{\varphi^0,\varphi^1\}\rangle| \le \|v\|_{L^2(\Gamma_1\times(0,T))}\|\{\varphi^0,\varphi^1\}\|_{F_T}.$$

It follows from (3.27) that $\{w'(T),-w(T)\}\in F_T'$ and that

$$\|\{w'(T),-w(T)\}\| \le \|v\|_{L^2(\Gamma_1\times(0,T))}.$$

Therefore $R_T\subset F_T'$ and the mapping $v\to\{w'(T),-w(T)\}$ is continuous in the appropriate topologies.

Conversely, suppose that $\{\overset{1}{w}_T,-\overset{0}{w}_T\}\in F_T'$. Then one may find $v\in L^2(\Gamma_1\times(0,T))$ such that $\{w'(T),-w(T)\}=\{\overset{1}{w}_T,-\overset{0}{w}_T\}$. Indeed, let $\{\varphi^0,\varphi^1\}\in F_T$, let φ solve (3.2)–(3.5), and set $v=-\varphi|_{\Gamma_1\times(0,T)}$. Then $\{w'(T),-w(T)\}$ depends linearly on $\{\varphi^0,\varphi^1\}$:

$$\{w'(T),-w(T)\} = \Lambda\{\varphi^0,\varphi^1\}.$$

From (3.26) we have

$$\langle\Lambda\{\varphi^0,\varphi^1\},\{\varphi^0,\varphi^1\}\rangle = \|\{\varphi^0,\varphi^1\}\|^2_{F_T}$$

and therefore Λ is an *isomorphism of* F_T *onto* F_T'. Thus, we need only choose $\{\varphi^0,\varphi^1\}=\Lambda^{-1}\{\overset{1}{w}_T,-\overset{0}{w}_T\}$.

Remark 3.4. The last argument (showing that $R_T\supset F_T'$) is the *Hilbert Uniqueness Method* introduced by J.L. LIONS [6].

Remark 3.5. It seems to be a difficult problem to characterize the reachable set R_T in terms of familiar spaces (e.g., Sobolev spaces). In addition, R_T will undoubtedly depend in a highly nontrivial way on the geometry of Ω. However, it seems *likely* that R_T is independent of T (for $T>T_0$). But even this has not been rigorously established. It is this apparently complicated structure of the reachable set which prevents us from obtaining the optimal control (whose existence was established in Section 2) as a linear feedback in $\{w,w'\}$. In the next section we consider a simpler situation in which the structure of R_T may be made precise.

4. The One–Dimensional Case. Here we consider the one–dimensional analog of (1.2)–(1.5), that is

(4.1) $$w'' + \frac{\partial^4 w}{\partial x^4} = 0, \quad 0 < x < \ell,\ t > 0,$$

(4.2) $$w = \frac{\partial w}{\partial x} = 0, \quad x = 0,\ t > 0,$$

(4.3) $$\frac{\partial^2 w}{\partial x^2} = 0,\ \frac{\partial^3 w}{\partial x^3} = v, \quad x = \ell,\ t > 0,$$

(4.4) $$w(0) = w^0\in V,\ w'(0) = w^1\in H$$

where

$$H = L^2(0,\ell),\ V = \{\chi: \chi\in H^2(0,\ell),\ \chi|_{x=0} = \frac{\partial\chi}{\partial x}|_{x=0} = 0\}.$$

(4.1)–(4.3) describes the motion of an Euler beam of length ℓ, clamped at the left end and controlled at the right end using a shear force v in the w–direction. (The time scale has been adjusted so that the elastic parameters do not explicitly appear in (4.1).)

The optimal control problem for (4.1)–(4.4) is

(4.5) $$\inf_v\{J(v): v\in U(w^0,w^1)\},\quad U(w^0,w^1) = \{v: v\in L^2(0,\infty),\ J(v) < \infty\},$$

where

(4.6) $$J(v) = \int_0^\infty \int_0^\ell [w'^2 + (\frac{\partial^2 w}{\partial x^2})^2]dxdt + N\int_0^\infty v^2 dt.$$

This problem admits a unique solution u, w(u).

Let us introduce the space

(4.7) $$F_T = \{\{\varphi^0,\varphi^1\}: \{\varphi^0,\varphi^1\}\in H\times V',\ \varphi|_{x=\ell}\in L^2(0,\ell)\},$$
$$\|\{\varphi^0,\varphi^1\}\|_{F_T} = [\int_0^T |\varphi(\ell,t)|^2 dt]^{\frac{1}{2}},$$

where φ is the solution of the problem

$$\varphi'' + \frac{\partial^4\varphi}{\partial x^4} = 0,\quad 0 < x < \ell,\ 0 < t < T,$$
$$\varphi = \frac{\partial\varphi}{\partial x} = 0,\quad x = 0,\ 0 < t < T,$$
$$\frac{\partial^2\varphi}{\partial x^2} = \frac{\partial^3\varphi}{\partial x^3} = 0,\quad x = \ell,\ 0 < t < T,$$
$$\varphi(0) = \varphi^0,\quad \varphi'(0) = \varphi^1.$$

The space F_T is characterized as follows.

Lemma 4.1. *If* $T>\ell^2$, *then*

(4.8) $$F_T = F = H\times V'$$

algebraically and topologically.

Remark 4.1. (4.8) is ***probably*** true for $T > 0$ (c.f. E. ZUAZUA [9]).

The proof of (4.8) is based on the use of multipliers as in Section 3. We start, for example, from the equation

$$\int_0^T \int_0^\ell (\psi'' + \frac{\partial^4\psi}{\partial x^4})x\frac{\partial\psi}{\partial x}dxdt = 0$$

in analogy with (3.13). However, unlike in the 2–dimensional situation, here we are able to obtain *two–sided estimates* on $\int_0^T |\varphi(\ell,t)|^2 dt$ of the form

(4.9) $$c(T-\ell^2)(\|\varphi^0\|_H^2 + \|\varphi^1\|_{V'}^2) \le \int_0^T |\varphi(\ell,t)|^2 dt \le CT(\|\varphi^0\|_H^2 + \|\varphi^1\|_{V'}^2).$$

We omit the details.

Remark 4.2. It follows from (4.8) and Theorem 3.1 that the reachable set for (4.1)–(4.3) is V×H. Thus, solutions with finite initial energy will have finite energy for all $t>0$ under the action of $L^2(0,\infty)$ controls v (c.f. Remark 1.1).

With the establishment of (4.8), we may now apply the formalism developed in J.L. LIONS [7], Chapter 8, to obtain the *optimality system* for the optimal control and the feedback operator giving the optimal control.

4.1. Optimality System. For $\epsilon>0$ we introduce the *penalized problem*

$$\inf_{v,f} J_\epsilon(v,f),\quad J_\epsilon(v,f) = \int_0^\infty\int_0^\ell [z'^2 + (\frac{\partial^2 z}{\partial x^2})^2]dxdt + N\int_0^\infty v^2 dt + \frac{1}{\epsilon}\int_0^\infty\int_0^\ell f^2 dxdt,$$

where

$$z'' + \frac{\partial^4 z}{\partial x^4} = f \in L^2((0,\infty);H),$$

$$z = \frac{\partial z}{\partial x} = 0,\quad x = 0,\ t>0,$$

$$\frac{\partial^2 z}{\partial x^2}\Big|_{x=\ell} = 0,\quad \frac{\partial^3 z}{\partial x^3}\Big|_{x=\ell} = v \in L^2(0,\infty),$$

$$z(0) = w^0,\quad z'(0) = w^1.$$

This problem admits a unique solution $(u_\epsilon, f_\epsilon, z_\epsilon)$. As $\epsilon\to 0$ one has

$$u_\epsilon \to u \text{ in } L^2(0,\infty),\quad f_\epsilon \to 0 \text{ in } L^2((0,\infty);H),$$

$$z_\epsilon \to w \text{ in } L^2((0,\infty);V),\quad z'_\epsilon \to w' \text{ in } L^2((0,\infty);H),$$

where $(u,w(u))$ is the optimal solution to the problem (4.1)–(4.6). Let us write the optimality system for $(u_\epsilon,f_\epsilon,z_\epsilon)$. We set $p_\epsilon = (1/\epsilon)f_\epsilon$. According to standard theory (see J.L. LIONS [5]), the optimal solution is characterized by

$$\int_0^\infty\int_0^\ell \left[z'_\epsilon\zeta' + \left[\frac{\partial^2 z_\epsilon}{\partial x^2}\right]\left[\frac{\partial^2\zeta}{\partial x^2}\right]\right]dxdt + N\int_0^\infty u_\epsilon v dt + \int_0^\infty\int_0^\ell p_\epsilon(\zeta'' + \frac{\partial^4\zeta}{\partial x^4})dxdt = 0 \tag{4.10}$$

for all ζ which satisfy

$$\zeta\in L^2((0,\infty);H),\quad \zeta'' + \frac{\partial^4\zeta}{\partial x^4} \in L^2((0,\infty);H),$$

$$\zeta = \frac{\partial\zeta}{\partial x} = 0,\quad x=0,\ t>0,$$

$$\frac{\partial^2\zeta}{\partial x^2}(\ell,\cdot) = 0,\quad \frac{\partial^3\zeta}{\partial x^3}(\ell,\cdot) = v\in L^2(0,\infty),$$

$$\zeta(0) = \zeta'(0) = 0,$$

and for which J_ϵ is finite. It follows from (4.10) that p_ϵ is a weak solution of the following system:

$$p''_\epsilon + \frac{\partial^4 p_\epsilon}{\partial x^4} - z''_\epsilon + \frac{\partial^4 z_\epsilon}{\partial x^4} = 0,\quad 0<x<\ell,\ t>0, \tag{4.11}$$

$$p_\epsilon = \frac{\partial p_\epsilon}{\partial x} = 0,\quad x=0,\ t>0, \tag{4.12}$$

$$\frac{\partial^2 p_\epsilon}{\partial x^2} = 0,\quad \frac{\partial^3 p_\epsilon}{\partial x^3} + u_\epsilon = 0,\quad p_\epsilon + Nu_\epsilon = 0,\quad x=\ell,\ t>0. \tag{4.13}$$

(Note that $p_\epsilon(0)$, $p'_\epsilon(0)$ are *not* given.) One may prove, by utilizing the estimate (4.8), that

$\{p_\epsilon(0), p'_\epsilon(0)\}$ *remains in a bounded set of* $H\times V'$.

This fact allows us to pass to the limit in (4.10)–(4.13). One obtains in the limit the *optimality system* for the optimal control problem.

THEOREM 4.1. *The optimal solution* u *of the problem* (4.1)–(4.6) *is characterized by the solution of the following optimality system:*

$$w'' + \frac{\partial^4 w}{\partial x^4} = 0,\quad p'' + \frac{\partial^4 p}{\partial x^4} - w'' + \frac{\partial^4 w}{\partial x^4} = 0,\quad 0<x<\ell,\ t>0, \tag{4.14}$$

$$w = \frac{\partial w}{\partial x} = p = \frac{\partial p}{\partial x} = 0,\quad x=0,\ t>0, \tag{4.15}$$

(4.16) $\frac{\partial^2 w}{\partial x^2} = \frac{\partial^2 p}{\partial x^2} = 0,\ N\frac{\partial^3 w}{\partial x^3} + p = 0,\ N\frac{\partial^3 p}{\partial x^3} - p = 0,\ x = \ell, t > 0,$

(4.17) $w(0) = w^0,\ w'(0) = w^1,$

(4.18) $\{w,w'\}\in C([0,\infty);V\times H)\cap L^2(0,\infty;V\times H),\ \{p,p'\}\in C([0,\infty);H\times V'),\ p|_{x=\ell}\in L^2(0,\infty).$

The optimal control is given by

(4.19) $u = -\frac{1}{N}p|_{x=\ell}.$

4.2. The Feedback Operator. Let $\{\hat{w}^0,\hat{w}^1\}$ be given in $V\times H$, let $\hat{v}\in U(\hat{w}^0,\hat{w}^1)$ and $\hat{w}$ be the corresponding state. Then from (4.14)–(4.18) one can obtain the following identity:

(4.20) $(p'(0) - w^1,\hat{w}^0) - (p(0),\hat{w}^1) = N\int_0^\infty u\hat{v}dt + \int_0^\infty\int_0^\ell [w'\hat{w}' + \frac{\partial^2 w}{\partial x^2}\frac{\partial^2 \hat{w}}{\partial x^2}]dxdt.$

In particular, if $\{\hat{w}^0,\hat{w}^1\}=\{w^0,w^1\}$ and $\hat{v}=u$, we obtain

(4.21) $(p'(0) - w^1,w^0) - (p(0),w^1) = J(u).$

We define an operator $P:H\times V \to H\times V'$ by setting

$$P\{w^1,w^0\} = \{-p(0),p'(0) - w^1\}.$$

Since the value of t which is selected for the time origin is immaterial, we have the *identity*

(4.22) $P\{w'(t),w(t)\} = \{-p(t),p'(t) - w'(t)\}$

for all $t\geq 0$. From (4.16), (4.17),

(4.23) $P\in\mathscr{L}(F',F),\ P = P^*,\ P\geq 0.$

Thus

$$P = \begin{bmatrix} P_1 & P_2 \\ P_3 & P_4 \end{bmatrix}$$

where $P_1\in\mathscr{L}(H)$, $P_2\in\mathscr{L}(V,H)$, $P_3\in\mathscr{L}(H,V')$, $P_4\in\mathscr{L}(V')$, $P_2^*=P_3$, $P_3^*=P_2$. Therefore, we may write

$$-p(t) = P_1w'(t) + P_2w(t).$$

Then, from (4.22), we see that *the optimal control* u *is given by the feedback law*

(4.24) $u = \frac{1}{N}(P_1w' + P_2w)|_{x=\ell}.$

Of course, in (4.24) we do not know P_1 and P_2 explicitly. However, it is possible to characterize P as a solution to an algebraic operator equation with a quadratic nonlinearity. To do so we proceed as follows (using the treatment given in [7] as our guide).

Let φ_1 and φ_2 be smooth functions on $[0,\ell]$. From (4.14) we have, for fixed t,

(4.25) $\int_0^\ell [p'' + p^{(4)} - w'' + w^{(4)}]\varphi_2 dx = 0,$

where $p^{(j)}=\partial^j p/\partial x^j$. Using (4.15) and (4.16), (4.25) may be written

(4.26) $(p'' - w'',\varphi_2) + (p^{(4)},\varphi_2) + (w^{(2)},\varphi_2^{(2)}) - \frac{1}{N}p(\ell)\varphi_2(\ell) = 0$

provided

$$\varphi_2\in H^2(0,\ell),\ \varphi_2(0) = \varphi_2'(0) = 0,$$

where $(f,g)=\int_0^\ell f(x)g(x)dx.$

Let us introduce

$$\tilde{w}(t) = \{w'(t), w(t)\}, \quad \tilde{p}(t) = \{-p(t), p'(t) - w'(t)\}, \quad \tilde{\varphi} = \{\varphi_1, \varphi_2\},$$

$$A = \begin{bmatrix} 0 & -I \\ (\frac{\partial}{\partial x})^4 & 0 \end{bmatrix} \quad B = \begin{bmatrix} I & 0 \\ 0 & (\frac{\partial}{\partial x})^2 \end{bmatrix} \quad J_1 = \begin{bmatrix} I & 0 \\ 0 & 0 \end{bmatrix} \quad J_2 = \begin{bmatrix} 0 & I \\ 0 & 0 \end{bmatrix}.$$

Then

$$\tilde{p}(t) = P\tilde{w}(t)$$

and (4.26) may be expressed in the form

(4.27) $$(\tilde{w}', P\tilde{\varphi}) - (AP\tilde{w}, \tilde{\varphi}) + (B\tilde{w}, B\tilde{\varphi}) + \frac{1}{N}(J_1P\tilde{w})(\ell)(J_2\tilde{\varphi})(\ell) = 0.$$

Next, we consider

$$\int_0^\ell [w'' + w^{(4)}]\varphi_1 dx = 0,$$

or, from (4.15), (4.16),

(4.28) $$(w'', \varphi_1) + (w, \varphi_1^{(4)}) - \frac{1}{N} p(\ell)\varphi_1(\ell) = 0$$

provided

$$\varphi_1 \in H^4(0,\ell), \quad \varphi_1(0) = \frac{\partial \varphi_1}{\partial x}(0) = \varphi_1^{(2)}(\ell) = \varphi_1^{(3)}(\ell) = 0.$$

(4.28) may be written

(4.29) $$(\tilde{w}', \tilde{\varphi}) + (\tilde{w}, A\tilde{\varphi}) + \frac{1}{N}(J_1P\tilde{w})(\ell)(J_1\tilde{\varphi})(\ell) = 0.$$

We replace $\tilde{\varphi}$ in (4.29) by $P\tilde{\varphi}$ and use (4.27) to replace the term $(\tilde{w}, P\tilde{\varphi})$. The result is

(4.30) $$(AP\tilde{w}, \tilde{\varphi}) + (\tilde{w}, AP\tilde{\varphi}) + \frac{1}{N}[(J_1P\tilde{w})(\ell)(J_2\tilde{\varphi})(\ell) + (J_1P\tilde{w})(\ell)(J_2P\tilde{\varphi})(\ell)] = -(B\tilde{w}, B\tilde{\varphi}).$$

(4.30) *is an identity which holds for every*

(4.31) $$\tilde{w} \in F', \quad \tilde{\varphi} \in F' \quad \textit{such that} \quad AP\tilde{\varphi} \in F, \quad AP\tilde{w} \in F,$$

and, together with (4.23), *characterizes* the feedback operator P.

THEOREM 4.2. *The optimal feedback operator* P *is characterized by* (4.23), (4.30) *and* (4.31), *and the optimal control* u *is given by* (4.24).

Acknowledgement. This research was supported by the Air Force Office of Scientific Research under grant AFOSR–86–0162.

REFERENCES

[1] R. DATKO, Extending a theorem of Liapunov to Hilbert spaces, J. Math. Anal. Appl., 32 (1970), pp. 610–616.

[2] G. DUVAUT and J.L. LIONS, *Inequalities in Mechanics and Physics*, Springer–Verlag, Berlin, 1976.

[3] J. LAGNESE, Uniform boundary stabilization of homogeneous, isotropic plates, Proc. 1986 Vorau Conference on Control of Distributed Parameter Systems, to appear.

[4] J. LAGNESE and J.L. LIONS, *Modelling, Analysis and Control of Thin Plates*, Lecture Notes of College de France, Masson, Paris, to appear.

[5] J.L. LIONS, *Optimal Control of Systems Governed by Partial Differential Equations*, Springer–Verlag, Berlin, 1971.

[6] J.L. LIONS, Contrôlabilité exacte des systèmes distribués, C. R. Acad. Sci. Paris, (302) 1986, pp. 471–475.

[7] J.L. LIONS, Exact controllability, stabilization and perturbations for distributed systems, The John Von Neumann Lecture, SIAM National Meeting, Boston, 1986. To appear in SIAM Review.

[8] J.L. LIONS and E. MAGENES, *Non–Homogeneous Boundary Value Problems and Applications*, Vol. I, Springer–Verlag, Berlin, 1972.

[9] E. ZUAZUA, Exact controllability of distributed systems for arbitrarily small time, Proc. of the 1987 IEEE Conference on Decision and Control, Los Angeles, CA, 1987.

ON BOUNDARY CONTROLLABILITY OF VISCOELASTIC SYSTEMS

G. Leugering
Fachbereich Mathematik
Technische Hochschule Darmstadt
D-6100 Darmstadt, Schlossgartenstr. 7

Abstract:

It is shown that a general isotropic viscoelastic solid with non vanishing Newtonian viscosity is never exactly controllable using L_2-boundary controls. For some models it is known that even spectral controllability does not hold. Here we show, thereby extending results obtained in Leugering and Schmidt [10], that the general model is approximatively controllable under some reasonable assumptions.

1. Introduction and Preliminaries:

Let Ω be a bounded domain in $\mathbb{R}^n$ with Lipschitzian boundary, e.g. $\Omega \in N^{0,1}$, see Necas [15]. Let L be a formally selfadjoint differential expression of even order

$$Lu = \sum_{\substack{|\alpha|\leq m \\ |\beta|\leq m}} (-1)^{|\beta|} D^{\beta}(a_{\alpha\beta} D^{\alpha} u)$$

with the familiar notation (α, β are multiindices) such that L is symmetric on $C_0^{\infty}(\Omega)$ as an operator in $L_2(\Omega)$. We suppose that L has a positive definite self-adjoint extension A in $L_2(\Omega)$ with a compact resolvent. Let J(u,v) be the positive, symmetric and closed sesquilinear form on V = D(J) associated with A:

$$D(A) = \{v \in V, \quad u \to J(u,v) \text{ continuous in } H = L_2(\Omega)\}$$

$$(Au,v) = (Lu,v) = J(u,v).$$

We suppose the existence of a boundary differential operator $B : H^{2m}(\Omega) \to L_2(\partial\Omega)^r$, $r \leq m$ such that

$$D(A) = \{v \in V = D(A^{1/2}), \quad B(v) = 0, \quad Lv \in H\}$$

The boundary conditions incorporated in $D(A^{1/2})$ are usually called geometric boundary conditions associated with Bv = 0 on $\partial\Omega$ are viewed as kinematic boundary

Without an essential restriction we set $f \equiv 0$. Since we are going to control the process u from the boundary we assume the boundary conditions

$$u(t) \in D(A^{1/2}) \tag{3}$$

$$c^2(Bu)(t) + a_0(Bu)_t + \int_0^t \dot{a}_1(t-s)(Bu)(s)ds = \phi(t) \text{ on } \partial\Omega \tag{4}$$

to hold for $t > 0$.

Let us consider, as a reference problem,

$$w_{tt}(t) + a_0 L w_t(t) + c^2 L w(t) = 0 \quad \text{on } \Omega \tag{5}$$

$$w(0) = u_0, \quad w_t(0) = v_0 \quad \text{on } \Omega \tag{6}$$

$$c^2(Bw)(t) + a_0(Bw)_t(t) = \phi(t) \quad \text{on } \partial\Omega$$
$$w(t) \in D(A^{1/2}). \tag{7}$$

In order to get an idea in which sense a solution can be expected if $u_0 \in D(A^{1/2})$, $v_0 \in H$ $\phi \in L_2(0,T,L_2(\partial\Omega)^r)$ we define the energy of u, u_t to be given by:

$$E(u)(t) := J(u(t), u(t)) + (u_t(t), u_t(t))$$

where $E(f,g) = \|A^{1/2}f\|^2 + \|g\|^2$ is a norm on $E := V \times H$.

We have the following energy estimate:

Proposition 1:

Let u be a classical solution of (5), (6), (7). Then for $T > 0$ we have with a constant $C > 0$:

$$E(u)(T) \leq C(E(u)(0) + \|\phi\|^2_{L_2(\Sigma_T)})$$

$L_2(\Sigma_T) = L_2(0,T,L_2(\partial\Omega)^r)$.

Proof:

The proof is an adaption of Prop. 0.1 in [10]. □

It makes, therefore, sense to define a continuous trajectory $t \to (u(t),u_t(t))$ in $E \subset V \times H$ as an appropriate mild solution.

conditions. Further, we assume the existence of a bounded sesquilinear form, b, on the boundary and a "dual" boundary operator D such that the abstract Lagrange-Identity holds:

$$(Lu,v) = J(u,v) + b(B(u),D(v)) = (A^{1/2}u,A^{1/2}v) + b(B(u),D(v))$$

$$\text{for all } u,\ v\in D(A^{1/2}) = D(J) = V, \quad D : V\to L_2(\partial\Omega)^r.$$

We assume further that D is relatively $A^{1/2}$-bounded. Let $0<\lambda_1^2\le\lambda_2^2\le\ldots\le\lambda_k^2\le\ldots$ $\lim_{j\to\infty}\lambda_j^2=\infty$ be the eigenvalues of A with the corresponding family of orthogonal projections P_k from H into the eigenspace H_k. We have in particular

$$Au = \sum_{k=1}^{\infty} \lambda_k^2 P_k u$$

$$H = \bigoplus_{k=1}^{\infty} H_k$$

$$\|u\| = \sum_{k=1}^{\infty} \|P_k u\|^2, \quad \|u\|_{A^{1/2}} = \sum_{k=1}^{\infty} \lambda_k^2 \|P_k u\|^2$$

$$D(A^{\alpha}) = \{u\in H \mid \sum_{k=1}^{\infty} \lambda_k^{4\alpha} \|P_k u\|^2<\infty\}.$$

2. The general viscoelastic solid:

Let u(t,x), we suppress the variable x in the sequel, be the deflection from an equilibrium configuration of a viscoelastic solid which is assumed to occupy the domain Ω is isotropic and isothermal. Then, if this material is assumed to be constituted by a Newtonian fluid part (short memory) and a relaxation part (long memory), under some simplifying assumptions the function u is governed by the following equation:

$$u_{tt}(t) + a_0Lu_t(t) + c^2Lu(t) + \int_0^t \dot{a}_1(t-s)Lu(s)ds = f(t) \quad \text{on } \Omega \tag{1}$$

with initial conditions

$$u(0) = u_0, \quad u_t(0) = v_0 \text{ on } \Omega \quad (u(t) \equiv 0 \text{ on } t<0). \tag{2}$$

Here, in (1), $a_0\ge0$, $c>0$ are real constants, while a_1 satisfies

A1 $a_1\in L_2(0,1)\cap C(0,\infty)$ is positive nonincreasing and log-convex.

The homogenity f can be considered as a cutting force, or distributed load.

Remark 1:

The system (5), (6), (7) with $\phi = 0$, that is with A instead of L in (5), has been studied by many authors, and in our special situation of a self-adjoint operator with compact resolvent (5), (6) can immediately been solved by Fouriers' method. The non homogeneous problem has been discussed in detail by Duvaut-Lions [4] and Gaiduk [5], for smoother boundary data and unit-step jumps, respectively. □

Just as in [10] we can show:

Lemma 1:

Given $(u_0,v_0) \in E$, $\phi \in L_2(\Sigma_T)$, there exists a unique mild solution u of (5), (6), (7) with $(u,u_t) \in C(0,T,E)$.

Proof:

Essentially Prop. 1.4 of [10].

Remark 2:

Being provided with a solution w of (5), (6), (7) it is a matter of applying the contraction principle to show that a unique mild solution of (5), (6) and (4) exists as well. We rename this solution by w. □

In order to achieve a solution, u, of the original system (1) to (4) we apply the Fattorini approach: decompose u as $u = v + z$, where z satisfies the quasi-static equation

$$Lz(t) = 0 \quad \text{in } Q_T \tag{8}$$

$$c^2Bz(t) + a_0(Bz)_t(t) + \int_0^t \dot{a}_1(t-s)(Bz)(s)ds = \phi(t) \text{ on } \Sigma_T,\ z(t) \in V \tag{9}$$

$$z(0) = 0 = z_t(0) \text{ in } \Omega. \tag{10}$$

Now, (9) is an integro-differential equation for the function Bz in $L_2(\Sigma_T)$. Thus, by classical results on Volterra-Integral-Resolvents in Hilbert spaces, for a very general approach see the paper by Prüss [17], we can represent Bz by the convolution with the resolvent, say S(t), associated with equation (9):

$$(Bz)(t) = \int_0^t S(t-s)\phi(s)ds =: \phi(t) \tag{11}$$

As a result, $\tilde{\Phi}$ is absolutely continuous. Therefore, solving (8) to (10) with $\phi \in L_2(\Sigma_T)$ is completely equivalent to solving (8), (10), (11) with $\tilde{\Phi}$ absolutely continuous in t and $L_2(\partial\Omega)^r$ in the space variable. The function v has to "solve" the following system:

$$v_{tt}(t) + a_0 Av_t(t) + c^2 Av(t) + \int_0^t \dot{a}_1(t-s)Av(s)ds = - z_{tt} \tag{12}$$

$$Bv(t) = 0, \quad v(t) \in V \tag{13}$$

$$v(0) = v_t(0) = 0. \tag{14}$$

Of course, (12) can only be true in a weak sense, if z_{tt} does not exist in a convenient space. However, z is given by a bounded operator G: $L_2(\Sigma_T) \to L_2(0,T,V)$, $z(t) = G\tilde{\Phi}(t)$, and is a solution in the variational sense, see [16]. We integrate (12) w.r.t t:

$$v_t(t) + a_0 Av(t) + \int_0^t (a_\infty + a_1(t-s))Av(s)ds = - G\tilde{\Phi}_t(t). \tag{15}$$

Now observe that $G\tilde{\Phi}_t \in L_2(0,T,V)$, and therefore we can apply the theory of Prüss [17] in case of assumption A1 and Lunardi [12] in case that a_1 is completely monotone, to the extend that v is given by a convolution with a resolvent operator associated with equation (15):

$$v(t) = - \int_0^t R(t-s)G\tilde{\Phi}_s(s)ds \tag{16}$$

and by (15), (16):

$$u_t(t) = - a_0 A \int_0^t R(t-s)G\tilde{\Phi}_s(s)ds - \int_0^t (a_\infty + a_1(t-s))A \int_0^s R(s-\tau)d\tau G\tilde{\Phi}_\tau(\tau)d\tau ds \tag{17}$$

where u is given by $u(t) = G\tilde{\Phi}(t) + v(t)$.

Theorem 1:

Under the assumption A1 there is a unique mild solution u of (1) - (4), where is given by u = v + z, z is solution of (8), (10), (11), ((9)), and v solves (12), (13), (14). We have

$$u, u_t \in C(0,T,V).$$

Proof:

By (16), (17), and Theorem 6, 7 and Corollary 2 in [17].

Remark 2:

The theory of Prüss covers also the case $a_0 = 0$, so that it is indeed possible to give a unifying existence theory even for inhomogeneous boundary value problems. If $a_0 = 0$ the dynamic of (1) is largely influenced by the behavior of $a_1(t)$ at the origin. As a result, if $a_1(0^+) < \infty$, $a_0 = 0$, and $\dot{a}_1(0^+) = -\infty$ one can have both propagation and C^∞-smoothing - a coexistence of genuinelly different phenomena. The reader is referred to [16], [17], [18]. Because of the limited space availlable, and with regard to the fact that, from a control theoretic point of view, the two cases are conceptually very different, we decided to treat only the case $a_0 > 0$, which has been considered by Baumeister [1] for the heat equation. For $a_0 = 0$ the reader is referred to [7], [8], [9].

3. Controllability:

Let us first consider the reference model, that is the Kelvin-Voigt solid, (5), (6), (7). If $\phi \equiv 0$ then the characteristics, the frequencies, are given by

$$\mu_k^{\pm} := \frac{1}{2} (- a_0\lambda_k^2 \pm (a_0^2\lambda_k^4 - 4c^2\lambda_k^2)^{-1}).$$

Now, all $\mu_k^{\pm}$ have negative real part. The sequence μ_k^+ is bounded, whereas for k large μ_k^- is real and tends to minus infinity. This spectral dychotomy leads to a direct decomposition of E into $E = E^+ \oplus E^-$. It was shown in [10] that the solution of (5), (6), (7) (for plate problem) with $\phi = 0$ is given by $(u,u_t)(t) = V(t)(u_0,v_0)$ where $V(t)$ is direct sum of a family $V^+(t)$ of isomorphisms and $V^-(t)$ consisting of compact operators. The same procedure is applicable to the more general material assumed here. We state, without proof, a lackness result the proof of which can be given along the lines of [10].

Theorem 2:

The system (5), (6), (7) is not exactly controllable in E with $L_2(\Sigma_T)$- controls ϕ.□

However, we believe, but this has not yet been proved, that the system (1), (2), (3), (4) is not exactly controllable as well. Although this appears to be fairly evident, one has to argue with perturbations on the boundary. We conclude the discussion of system (5), (6), (7) with the remark that in some cases this system is not even spectrally controllable, this was pointed out recently by Russell [19]. In this light the extension of the main result in [10] to (5), (6), (7) is of interest:

Theorem 3:

Assume that for $u \in H$, $DP_k u = 0$ for all $k \in \mathbb{N}$ implies $u \equiv 0$ on Ω. Then the system (5), (6), (7) is approximately controllable in E with $L_2(\Sigma_T)$- controls ϕ.

We would like to show that Theorem 3 holds for (1) to (4). Now, given the solution u by Theorem 1 together with its derivative u_t given by (16), (17), evaluated at the given final time $T>0$, with $\tilde{\Phi} = \xi \cdot \eta$, $\xi \in L_2(\partial\Omega)^r$ $\eta \in H_0^1(0,T) = \{\varphi$ is absolutely continuous, $\varphi(0) = \varphi(T) = 0\}$ we have:

$$u(T) = -\int_0^T R(T-s)G\xi\eta_s(s)ds$$

$$u_t(T) = -a_0A\int_0^T R(T-s)G\xi\eta_s(s)ds - \int_0^T (a_\infty+a_1(T-s))A\int_0^s R(s-\tau)G\xi\eta_\tau(\tau)d\tau ds \qquad (18)$$

Assume we have a pair $(v^0,v^1) \in E = V \times H$ being orthogonal to the range of all phase points $(U(T),U_t(T))$ given by (18), u satisfying (1) - (4).

$$0 = <(v^0,v^1),\ (u(T),u_t(T))>_E = <(v^0,v^1),\ (\int_0^T R(T-s)\xi\eta_s(s)ds,$$

$$A\int_0^T R(T-s)(a_0G\xi\eta_s(s) + \int_0^s (a_\infty+a_1(t-s)G\xi\eta_r(r)ds)>_E =$$

$$= \int_0^T <(v^0,v^1),\ (R(T-s)G\xi,\ (\int_s^T (a_\infty+a_1(r-s))AR(T-r)dr + a_0AR(T-s)G\xi)>_E\eta_s(s)ds \quad (19)$$

with

$$h_s(s) := <(v^0,v^1),\ (R(T-s)G\xi,\ (\int_s^T(a_\infty+a_1(r-s)) \cdot AR(T-s)dr + a_0AR(T-s)G\xi)>_E \qquad (20)$$

(19) is an orthogonality relation in $H_0^1(0,T)$:

$$\int_0^T h_t(t)\eta_t(t)ds = 0 \quad \forall \eta \in H_0^1(0,T) \qquad (21)$$

implying

$$q(t) := <(v^0,v^1),\ (R(t)G\xi,\ a_0AR(t)G\xi + \int_0^t (a_\infty+a_1(t-r)AR(r)dr\ G\xi)>_E = 0 \quad \text{on } (0,T) \qquad (22)$$

In order to exploit the condition (22) we need more information on q(t). This is precisely the point where a conceptual difference between the two cases $a_0 = 0$, $a_0 > 0$ comes into play.

However, if $a_0 > 0$ (as assumed always) we can use the results by Da Prato and Ianelli, Lunardi [2], [3], [12] if we assume that the kernel a_1 is completely monotone, when applied to the system in the phase space (not to (15)).

A2 a_1 is completely monotone, that is $(-1)^n\, a_1{}^{(n)}(t) > 0$ or equivalently $a_1(t) = \int_0^\infty e^{-t\tau}d\alpha(\tau)$ with a positive Borel measure α.

<u>Theorem 4:</u>

If a_1 satisfies A2 then R(t) is real analytic on $(0,\infty)$ with values in B(H). □

This is definitely not true in case $a_0 = 0$. Now, a_1 itself, as Laplace-Stieltjes transform is real analytic on $(0,\infty)$, hence the convolution of R and a_1 is real analytic on $(0,\infty)$. That is, $q(t)$ is a real analytic function on $(0,\infty)$ being equivalent to zero on $(0,T)$ according to (22). We conclude

$$q(t) = 0 \quad \text{on } (0,\infty). \tag{23}$$

Let $R(t)$ be given in its spectral realisation

$$R(t) = \Sigma\, r_k(t) P_k. \tag{24}$$

Then $q(t)$ is given by the series:

$$\begin{aligned} q(t) = \sum_{k=1}^{\infty} \{ & r_k(t)\lambda_k^2 \langle v^0, P_k\, G\xi\rangle + [a_0\, \lambda_k^2 r_k(t) + \\ & + \int_0^t (a_\infty + a_1(t-s)) r_k(s) ds \lambda_k^2] \langle v^1, P_k G\xi\rangle\} =: \sum_{k=1}^{\infty} q_k(t). \end{aligned} \tag{25}$$

Since $q(t) \equiv 0$ on $(0,\infty)$ its Laplace-Transform vanishes everywhere. We wish to compute the Laplace-transform $\hat{q}(s)$ of q. Now,

$$\hat{R}(s) = \frac{1}{s}\,(I + \hat{a}(s)A)^{-1}, \quad a(t) = a_0 + a_\infty t + \int_0^t a_1(s) ds$$

that is

$$\hat{R}(s) = \frac{1}{s} \sum_{k=1}^{\infty} (1 + \hat{a}(s)\lambda_k^2)^{-1} P_k \tag{26}$$

Lemma 2:

There is a $c > 0$ such that for all $\lambda \in \mathbb{R}$, $s \in C^+$

$$\left| \frac{1}{\hat{a}(s)} + \lambda^2 \right| \geq \frac{1}{c}\,(\lambda^2 + |s|) \tag{27}$$

Proof:

The proof follows the line of Miller and Wheeler [14].|| By Lemma 2 one concludes that

$$\sum_{k=1}^{\infty} \hat{q}_k(s) = \sum_{k=1}^{\infty} \frac{\lambda_k^2}{\frac{1}{\hat{a}(s)} + \lambda_k^2} \left(\frac{1}{s\hat{a}(s)} \langle v^0, P_k\, G\xi\rangle + \langle v^1, P_k G\xi\rangle \right) \tag{28}$$

converges uniformly and absolutely in $s \in C^+$. Hence the Laplace-transform of $q(t)$ can indeed be obtained term by term:

$$\hat{q}(s) = \sum_{k=1}^{\infty} \hat{q}_k(s).$$

Hence, (22) amounts to

$$\begin{aligned} 0 = \sum_{k=1}^{\infty} \hat{q}_k(s) = & \frac{1}{s} \sum_{j=0}^{\infty} (-1)^j \left(\sum_{k=1}^{\infty} \frac{1}{\lambda_k^{2j}} \langle v^0, P_k\, G\xi\rangle \right) \left(\frac{1}{\hat{a}(s)} \right)^{j+1} + \\ & + \sum_{j=0}^{\infty} (-1)^j \left(\sum_{k=1}^{\infty} \frac{1}{\lambda_k^{2j}} \langle v^1, P_k G\xi\rangle \right) \left(\frac{1}{\hat{a}(s)} \right)^j. \end{aligned}$$

Define

$$a_{j-1} = \sum_{k=1}^{\infty} \frac{1}{\lambda_k^{2j}} <v^0, P_k\, G\xi>(-1)^{j-1}$$

$$b_j = \sum_{k=1}^{\infty} \frac{1}{\lambda_k^{2j}} <v^1,\, P_k\, G\xi>\, (-1)^j$$

both having a positive radius of convergence. Now $\hat{a}(s)$ is analytic in the slit complex plane $C \setminus \{s \mid \mathrm{Re}\, s \le 0\}$ and $1/\hat{a}(s)$ is analytic in an open set with 0 at least on the boundary, $\frac{1}{\hat{a}(0)} = 0$. By the open mapping theorem the range of $1/\hat{a}(s)$ is an open set which must include, at least at the boundary, the point 0. In fact, if $a_1(t)$ decays exponentially at $t = \infty$, 0 is an interior point. Therefore the series

$$\sum_{j=1}^{\infty} a_{j-1} \left(\frac{1}{\hat{a}(s)}\right)^j, \quad \sum_{j=0}^{\infty} b_j \left(\frac{1}{\hat{a}(s)}\right)^j$$

are in fact power series with

$$\sum_{j=1}^{\infty} a_{j-1} \left(\frac{1}{\hat{a}(s)}\right)^j = - s \sum_{j=0}^{\infty} b_j \left(\frac{1}{\hat{a}(s)}\right)^j \tag{29}$$

$$1/\hat{a}(s) = s^2/(sa_0 + a_\infty + sa_1(s)).$$

Since $s\hat{a}(s) \to a_1(\infty) = 0$, because a_1 is completely monotonic, the denominator tends to a_∞ for $s \to 0$. Therefore, $1/\hat{a}(s)$ tends to zero for $s \to 0$ like a square, and this implies that s cannot been developed into a power series at zero in terms of $1/\hat{a}(s)$. From this we infer

$$a_j = b_j = 0 \quad \text{for all } j \in \mathbb{N}$$

or

$$\begin{aligned} &\sum_{k=1}^{\infty} \frac{1}{\lambda_k^{2j}} <v^0, P_k\, G\xi>(-1)^j = 0 \quad j \in \mathbb{N}, \\ &\sum_{k=1}^{\infty} \frac{1}{\lambda_k^{2j}} <v^1, P_k\, G\xi>(-1)^j = 0 \quad j \in \mathbb{N}. \end{aligned} \tag{30}$$

Let p(s) be an arbitrary polynomial, then (30) implies

$$\begin{aligned} &\sum_{k=1}^{\infty} <v^0, P_k\, G\xi> P\left(\frac{1}{\lambda_k^2}\right) = 0 \\ &\sum_{k=1}^{\infty} <v^1, P_k\, G\xi> P\left(\frac{1}{\lambda_k^2}\right) = 0. \end{aligned} \tag{32}$$

Let f be a continuous function vanishing on all numbers $1/\lambda_k^2$ except at $1/\lambda_\ell^2$, where it is 1. The domain of f is the interval $(0,\, 1/\lambda_1^2)$. By the Weierstraß' theorem this function is the limit of polynomials. Therefore, using the uniform convergence of (30), (31) we conclude

$$<v^0, P_k G> = <v^1, P_k G> = 0 \tag{32}$$

for all $k \in \mathbb{N}$.

Now

$$<v^0, P_k G\xi> = -\frac{1}{\lambda_k^2} b(\xi, D(P_k v^0)) = 0,$$

and

$$<v^1, P_k G\xi> = -\frac{1}{\lambda_k^1} b(\xi, D(P_k v^1)) = 0.$$

Since ξ can vary in all of $L_2(\partial\Omega)^r$ we are left with

$$D(P_k v^0) = D(P_k v^1) = 0.. \tag{33}$$

Since $P_k v^0 \in D(A)$, $P_k v^1 \in D(A)$, the condition (34) requires additional boundary conditions to hold for $P_k v^0$, $P_k v^1$ for all k. This is usually not possible without v^0, v^1 being itself zero. We therefore have the main result:

Theorem 5:

If for $v \in H$, $DP_k v = 0$ for all k implies $v = 0$, then the system (1) to (4) are approximately controllable with $L_2(\Sigma_T)$ - controls.

Remark 3:

The assumption in Theorems 3, 5 are met in case of second order operators, see [13] and [20] for details and additional references. For the general case such a uniqueness result for the boundary value problem concerning L, namely whether

$$Au = \lambda u, \quad u \in D(A), \quad D(u) = 0$$

implies $u \equiv 0$, does not seem to exist, even though it is very likely to hold in any particular situation of second or fourth order operators. If, however, the conditions $u \in D(A)$, $D(u) = 0$ amount to a set zero Cauchy data on a non-characteristic part of the boundary, then one can confirm the hypothesis on the base of the Cauchy-Kowalewski-Theorem, see [6]. For one-dimensional models there are now problems at all.

Remark 4:

Applying the (F,F',Λ)-principle of Lions [11] to (1) - (4) or (5) - (7), the Theorem 3, 5 give rise to uniqueness results concerning the homogeneous problems. Thus by [11] there is a possibly very small space $F' \subset V \times H$ which can be exactly controlled in the sense above. This can be seen as follows:

Consider ψ, φ as classical solutions of

$$\begin{cases} \psi_{tt}(t) + a_0 L\psi_t(t) + c^2 L\psi(t) + \int_0^t \dot{a}_1(t-s)L\psi(s)ds = 0 \\ \psi(0) = \psi_t(0) = 0 \\ \psi(t) \in V,\ a_0(R\ )_t(t) + c^2 R\psi(t) + \int_0^t \dot{a}_1(t-s)(R\psi)(s)ds = \phi(t) \end{cases}$$

and

$$\begin{cases} \varphi_{tt}(t) - a_0 L\varphi_t(t) + c^2 L\varphi(t) + \int_t^T \dot{a}_1(s-t)L\varphi(s)ds = 0 \\ \varphi(T) = \varphi_0,\ \varphi_t(T) = \varphi_1 \\ \varphi(t) \in V,\ a_0(R\varphi)_t(t) - c^2(R\varphi)(t) - \int_t^T \dot{a}_1(s-t)L\varphi(s)ds = 0 \end{cases}$$

respectively, then defining $\phi(t) := -\ D\varphi(t)$

$$\Lambda(\varphi_0,\varphi_1) := (\psi_t(T) + a_0 L\psi(T)\ -\ \psi(T))$$

one gets

$$<\Lambda(\varphi_0,\varphi_1),\ (\varphi_0,\varphi_1)> = \int_0^T b(D\varphi(t))^2 dt. \tag{34}$$

Now, on choosing $\varphi_0 = A^{-1}\xi_1$, $\varphi_1 = -\ \xi_0 + a_0\xi_1$, we define $U : X \times V^* \to V \times V^*$ by $U(\xi_0,\xi_1) = (A^{-1}\xi_1,\ a_0\xi_1 - \xi_0)$ and rewrite (34) as

$$<(\psi(T),\psi_t(T)),\ (\xi_0,\xi_1)>_{H\times V^*} = <U^*\Lambda U(\xi_0,\xi_1)(\xi_0,\xi_1)> = \tag{35}$$

$$= <\Lambda(\varphi_0,\varphi_1),(\varphi_0,\varphi_1)> = \int_0^T b(D\varphi(t))^2 dt.$$

Under the assumptions of Theorem 4 the r.h.s of (35) defines a norm on the initial data and the Lions principle applies, and among other facts, we have a uniqueness result concerning φ, i.e. if φ satisfies in addition $D\varphi = 0$ on $\partial\Omega$ then $\varphi \equiv 0$. □

Remark 5:

Thm5 and related results in [8],[9] for systems with a_0=o and exact instead of appr. controllability are concerned with "point control" that is disregarding the history as a state to be controlled. However, taking into account the history the classical control problems cannot be solved in finite time at all, hence a new concept has to be developed, where one does not insist on being able to hold a particular phase-point for all future time but to stay in a small neighbourhood of it.

References:

[1] Baumeister J (1983) Boundary control of an integrodifferential equation. J Math Anal Appl 93: 550-570

[2] Da Prato G, Ianelli M (1980) Linear integro-differential equations in Banach spaces. Red Sem Mat Univ Padova 62: 207-219

[3] Da Prato G, Lunardi A (1987) Solvability of the real line of a class of linear Volterra integrodifferential equations of parabolic type. Universita Degli Studi di Pisa Preprint Nr 178

[4] Duvaut G, Lions JL (1972) Les Inéquations en Mécanique et en Physique. Dunod Paris

[5] Gaiduk SI (1977) Some problems related to the theory of the action of transverse impulse on a rod. Differential equations 13/II: 854-861

[6] Hörmander L (1963) Linear Partial Differential Operators. Springer Verlag

[7] Leugering G (1986) Boundary controllability of a viscoelastic beam. Applicable Analysis 23: 119-137

[8] Leugering G (1987) Exact boundary controllability of an integrodifferential equation. Appl Math Optim 15: 223-250

[9] Leugering G (1987) On boundary controllability of Volterra integrodifferential equations in Hilbert spaces. To appear in the proceedings of the Conference on Control of Distributed Parameter Systems, Vorau, Austria 1986

[10] Leugering G, Schmidt EJPG (1987) Boundary control of a vibrating plate with internal damping. Submitted

[11] Lions JL (1986) Exact controllability, stabilization and perturbations for distributed systems. The John von Neumann Lecture, SIAM National Meeting, Boston, USA 1986

[12] Lunardi A (1985) Laplace transforms methods in integrodifferential equations. J Integral Eq 10 (1-3) suppl 185-211

[13] MacCamy RC, Mizel VJ, Seidman TI (1968) Approximate boundary controllability for the heat equation. J Math Anal Appl 23: 699-703

[14] Miller RK, Wheeler RL (1977) Asymptotic behavior for a linear Volterra integral equation in Hilbert space. J Diff Eq 23: 270-284

[15] Nečas J (1967) Les méthodes directes en théorie des équation elliptiques. Masson et Cie, Paris

[16] Nerain A, Joseph DD (1982) Linearized dynamics for step jumps of velocity and displacement of shearing flows of a simple fluid. Rheol Acta 21: 228-250

[17] Prüss J (1987) Positivity and regularity of hyperbolic Volterra equations in Banach spaces. Math Annalen to appear

[18] Renardy M (1982) Some remarks on propagation and nonpropagation of discontinuities in linearly viscoelastic liquids. Rheol Acta 21: 251-254

[19] Russell DL (1985) Mathematical models for the elastic beam and their control theoretic implications. Semigroup Theory and Applications (Brezis et at Eds) Longman New York

[20] Schmidt EJPG, Weck N (1978) On the boundary behavior of solutions to elliptic and parabolic equations - with applications to boundary control for parabolic equations. SIAM Control Optim 4: 593-598

REMARKS ON EXACT CONTROLLABILITY AND STABILIZATION OF A HYBRID SYSTEM IN ELASTICITY THROUGH BOUNDARY DAMPING

Walter Littman and Lawrence Markus
School of Mathematics, University of Minnesota,
Minneapolis, MN, 55455 U.S.A.

1. Introduction.

In this note we discuss some recent work [4] [5] on the exact controllability and strong stabilization of a hybrid system consisting of an elastic beam attached at its ends to two rigid bodies. We will review some of the main results of these papers as well as make some new observations and amplifications of the results and methods not made in the original papers.

We begin with a brief discussion of the reduced (two dimensional) "SCOLE" model. Among the variety of large flexible space structures currently under consideration, one prototype model discussed by NASA is the "Spacecraft Control Laboratory Experiment (SCOLE) [1]. In this model a large rigid body, the Space Shuttle Orbiter (S) is joined to a small rigid body - the antenna reflector (A) via a long flexible mast (M). The motions of the two rigid bodies (S) and (A) each are governed by ordinary differential equations, while the motion of the flexible mast must satisfy a partial differential equation with boundary conditions imposed at the two ends by the control forces and torques acting on (S) and (A). These control forces and torques regulate the coupled PDE and ODE's as a unified hybrid dynamical system. In [4] and [5] we have investigated a two dimensional reduced SCOLE model. This has the advantage of allowing a clear and much more transparent treatment, while at the same time attacking the essential mathematical difficulties.

2. The Physical Problem.

The physical problem requires the shuttle (S) to slew rapidly in orientation, while controlling the corresponding motion and structural vibrations of the mast (M) so that the antenna (A) points accurately in a given direction. To isolate the difficulties, we take the view that a definite thrust-torque program has been established for reorienting (S) about its fixed centroid to its final specified attitude. Once such a control problem for (S) has been fixed, this is carried through regardless of (M) and (A). This is a plausible approach since (S) is much more massive than either (A) or (M). Of course, the rotational influence on (A) and (M) must be taken into account.

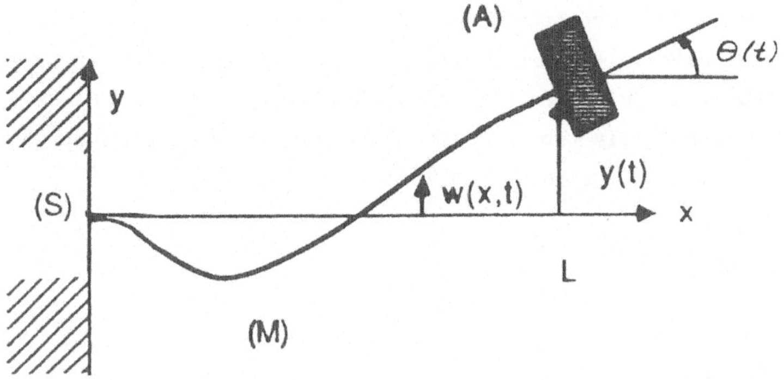

Figure 1

In figure 1 we indicate the two dimensional simplified version of SCOLE, namely an elastic beam (M) linked at its right end $(x = L)$ to the rigid body of the antenna (A), while the left end $(x = 0)$ of the mast is clamped to the massive shuttle (S) which is stationary in the (x,y) coordinate system, but which is truly rotating about the centroid K of the system (approximately the centroid of (S)).

3. Statement of Results.

Once the shuttle has been oriented we may use the Euler-Bernoulli equation for the beam with respect to the coordinates fixed in the shuttle. By appropriate scaling we may take the mast to extend over the interval $0 \leq x \leq 1$ and thus the vibrations of the mast are described by the transverse displacement $w(x,t)$ which satisfies the Euler-Bernoulli PDE of linear elasticity theory:

$$w_{tt} + w_{xxxx} = 0 \qquad 0 \leq x \leq 1 \ , \quad t \geq 0 \ ,$$

with clamped end conditions (Dirichlet) conditions at $x = 0$:

$$w(0,t) \equiv 0 \ , \quad w_x(0,t) \equiv 0 \, .$$

The linked end conditions, at the antenna end, take the form

$$(B_1w)(t) = w_{tt}(1,t) - \beta_1 w_{xxx}(1,t) = f_1(t)$$
$$(B_2w)(t) = w_{xtt}(1,t) + \beta_2 w_{xx}(1,t) = f_2(t)$$

with $\beta_1 > 0$ and $\beta_2 > 0$ constants depending on the antenna (A), $f_1(t)$ and $f_2(t)$ representing the external control forces and torques applied at the right (antenna) end with a view of bringing the vibration displacements w of the beam to rest.

There are essentially two problems here. One is the open loop control problem, where given an initial disturbance

$$w(x,0), \quad w_t(x,0)$$

in an appropriate function space, one is to find controllers $f_1(t)$ and $f_2(t)$ that will bring $w(x,t)$ to rest in finite time. It is usual to demand that $f_1(t)$ and $f_2(t)$ be L^1 or L^2. The other problem is the one of closed loop stabilization - usually viewed as the one of greater practical interest. Here the object is to choose f_1 and f_2 as functions of the state of the antenna so as to stabilize w - i.e., to make the energy of (M) + (A) approach zero as $t \to \infty$.

For the purpose of brevity we will use physical language to state our results, with reference to [4] and [5] for greater precision.

Theorem I (theorem 4 of [4]) The mast (M) can be brought to complete rest within an arbitrarily small time interval by open loop controllers $f_1(t)$ and $f_2(t)$ which are continuous on $t \geq 0$ and are C^∞ for $t > 0$ - provided the initial disturbance $w(x,0)$, $w_t(x,0)$ has a certain amount of smoothness $(H^6 \times H^4)$ and the obvious compatibility conditions hold at the clamped end. It should be pointed out that the means of obtaining this result are constructive: f_1 and f_2 are established as convergent series involving known functions.

Theorem II (theorem 2 of [5]) For each initial disturbance of finite energy, given "strictly dissipative" feedback laws

$$f_1 = L_1 w \quad \text{and} \quad f_2 = L_2 w ,$$

the energy of the system is driven to zero asymptotically. In particular the choices

$$f_1 = -w_t(1,t) \quad \text{and} \quad f_2 = -w_{xt}(1,t)$$

accomplish this feedback stabilization.

Let us note that for this discussion, the energy of the system is defined by means of the formula

$$E(t) = \frac{1}{2}\int_0^1 [w_t(x,t)^2 + w_{xx}(x,t)^2]dx + \frac{1}{2}[\mu_1 w_t(1,t)^2 + \mu_2 w_{xt}(1,t)^2] ,$$

where μ_1 and μ_2 represent positive physical constants of the antenna.

4. Questions Concerning Open Loop Control.

Several questions arise naturally in connection with Theorems I and II. With respect to the first, it would seem that the additional smoothness imposed on the initial data may be due to a shortcoming of the proof rather than be inherent in the problem. Thus a reasonable conjecture would be that finite initial energy should suffice for open loop controllability by L^1 controllers. However this conjecture is false:

Theorem III: An initial disturbance of the mast (M) having finite energy cannot always be steered to zero in finite time by open loop L^1 controllers $f_1(t)$ and $f_2(t)$.

The proof of this theorem rests principally on a negative result of R. Triggiani [8] which states:

Let X be a complex separable infinite dimensional Banach space; B a bounded linear operator from a Banach space U into X . Let A be a closed linear operator on X with dense domain, generator of a strongly continuous semigroup $S(t)$, $t > 0$, on X . Then the autonomous system

$$\frac{dx}{dt} = Ax + Bu$$

can never be exactly controllable on any finite interval by using locally L^1 controls $u(t)$, if the operator B is of finite dimensional range. (The last condition can be replaced by compactness of B , provided X has a Schauder basis.)

To apply Triggiani's result, we must, of course, express the control dynamics of our model as such a system. There are many ways to do this but for the present purpose perhaps the most convenient is that used in [6]. If the role of A and B in Triggiani's result are assigned to the operators designated in [6] by $\mathcal{A}$ and $\mathcal{B}$, the conditions of Triggiani's theorem will be satisfied and the conclusions will be valid.

While this answers the question of the exact controllability of a finite energy initial disturbance with locally L^1 controllers, it leaves open the exact degree of smoothness (in terms of Sobolev spaces, for example) required of the initial data to achieve the same end - or with L^2 controllers.

It may be asked what the physical significance of a "somewhat smooth" initial disturbance is for the SCOLE model. In other words, how can we be sure that the required degree of smoothness is achieved? One answer is that to the extent vibrations in the mast are caused by orientation maneuvers of the shuttle (S), smooth maneuvers of the latter give rise to "smooth" vibrations in the former. Thus, vibrations set up in the mast by orientation maneuvers having a certain degree of smoothness are exactly controllable by open loop continuous controls.

In connection with the nature of the controls, it should be noted that even if the initial data only has finite energy - or is in appropriate spaces of distributions - the method utilized in [4] gives controls which are C^∞ for $t > 0$. It is only to get continuity up to $t = 0$ that the additional smoothness of the initial data was hypothesized. Without this smoothness one can still say something of the nature of the controllers f_1 and f_2 near zero. They are distributions which are represented near zero by functions vanishing for $t < 0$, smooth on $t > 0$, and possibly oscillating with ever increasing rapidity and magnitude as t approaches zero through positive values.

Another question that may be asked in connection with open loop controllability is to what extent do the results continue to hold without the assumption that the shuttle (S) has been brought to rest before attempts at open loop control are initiated. For example, are the results of Theorem I still valid if (S) moves at a constant angular velocity relative to some point fixed in the shuttle. An affirmative answer to this question has been given recently by a graduate student at the University of Minnesota, Steve Taylor [7].

The method of [5] yields (without additional effort) exact controllability results for other boundary conditions at the right end (keeping zero Dirichlet conditions at the left end). In that case the method gives controllers f_1 and f_2 which are continuous for $t \geq 0$ and C^∞ for $t > 0$, even if the initial data is only of finite energy. This may be compared to the H.U.M. method of J.L. Lions, where the initial data is allowed in the "wilder" space $L^2 \times H^{-2}$, but where the controls are achieved in L^2.

5. Comments Concerning Closed Loop Control.

In [3] Chen et al. show that under the assumption that the antenna (A) has zero mass and moment of inertia (dropping the terms w_{tt} and w_{xtt} in the operators B_1 and B_2), there occurs exponential stabilization via closed loop control. In contrast, it is shown in [5] that if the mass and moment of inertia of the antenna are not neglected, closed loop stabilization (as described by Theorem II) with f_1 and f_2 depending linearly in the state of (A) cannot be exponential (see also [6]). However the situation is markedly worse [5].

Theorem IV: Given any positive function of t which approaches zero monotonically as $t \to \infty$, if one starts with an appropriately chosen finite energy initial disturbance, the energy of the closed loop stabilized system will exceed this given function at an infinite sequence of times that approach ∞.

The physical implications seem to be that the closed loop stabilizability is enhanced by making the mass and moment of inertia of the antenna (A) small and preferably negligible relative to the mast (M).

Acknowledgement: This research was partially supported by NSF Grant DMS 86-07687 and AFOSR-ISSA-860088 and the second author who received support from SERC.

References

[1] Balakrishnan, A.V., and Taylor, L., The SCOLE Design Challenge, 3rd Annual NASA SCOLE Workshop (1986).

[2] Balakrishnan, A.V., On large scale space structure control problems, Proc. IFIP Conference, Gainesville, 1986 (to appear).

[3] Chen, G., Delfour, M., Krall, A., and Payre, G., Modeling, stabilization and control of serially connected beams, SIAM J. Control and Optimization, to appear.

[4] Littman, W., and Markus, L., Exact boundary controllability of a hybrid system of elasticity, Mathematics Report #86-147, 1987, University of Minnesota, to appear, Archive for Rational Mechanics and Analysis.

[5] Littman, W., and Markus, L., Stabilization of a hybrid system of elasticity by feedback boundary damping, Mathematics Report #86-135, 1987, University of Minnesota.

[6] Littman, W., Markus, L., and You, Y.C., A note on stabilization and controllability of a hybrid elastic system with boundary control, Mathematical Report #103, 1987, University of Minnesota.

[7] Taylor, Steve, Boundary control of a vibrating beam, Manuscript, 1987.

[8] Triggiani, R., Controllability and Observability in Banach space with Bounded Operators, SIAM J. Control, pp. 462-491 (1975).

UN PROBLEME DE CONTROLE AVEC CONTRAINTES SUR L'ETAT

(Eric Lunéville, Fulbert Mignot)
Ensta Centre de l'Yvette Chemin de la Huniére
91120 Palaiseau

Introduction

Il s'agit d'une modélisation mathématique très simplifiée de la trempe d'un acier par faisceau laser et de son contrôle.Brièvement l'éclairage bref d'une pièce de métal par un laser provoque une trempe de la zone superficielle : si la température du matériau est supérieure à T_C (de l'ordre de 1400^0)il y a un changement de phase austhénite-martensite qui se conserve à température ambiante si le refroidissement est suffisamment rapide : c'est ce qui se passe dès que l'éclairage laser a cessé.

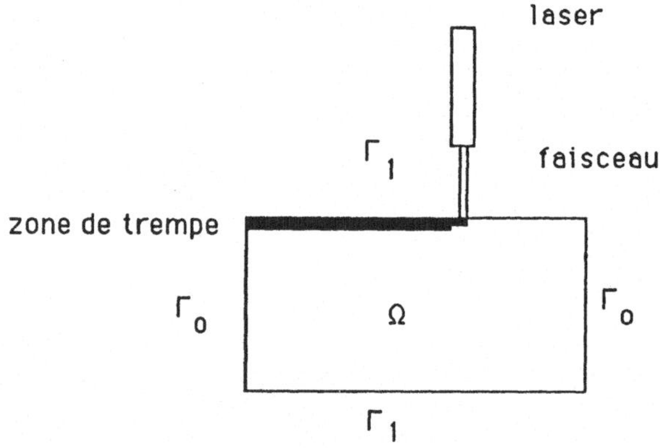

Schéma du dispositif

Le problème initial est d'optimiser la vitesse v de déplacement du faisceau et I(t) l'énergie fournie par le laser pour que la zone trempée Z ait une épaisseur donnée e.On modélise cet objectif en imposant la contrainte sur la température $\{\ \forall\ x \in Z\ ,\text{mesure}\{t,y\ (x,t) \geq T_c\ \} \geq t_0\ \}$ (t_0 est fixé) . Cette température y définit l'état du système et est solution de l'équation parabolique doublement non-linéaire:

$$\begin{aligned} & c(y)\partial_t y - \text{div}(k(y)\text{grad } y) = 0 \quad \text{dans } \Omega \\ & k(y)\partial_n y = h(y-y_{ex}) + \beta(y)(y^4 - y^4_{ex}) - \alpha(y)I(x,t,v) \ \text{ sur } \Gamma_1 \qquad (0) \\ & k(y)\partial_n y = h(y-y_{ex}) \qquad \text{sur } \Gamma_0 \\ & y(0)=y_0 \end{aligned}$$

On cherche à minimiser la fonctionnelle $\int l^2(x,t,v)dx$ oû à maximiser v avec la contrainte précédente sur l'état et $l \in U_{ad}$ convexe fermé de $L^2(\Gamma_1)$.

Dans TALBOT (9) on trouvera une analyse du problème (0) (théorème d'existence,cas d'évolution ,cas stationnaire,approximation numérique).

Ici on résoud un problème de contrôle simplifié par rapport au problème précédent:on se limite au cas stationnaire et l'équation d'état est linéarisée.Les résultats obtenus sur le plan théorique recoupent ceux de BONNANS(2) BONNANS-CASAS(3) ,CASAS(4) ,et MACKENROTH(7).

Position du problème

Soit Ω un ouvert régulier de R^2 de frontière Γ et $\psi_d \in C^o(\Omega)$. L'état du système est la solution du problème de Neuman

$$\begin{aligned} &-\Delta y + v\partial_x y = 0 \quad \text{dans } \Omega \\ &\partial_n y = v \quad \text{sur } \Gamma_1 \\ &y = 0 \quad \text{sur } \Gamma_0 \end{aligned} \tag{1}$$

Le contrôle v appartient à U_{ad} ,oû U_{ad} est un convexe fermé de $L^2(\Gamma_1)$ On cherche à minimiser

$$J(v) = 1/2 \int_\Gamma v^2 d\Gamma \tag{2}$$

sur l'ensemble:

{$v \in U_{ad}$, l'état vérifiant la contrainte $\psi_d \leqslant y(v)$ sur K oû K est un compact inclus dans $\Omega, \psi_d \in C^o(K)$.}

Existence

Proposition:Le problème précédent admet une solution uniquePreuve.La fonction J est convexe s.c.i., sur un ensemble convexe fermé car la contrainte sur l'état définit un sous ensemble convexe fermé de U_{ad}.(Ceci est vrai quelle que soit la dimension).

Conditions du premier ordre.

Théorème La solution u du problème de minimisation ( ) est caractérisée par:il existe une mesure de Radon négative ξ portée par K telle que si $\eta \in W^{1,(p/p-1)-\varepsilon}(\Omega)$ est solution du problème:

$$\begin{aligned} &-\Delta\eta - \partial_x\eta = \xi.1_\Omega \\ &\partial_n\eta = \xi.1_\Omega \quad \text{sur } \Gamma_1 \\ &\eta = 0 \quad \text{sur } \Gamma_0 \end{aligned} \tag{3}$$

on ait:

$$\int_K (\varphi - y(u))d\xi \leqslant 0 , \quad \forall \varphi, \varphi \in C^0(K) , \varphi \geqslant \psi \text{ sur } K \tag{4}$$

$$\int_\Gamma (u+\eta)(v-u)d\Gamma \geqslant 0 , \quad \forall v \in U_{ad.} \tag{5}$$

Schéma de la démonstration.(Elle repose sur le calcul du sous différentiel d'une somme de fonctions convexes)

Soit $\psi_K : C^0(K)$ _____ $[o,\infty[$,la fonction caractéristique du cone $\mathcal{K} = \{ \varphi$,

$\varphi \in C^0(K)$, $\varphi \geqslant \psi$ sur K : $\psi_K(\varphi)=0$ si $\varphi \geqslant \psi$,$+\infty$ sinon} et ψ_U la fonction caractéristique de U_{ad} dans $L^2(\Gamma_1)$.

On définit l'opérateur $\Lambda : L^2(\Gamma) \dashrightarrow C^0(K)$

$v \dashrightarrow y(v)|_K$ (y(v) solution de (1),Λ est bien défini car $n \leqslant 2$ et y(v) estcontinu sur Ω.).

Le problème de contrôle s'écrit alors:

Minimiser $\Phi(v) = 1/2 \int_\Gamma v^2 d\Gamma + \psi_K(y(v)) + \psi_U(v)$ pour v décrivant $L^2(\Gamma_1)$. (6)

La solution du problème de contrôle (2) est la solution du problème(6).Il s'agit maintenant de calculer le système d'optimalité.Comme le cône $\mathcal{K}$est d'intérieur non vide dans $C^0(K)$on peut écrire (prop. 5.6 et5.7 (p.26) de EKELAND TEMAM (5)) :

$$\partial(1/2\int u^2 d\Gamma + \psi_K(\Lambda u) + \psi_U(u)) = \partial(1/2\int u^2 d\Gamma + \psi_K(\Lambda u)) + \partial\psi_U(u)$$
$$= u + \Lambda^* \partial\psi_K(\Lambda u) + \partial\psi_U(u) \ni 0.$$

Il existe donc une mesure $\xi \in \mathcal{M}(K) = (C^0(K))'$,$\xi \in \partial\psi_K(\Lambda u)$ vérifiant:

$$\int_K (\varphi - y(u)) d\xi \leqslant 0 \;\; \forall \varphi \in C^0(K), \varphi \geqslant \psi \text{ et tel que } 0 \in u + \Lambda^* \xi + \partial\psi_U(u). \quad (7)$$

Pour le calcul de $\Lambda^* \partial\psi_K(\Lambda u)$ on introduit un état adjoint η donné par (3):tout d'abord la mesure ξ se décompose, $\xi = \xi_1 + \xi_2$ avec $\xi_1 = \xi . 1_\Omega$, $\xi_2 = \xi . 1_\Gamma$. ξ_1 et ξ_2 sont deux mesures de Radon sur $\overline{\Omega}$,ξ_2 ayant son support inclus dans Γ) ,le problème (3) admet une solution dans $W^{1,p}(\Omega)$ pour tout $p \in [1, \infty[$.(R. DAUTRAY,P.L. LIONS (4) page 585)(La démonstation se fait en deux étapes ,1.existence de solutions avec données L^1grâce à la formule de Green ,2. passage aux données mesures par densité.)Cette solution est caractérisée par sa formulation faible:

$$\forall \varphi \in C^2(\Omega), \; \varphi\big|\Gamma_1 = 0 :$$
$$\int_\Omega (-\Delta\varphi + \partial_x\varphi) d\Omega + \int_\Gamma \partial_n\varphi . \eta d\Gamma = \int_\Omega \varphi d\xi_1 + \int_\Gamma \varphi d\xi_2 \quad (8)$$
$$\eta \in W^{1,1}(\Omega),$$

Alors la formule de Green montre que $\Lambda^* \xi = \eta$ et que :

$$\langle \Lambda(v), \xi \rangle = \int_K y(v) d\xi_1 = \int_\Omega (-\Delta\eta - \partial_x\eta) y(v) d\Omega + \int_\Gamma y(v) d\xi_2 \quad (9)$$
$$= \int_\Omega (-\Delta y + \partial_x y)\eta d\Omega - \int_\Gamma \partial_n \eta . y + \int_\Gamma \eta \partial_n y \, d\Gamma = \int_\Gamma \eta v d\Gamma$$

Les relations (7) et (9) impliquent alors le système d'optimalité.

Remarques.

1.La restriction $n \leq 2$ permet d'avoir un convexe de contraintes d'intérieur non vide. Si $n \geq 3$ et si $K \subset\subset \Omega$ la même méthode s'applique car $y(v)|_K \in C^0(K)$.Dans le cas ou $K \cap \Gamma_1 \neq \emptyset$ il faut travailler dans $(L^\infty(\Omega))'$.

2.Si l'opérateur $-\Delta+\partial_x$ est remplacé par un opérateur elliptique non linéaire on peut obtenir dans certains cas un système d'optimalité comparable , l'opérateur $-\Delta-\partial_x$ étant remplacé par l'adjoint de l'opérateur dérivé en utlisant la notion de gradient généralisé au sens de CLARKE. (BONNANS-CASAS (3)).

3. Le problème du contrôle par une condition au bord de type Dirichlet a été étudié par MACKENROTH (7).

4 Sur des problèmes analogues , MOSSINO(8), ABERGEL (1) ont appliqué les techniques de dualité .

Formulation lagrangienne

On introduit un lagrangien ayant pour solution (u,ξ),ceci permet de dualiser la contrainte sur l'état:

$$\mathcal{L}(v,\zeta)=1/2\int_\Gamma v^2 d\Gamma+\int_K (y(v)-\psi_d)d\zeta \;,v\in U_{ad}\subset L^2(\Gamma_1)\;,\zeta\in\mathcal{M}(K)\;,\zeta\leq 0\,.$$

On a : (LUNEVILLE(6))

Théorème : La solution (u,ξ) du problème de contrôle optimal ( ) est l'unique point-selle de $\mathcal{L}$:

$$\mathcal{L}(u,\zeta)\leq\mathcal{L}(u,\xi)\leq\mathcal{L}(v,\xi)\,,\qquad\forall\,(v,\zeta)\in(U_{ad},\mathcal{M}(K))\;,\zeta\leq 0\,.$$

$$\mathcal{L}(u,\xi)=\inf_{\{v\in U_{ad}}\;\sup_{\zeta\in\mathcal{M},\zeta\leq 0\}}\mathcal{L}(v,\zeta)$$

Cette formulation lagrangienne discrètisée permet la mise en oeuvre de l'algorithme d'UZAWA pour l'approximation numérique du contrôle optimal.Ontrouvera plusieurs exemples dans LUNEVILLE (6).

Bibliographie

(1)F.ABERGEL :Problèmes de contrôle mal posés .Dualité généralisée. Thèse PARIS 11 (1986)

(2)F.BONNANS-E. CASAS:Contrôle de systèmes non linéaires comportant des contraintes distribuées sur l'état. Rapport INRIA n°300 (1984).

(3)E. CASAS :Control of an elliptic problèm with pointwise state constraints .S.I.A.M. Optimization (à paraitre).

(4)R. DAUTRAY-J.L. LIONS: Analyse mathématique et calcul numérique pour les sciences et techniques. tome 1, MASSON (1985)

(5)I.EKELAND-R. TEMAM :Analyse convexe et problèmes variationnels.DUNOD(1974)

(6)E. LUNEVILLE Preprint ENSTA (1987)

(7)U.MACKENROTH: Convex parabolic boundary control problem with pointwise state constraints,J.Math.An.and Appli. 87,256,277 (1982)

(8)J.MOSSINO: An application of duality to distribued optimal control problèm with constraints on the control and state,J. of Math. and Appl. 50,223-243 (1975).

(9) J.M. TALBOT:Modélisation,étude mathématique et numérique du traitement de surface par laser. Thèse 3^{éme} cycle. Université Paris 6 (1984).

A VARIATIONAL INEQUALITY APPROACH TO THE PROBLEM OF THE DESIGN OF THE OPTIMAL COVERING OF AN OBSTACLE

P. Neittaanmäki[1], D. Tiba[2] and R. Mäkinen[1]

[1]Department of Mathematics, University of Jyväskylä,
Seminaarinkatu 15, SF–40100 Jyväskylä, Finland
[2]Department of Mathematics, INCREST, Bd. Păcii 220,
R–79622 Bucaresti, Romania

1. The problem

Consider a membrane $\Omega(\alpha)$ possibly in contact with a rigid obstacle G. Let φ describe the shape of the obstacle and let $\Omega(\alpha)$ be given by

$$\Omega(\alpha) = \{(x_1, x_2) \in \mathbf{R}^2 \mid x_2 \in]0,1[, \ 0 < x_1 < \alpha(x_2)\} ,$$

where $\alpha \in U_{ad}$ is the (control) function describing the moving part $\Gamma(\alpha)$ of the boundary $\partial\Omega(\alpha)$:

$$\Gamma(\alpha) = \{(x_1, x_2) \mid x_1 = \alpha(x_2), \ x_2 \in]0,1[\} ,$$
$$U_{ad} = \{\alpha \in W^{1,\infty}(]0,1[) \mid a \le \alpha \le b, \ |\alpha'| \le c\} ,$$

with a, b, c positive constants such that $U_{ad} \neq \emptyset$ (see Figure 1).

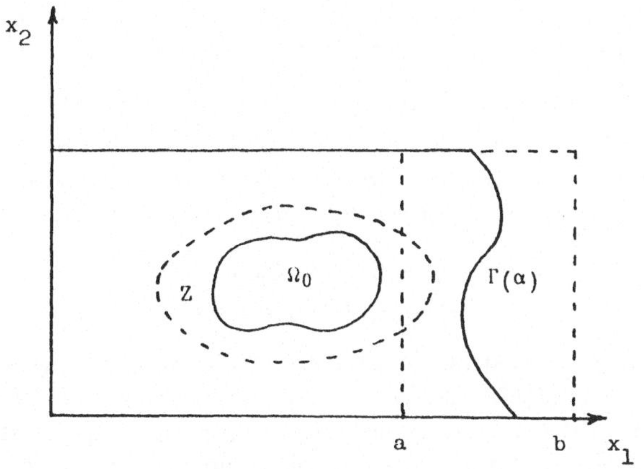

Figure 1

On any $\Omega(\alpha)$, $\alpha \in U_{ad}$, we introduce the following variational inequality: find $u(\alpha) \in K(\Omega(\alpha)) = \{v \in H_0^1(\Omega(\alpha)) \mid v \ge \varphi \text{ a.e. in } \Omega(\alpha)\}$ such that

$$\big(\operatorname{grad} u(\alpha), \operatorname{grad}(v - u(\alpha))\big)_{L^2(\Omega(\alpha))} \ge \big(f, v - u(\alpha)\big)_{L^2(\Omega(\alpha))} \qquad \forall v \in K(\Omega(\alpha)) . \tag{1.1}$$

We assume that $f \in L^2(\hat{\Omega})$, $\hat{\Omega} =]0, b[\times]0,1[$, and $\varphi \in H^1(\hat{\Omega})$, $\varphi \le 0$ on $\partial\Omega(\alpha)$ and in $[a,b]\times[0,1]$.

The equation (1.1) describes the vertical displacement $u(\alpha)$ of the membrane $\Omega(\alpha)$ (the equilibrium position) under the load f and in contact with the obstacle G. Formally, (1.1) may be rewritten as:

$$\begin{aligned} &-\Delta u(\alpha) \geq f && \text{in } \Omega(\alpha)\,, \\ &u(\alpha) \geq \varphi && \text{in } \Omega(\alpha)\,, \\ &(-\Delta u(\alpha) - f)(u - \varphi) = 0 && \text{in } \Omega(\alpha)\,, \\ &u(\alpha) = 0 && \text{on } \partial\Omega(\alpha)\,. \end{aligned} \tag{1.2}$$

We denote by

$$Z(u(\alpha)) = \{x \in \Omega(\alpha) \mid u(\alpha)(x) = \varphi(x)\}$$

the contact region, which is also called the coincidence set.

The ***packaging problem*** introduced in [1] consists of minimizing the area of $\Omega(\alpha)$ such that the *contact region* $Z(u(\alpha))$ of the solution $u(\alpha)$ of (1.1) *contains a given subset* $\Omega_0 \subset \hat{\Omega}$. That is, we consider the optimization problem

$$\text{Minimize } \{J(\alpha) \equiv \int_0^1 \alpha(x)\,dx\} \tag{1.3}$$

for $\alpha \in U_{ad}$ and such that $u(\alpha)$, the solution of (1.1) corresponding to α, satisfies the constraint

$$\Omega_0 \subset Z(u(\alpha))\,. \tag{1.4}$$

We suppose that the problem (1.1), (1.3), (1.4) has at least one admissible pair $\{u(\hat{\alpha}), \hat{\alpha}\}$. The following continuity result of Haslinger and Neittaanmäki ([3]) plays an important role in the sequel.

THEOREM 1.1. *Let $\alpha_n \to \alpha$ uniformly in $[0,1]$ and let $u_n = u(\alpha_n)$ be the solutions of (1.1). Then, there exists a subsequence, denoted again u_n, such that $\tilde{u}_n \to U$ strongly in $H^1(\hat{\Omega})$ and $U|_{\Omega(\alpha)}$ is the solution of (1.1) corresponding to α and $\tilde{u}$ is the extension of $u(\alpha)$ to $\hat{\Omega}$ by zero.*

PROOF: One may easily infer that $\|\tilde{u}_n\|_{H_0^1(\hat{\Omega})} \leq C$ and, by taking subsequences, that $\tilde{u}_n \to U$ weakly in $H_0^1(\hat{\Omega})$. Since $\alpha_n \to \alpha$ uniformly in $[0,1]$ we see that $U|_{\hat{\Omega}-\Omega(\alpha)} = 0$, so $U|_{\Omega(\alpha)} \in H_0^1(\Omega(\alpha))$. Moreover $\tilde{u}_n \geq \varphi$ a.e. in $\hat{\Omega}$ and we obtain $U \geq \varphi$ a.e. in $\hat{\Omega}$, that is $U \in K(\Omega(\alpha))$.

For any $v \in K(\Omega(\alpha))$ there exist a subsequence $\{\alpha_n\}$ and a sequence $\{v_n\} \subset H_0^1(\hat{\Omega})$ such that

$$v_n \to \tilde{v} \quad \text{in } H_0^1(\hat{\Omega})\,, \tag{1.5}$$

$$v_n|_{\Omega(\alpha_n)} \in K(\Omega(\alpha_n))\,. \tag{1.6}$$

The sequence $\{v_n\}$ may be constructed as follows. There is $\{\omega_k\} \subset \mathcal{D}(\Omega(\alpha))$ such that $\omega_k \to v$ in $H_0^1(\Omega(\alpha))$ and $\tilde{\omega}_k \to \tilde{v}$ in $H_0^1(\hat{\Omega})$. Let $v_k = \sup(\tilde{\omega}_k, \varphi)$. Obviously $v_k \in H_0^1(\hat{\Omega})$ and $v_k \geq \varphi$ a.e. in $\hat{\Omega}$. By the continuity of the $\sup(\cdot,\cdot)$ application with respect to the $H^1(\hat{\Omega})$ norm, we get (1.5).

Let k_0 be fixed and $G_{k_0} = \operatorname{supp} \omega_{k_0}$. As $\alpha_n \to \alpha$ uniformly, there is $n_0 = n(k_0)$ such that $\Omega(\alpha_{n_0}) \supset G_{k_0}$ and $v_{k_0} = 0$ on $\hat{\Omega} - \Omega(\alpha_{n_0})$. Hence $v_{k_0}|_{\Omega(\alpha_{n_0})} \in K(\Omega(\alpha_{n_0}))$ and (1.6) follows too.

From the definition of $u(\alpha_n)$ as the solution of (1.1) and from (1.6) we obtain

$$(\operatorname{grad} u_n, \operatorname{grad}(v_n - u_n))_{L^2(\Omega(\alpha_n))} \geq (f, v_n - u_n)_{L^2(\Omega(\alpha_n))}\,,$$

$$(\operatorname{grad} \tilde{u}_n, \operatorname{grad}(v_n - \tilde{u}_n))_{L^2(\hat{\Omega})} \geq (f, v_n - \tilde{u}_n)_{L^2(\hat{\Omega})}\,.$$

One may pass to the limit and deduce:

$$(\operatorname{grad} U, \operatorname{grad}(\tilde{v} - U))_{L^2(\hat{\Omega})} \geq (f, \tilde{v} - U)_{L^2(\hat{\Omega})}\,,$$

$$(\operatorname{grad} u, \operatorname{grad}(v - u))_{L^2(\Omega(\alpha))} \geq (f, v - u)_{L^2(\Omega(\alpha))}\,,$$

where $u = U|_{\Omega(\alpha)}$. Because $v \in K(\Omega(\alpha))$ is arbitrary, we see that u solves (1.1), so $u = u(\alpha)$.

Finally, we show that for a *subsequence* $\{\tilde{u}_n\}$ we have $\tilde{u}_n \to U$ in the norm of $H^1(\hat{\Omega})$. We apply (1.5), (1.6) to U and we denote w_n the obtained sequence. We have with some constant $C > 0$

$$
\begin{aligned}
0 &\le C\|U - \tilde{u}_n\|^2_{H^1(\hat{\Omega})} \\
&\le (\operatorname{grad}(U - \tilde{u}_n), \operatorname{grad}(U - \tilde{u}_n))_{L^2(\hat{\Omega})} \\
&= (\operatorname{grad} U, \operatorname{grad}(U - \tilde{u}_n))_{L^2(\hat{\Omega})} - (\operatorname{grad} \tilde{u}_n, \operatorname{grad}(U - w_n))_{L^2(\hat{\Omega})} \\
&\quad - (\operatorname{grad} \tilde{u}_n, \operatorname{grad}(w_n - \tilde{u}_n))_{L^2(\hat{\Omega})} \\
&\le (\operatorname{grad} U, \operatorname{grad}(U - \tilde{u}_n))_{L^2(\hat{\Omega})} - (\operatorname{grad} \tilde{u}_n, \operatorname{grad}(U - w_n))_{L^2(\hat{\Omega})} \\
&\quad - (f, w_n - \tilde{u}_n)_{L^2(\hat{\Omega})} \to 0 \quad \text{for } n \to \infty .
\end{aligned}
$$

□

REMARK 1.2. Since the solution of (1.1) is unique, we obtain the convergence of the whole sequence $\{u_n\}$ in Theorem 1.1. □

REMARK 1.3. As an easy consequence of Theorem 1.1 and of the compactness of U_{ad}, one may deduce the existence of at least one optimal control α^* for the problem (1.1), (1.3), (1.4). □

One of the main difficulties in the treatment of the problem (1.1), (1.3), (1.4) consists in the variable character of the domain $\Omega(\alpha)$ on which the problem is given. By scaling the domain $\Omega(\alpha)$ such that it becomes fixed, the optimization parameter α appears as a coefficient in the state system. Optimal design problems of this type were discussed in the work [**4**].

The second difficulty in the numerical realization of (1.3) is the presence of the state constraint $u(\alpha) = \varphi$ in Ω_0. One straightforward method to overcome this difficulty would be the penalty approach (see [**1**]): Let

$$J_\varepsilon(\alpha) = J(\alpha) + \frac{1}{\varepsilon}\int_{\Omega_0} (u(\alpha) - \varphi)\, dx \,, \quad (\varepsilon > 0, \text{ a penalty parameter})$$

be the modified cost functional, where $u(\alpha)$ denotes the solution of (1.1).

The penalty form of (1.3) reads

$$\text{Minimize } J_\varepsilon(\alpha) \tag{1.7}$$

for $\alpha \in U_{ad}$ with the corresponding solution $u(\alpha)$ of the state problem (1.1). In [**1**] it is shown that the mapping $\alpha \mapsto J_\varepsilon(\alpha)$ is differentiable. Therefore it seems that one could solve easily the problem (1.7) numerically. It was shown in [**3, 5**] that in the discrete case the mapping $\alpha \mapsto x(\alpha)$ (and therefore the mapping $\alpha \mapsto J_\varepsilon(\alpha)$) may be nondifferentiable. Here we denote by $x(\alpha)$ the vector of nodal values of the solution of the discrete state problem.

We illustrate the situation with an example.

EXAMPLE 1.4. Let us consider the problem:

$$\begin{cases} -u^t(x)'' = -1 \,, & x \in (0,t) \\ u^t(x) \ge 0 \,, & x \in (0,t) \\ u^t(0) = 0 \,, \ u^t(t) = 1 \,. \end{cases} \tag{1.8}$$

The solution of (1.8) is

$$u^t(x) = \left(\frac{1}{t} - \frac{t}{2}\right)x + \frac{1}{2}x^2 \quad \text{for } t \le \sqrt{2}$$

$$u^t(x) = \begin{cases} \frac{1}{2}(x - t + \sqrt{2})^2 & \text{, if } x \ge t - \sqrt{2} \\ 0 & \text{, if } x \le t - \sqrt{2} \end{cases} \quad \text{for } t \ge \sqrt{2} \,.$$

The derivative of u^t with respect to the design parameter t is

$$\frac{\partial}{\partial t}u^t(x) = \left(-\frac{1}{t^2} - \frac{1}{2}\right)x \quad \text{for } t \le \sqrt{2}$$

$$\frac{\partial}{\partial t}u^t(x) = \begin{cases} t - x - \sqrt{2} & \text{, if } x \ge t - \sqrt{2} \\ 0 & \text{, if } x \le t - \sqrt{2} \end{cases} \quad \text{for } t \ge \sqrt{2}\,.$$

For $t = \sqrt{2}$ both expressions reduce to $-x$. Thus u^t is continuously differentiable in t.

Let $\{0, \frac{t}{3}, \frac{2t}{3}, t\}$ be a partition of $[0, t]$. The discrete state inequality reads:

$$\begin{bmatrix} -2 & 1 \\ 1 & -2 \end{bmatrix} \begin{bmatrix} x_1 \\ x_2 \end{bmatrix} \ge \frac{t^2}{18}\begin{bmatrix} 1 \\ 1 \end{bmatrix} - \begin{bmatrix} 0 \\ 1 \end{bmatrix}, \qquad x_1, x_2 \ge 0\,.$$

For $t \le \sqrt{6}$ we have

$$x_1 = \frac{1}{3}\left(1 - \frac{t^2}{6}\right),$$

$$x_2 = \frac{1}{3}\left(2 - \frac{t^2}{6}\right),$$

i.e. the contact condition $x_i \ge 0$ is not active.

For $t \ge \sqrt{6}$ we have

$$x_1 = 0\,,$$

$$x_2 = \frac{1}{2}\left(1 - \frac{t^2}{18}\right).$$

Thus it can be seen that neither $x_1(t)$ nor $x_2(t)$ are differentiable in t at $t = \sqrt{6}$. Consequently, we can note that also $\sum_i x_i(t)$ is nondifferentiable. □

We can say in general that if $x(\alpha)$ is the FE-solution of some obstacle problem with the obstacle $\varphi(\alpha)$, then $\sum_{i \in I}(x_i(\alpha) - \varphi_i(\alpha))$ is not differentiable in α.

This example shows that the straightforward application of the penalty method is not enough to make it possible to apply standard nonlinear programming methods for minimizing J_ε. Therefore non-smooth programming methods or a regularization of the state problem are needed. Instead of the state inequality (1.2) one can solve the state problem

$$\begin{cases} -\Delta u(\alpha) + \beta_\varrho(u(\alpha) - \varphi) = f & \text{in } \Omega(\alpha) \\ u(\alpha)|_{\partial\Omega(\alpha)} = 0\,, \end{cases} \tag{1.9}$$

where $\beta_\varrho(v) = -\frac{1}{\varrho}(v^-)^2$, $\varrho > 0$ is a penalty term corresponding to the nonpenetrating condition $u \ge \varphi$ on $\Omega(\alpha)$ (see [5]).

If one minimizes J_ε over U_{ad} with (1.9), one gets an approximation for the original solution α^* of (1.3).

2. A VARIATIONAL INEQUALITY METHOD

In this chapter we show how to use the so called variational inequality method ([4, 6, 7]) directly in the problem (1.1), (1.3), (1.4) and how to obtain precise approximation results. In the variational inequality method the state constraint $\Omega_0 \subset Z(u(\alpha))$ is forced to be satisfied by modifying the state problem (1.1) as well as the criterion function.

We start with a relaxation of the state constraints. Let $\{\psi_n^\delta\}$ be a family of smooth functions on $\hat{\Omega}$ satisfying the conditions

$$\psi_n^\delta \leq \frac{1}{n} + \varphi \quad \text{in } \Omega_{0,n} , \tag{2.1}$$

$$\psi_n^\delta \geq n \quad \text{in } \Omega(\alpha) - \Omega_0^n , \tag{2.2}$$

$$\psi_n^\delta \leq \delta + \varphi \quad \text{in } \Omega_0 , \quad \psi_n^\delta = \delta + \varphi \quad \text{in } \partial\Omega_0 , \tag{2.3}$$

$$\psi_n^\delta \geq \psi_{n-1}^\delta \quad \text{in } \Omega(\alpha) - \Omega_0 , \tag{2.4}$$

$$\begin{aligned}&\text{for any continuous function } u \text{ on } \Omega(\alpha), \text{ with } u|_{\Omega_0} = \varphi,\\ &\text{there exists } n_0 \in N \text{ such that } \psi_n^\delta \geq u \text{ in } \Omega(\alpha) \text{ for } n \geq n_0.\end{aligned} \tag{2.5}$$

Above we denote

$$\Omega_{0,n} = \left\{ x \in \Omega_0 \mid \text{dist}(x, \partial\Omega_0) \geq \frac{1}{n} \right\} , \tag{2.6}$$

$$\Omega_0^n = \left\{ x \in \Omega(\alpha) \mid \text{dist}(x, \Omega_0) \leq \frac{1}{n} \right\} \tag{2.7}$$

and we assume that $n \geq 1/\delta$.

Roughly speaking, the family $\{\psi_n^\delta\}$ is an approximation of the indicator function of Ω_0 plus φ and the conditions (2.1)–(2.5) may be viewed as regularity assumptions on Ω_0. For an effective construction of such a family see the end of this section.

For the sake of simplicity, we put $\varphi = 0$.

LEMMA 2.1. *For any admissible pair $\{u(\hat{\alpha}), \hat{\alpha}\}$ of the problem (1.1), (1.3), (1.4), there exists $\bar{n}$ such that for $n \geq \bar{n}$, $\{u(\hat{\alpha}), \hat{\alpha}\}$ is an admissible pair for the approximating problem (1.1), (1.3) and*

$$u \leq \psi_n^\delta \quad \text{a.e. in } \Omega(\alpha) . \tag{2.8}$$

PROOF: It is wellknown that in two dimensional case the solution of the variational inequality (1.1) is continuous in $\Omega(\hat{\alpha})$ for $f \in L^2(\hat{\Omega})$. As $\{u(\hat{\alpha}), \hat{\alpha}\}$ is an admissible pair, we have $u = 0$ on Ω_0 and, by (2.5), there exists $\bar{n} \in N$ such that $u(\hat{\alpha}) \leq \psi_n^\delta$ for $n \geq \bar{n}$ and (2.8) is fulfilled. □

REMARK 2.2. In particular, Lemma 2.1 is valid for any optimal pair $\{u(\alpha^*), \alpha^*\}$. □

REMARK 2.3. Using again Theorem 1.1 and the compactness of U_{ad}, one may easily establish the existence of at least one optimal pair for the approximating problem, which we denote by $\{u(\alpha_n^\delta), \alpha_n^\delta\}$. □

THEOREM 2.4. *For $n \to \infty$, on a subsequence, we have $\alpha_n^\delta \to \alpha^\delta$ uniformly and α^δ is an optimal control for the problem (1.1), (1.3), (1.4).*

PROOF: Let J denote the cost functional of (1.3). By Lemma 2.1 α^* is adnissible for the approximating problem for $n \geq \bar{n}$ and we get

$$J(\alpha_n^\delta) \leq J(\alpha^*) . \tag{2.9}$$

Moreover, as U_{ad} is compact in $C(]0,1[)$, we may assume that, for a subsequence, $\alpha_n^\delta \to \alpha^\delta \in U_{ad}$, uniformly in $[0,1]$ and (2.9) gives

$$J(\alpha^\delta) \leq J(\alpha^*) . \tag{2.10}$$

Then, Theorem 1.1 implies that $\tilde{u}(\alpha_n^\delta) \to \tilde{u}(\alpha^\delta)$ strongly in $H_0^1(\hat{\Omega})$. By (2.1) we see that $\tilde{u}(\alpha^\delta) = 0$ a.e. in Ω_0, that is the pair $\{u(\alpha^\delta), \alpha^\delta\}$ is admissible for the problem (1.1), (1.3), (1.4). From (2.10) we see that it is optimal. □

REMARK 2.5. For n sufficiently large, the pairs $\{u(\alpha_n^\delta), \alpha_n^\delta\}$ satisfy:

- $J(\alpha_n^\delta) \leq J(\alpha^*)$,
- $u(\alpha_n^\delta)$ satisfies the state problem (1.1),
- $\alpha_n^\delta \in U_{ad}$,
- $0 \leq u(\alpha_n^\delta) \leq \delta$ in Ω_0 (pointwise estimate).

Therefore they are suboptimal for the problem (1.1), (1.3), (1.4). □

In order to remove the state constraint (2.8) we apply the variational inequality technique. We consider the problem without state constraints:

$$\text{Minimize } \left\{ J(\alpha) + \frac{1}{\varepsilon}\|\gamma_\varepsilon(u - \psi_n^\delta)\|^2_{L^2(\Omega(\alpha))} \right\}, \tag{2.11}$$

$$-\Delta u + \beta(u) + \gamma_\varepsilon(u - \psi_n^\delta) \ni f , \tag{2.12}$$

$$u|_{\partial\Omega(\alpha)} = 0 , \tag{2.13}$$

$$\alpha \in U_{ad} . \tag{2.14}$$

Here γ_ε is the Yosida approximation of the maximal monotone graph

$$\gamma(y) = \begin{cases} 0 & , \ y < 0 \\ [0, \infty[& , \ y = 0 \\ \emptyset & , \ y > 0 \end{cases}$$

and

$$\beta(y) = \begin{cases} 0 & , \ y > 0 \\ ]-\infty, 0] & , \ y = 0 \\ \emptyset & , \ y < 0 . \end{cases}$$

LEMMA 2.6. *Assume that $f \in L^\infty(\hat{\Omega})$, then*

$$\|\gamma_\varepsilon(u - \psi_n^\delta)\|_{C(\overline{\Omega(\alpha)})} \le \|f + \Delta\psi_n^\delta\|_{L^\infty(\Omega(\alpha))} , \tag{2.15}$$

where u is the solution of (2.12), (2.13).

PROOF: By the regularity of u we know that $\gamma_\varepsilon(u - \psi_n^\delta) \in C(\overline{\Omega(\alpha)})$ since γ_ε is Lipschitzian. Moreover, as $\gamma_\varepsilon(y) = 0$ for $y \le 0$, we get

$$\beta(u)\gamma_\varepsilon(u - \psi_n^\delta) = 0 \quad \text{in } \Omega(\alpha) . \tag{2.16}$$

Multiplying (2.12) by $\gamma_\varepsilon^{p-1}(u - \psi_n^\delta)$, $p > 2$ even we have:

$$\begin{aligned} &- \int_{\Omega(\alpha)} \Delta u \gamma_\varepsilon^{p-1}(u - \psi_n^\delta)\, dx \\ &= - \int_{\Omega(\alpha)} \Delta(u - \psi_n^\delta)\gamma_\varepsilon^{p-1}(u - \psi_n^\delta)\, dx - \int_{\Omega(\alpha)} \Delta\psi_n^\delta \gamma_\varepsilon^{p-1}(u - \psi_n^\delta)\, dx \\ &= \int_{\Omega(\alpha)} \operatorname{grad}(u - \psi_n^\delta)\operatorname{grad}\gamma_\varepsilon^{p-1}(u - \psi_n^\delta)\, dx - \int_{\partial\Omega(\alpha)} \frac{\partial}{\partial n}(u - \psi_n^\delta)\gamma_\varepsilon^{p-1}(-\psi_n^\delta)\, d\tau \\ &\quad - \int_{\Omega(\alpha)} \Delta\psi_n^\delta \gamma_\varepsilon^{p-1}(u - \psi_n^\delta)\, dx \\ &\ge - \int_{\Omega(\alpha)} \Delta\psi_n^\delta \gamma_\varepsilon^{p-1}(u - \psi_n^\delta)\, dx . \end{aligned}$$

Combining this with (2.16) we have the estimate

$$\int_{\Omega(\alpha)} |\gamma_\varepsilon(u - \psi_n^\delta)|^p\, dx \le \int_{\Omega(\alpha)} |f + \Delta\psi_n^\delta||\gamma_\varepsilon(u - \psi_n^\delta)|^{p-1}\, dx$$

We close this section with an example of a family $\{\psi_n^\delta\}$ satisfying (2.1)–(2.5). We discuss in the details only the case $\Omega(\alpha) \subset \mathbf{R}$. So, let $\Omega(\alpha)$ be an interval of $\mathbf{R}$ and Ω_0 be some subinterval of positive length. The situation when Ω_0 is the union of some disjoint subintervals may be studied similarly.

For $n \geq 1/\delta$ we define the sequence $\{\psi_n^\delta\}$ such that it satisfies the conditions (2.1)–(2.4) and

$$|\psi_n'| \geq n \quad \text{on } \Omega_0^n - \Omega_0 \ . \tag{2.5'}$$

We have to show that $\{\psi_n^\delta\}$ satisfies (2.5) too.

Let u be any continuous function on $\overline{\Omega(\alpha)}$, $u|_{\partial\Omega(\alpha)} = 0$, $u|_{\Omega_0} = 0$ (we have fixed as before $\varphi = 0$). Let u_λ be defined by

$$u_\lambda(x) = \begin{cases} \int_{-\infty}^{\infty} u(x + \lambda - \lambda\tau)\varrho(\tau)\,d\tau & x \leq x_0 \ , \\ \int_{-\infty}^{\infty} u(x - \lambda - \lambda\tau)\varrho(\tau)\,d\tau & x \geq x_0 \ , \end{cases}$$

where x_0 is the middle of Ω_0 and ϱ is Friedrich's mollifier, that is $\varrho \in C_0^\infty(\mathbf{R})$, $\varrho(x) = 0$ for $|x| \geq 1$, $\varrho(-s) = \varrho(s)$, $\varrho(s) \geq 0$, $\int_{-\infty}^{\infty} \varrho(s)\,ds = 1$. Since $u|_{\Omega_0} = 0$, we get $u_\lambda \in C^\infty(\overline{\Omega(\alpha)})$, $u_\lambda|_{\Omega_0} = 0$ and, for sufficiently small λ, we have

$$u_\lambda(x) \geq u(x) - \delta \ , \tag{2.19}$$

where $\delta > 0$ is fixed.

As u is bounded on $\overline{\Omega(\alpha)}$, $\{u_\lambda(x)\}$ is uniformly bounded on $\overline{\Omega(\alpha)}$. There is $n_0 \in N$ such that

$$\begin{aligned} \psi_n^\delta &\geq u + \delta \quad \text{in } \Omega(\alpha) - \Omega_0^n \ , \quad n \geq n_0 \ , \\ \psi_n^\delta &\geq u \qquad\quad \text{in } \Omega(\alpha) - \Omega_0^n \ , \quad n \geq n_0 \ , \end{aligned}$$

(n_0 depends only on u).

As $u_\lambda \in C^\infty(\overline{\Omega(\alpha)})$, there exists $m_\lambda = \sup |u_\lambda'|$ in $\overline{\Omega(\alpha)}$. We can find n_λ (which depends on λ) such that

$$|(\psi_n^\delta)'| \geq |u_\lambda'| \quad \text{on } \Omega_0^n - \Omega_0 \ . \tag{2.20}$$

By (2.3) we have $\psi_n^\delta|_{\partial\Omega_0} = \delta \geq \delta + u_\lambda|_{\partial\Omega_0}$. This, combined with (2.20), gives

$$\psi_n^\delta \geq \delta + u_\lambda \quad \text{on } \Omega_0^n - \Omega_0 \ .$$

Of course, (2.20) should be understood in the correct way without the modulus. Finally, (2.19) shows $\psi_n^\delta \geq u$ on $\Omega_0^n - \Omega_0$ and we conclude that $\psi_n^\delta \geq u$ in $\Omega(\alpha)$ for $n \geq n_\lambda$.

The above construction may be extended directly to $\Omega(\alpha) \subset \mathbf{R}^2$, when Ω_0 is a disc in $\Omega(\alpha)$.

3. The variational inequality method in finite dimensional case

In this section we briefly describe how to apply the variational inequality method in the finite dimensional case.

Let $\alpha = (\alpha_0, \ldots, \alpha_{N(h)})$ be the vector of control parameters (x_2-coordinates of design nodes $A_i = (ih, \alpha_i)$, $i = 0, \ldots, N(h)$),

$$\begin{aligned} \alpha \in \mathcal{U} = \{\alpha \in \mathbf{R}^{N(h)+1} \,|\, & a \leq \alpha_i \leq b, \ i = 0, \ldots, N(h), \\ & -ch \leq \alpha_{i+1} - \alpha_i \leq ch, \ i = 0, \ldots, N(h)\} \ . \end{aligned}$$

The (polygonal) domain $\Omega(\alpha)$ is defined by design nodes A_i in an usual way. Let $T_h(\alpha)$ be a regular triangular of $\Omega(\alpha)$.

for all $p > 2$ even. Then

$$\|\gamma_\varepsilon(u - \psi_n^\delta)\|_{L^p(\Omega(\alpha))} \le \|f + \Delta\psi_n^\delta\|_{L^p(\Omega(\alpha))}$$

and passing to the limit $p \to \infty$, we obtain (2.15). □

COROLLARY 2.7. *We have*

$$\sup_{x\in\hat{\Omega}_0} (u - \psi_n^\delta)_+ \le \varepsilon\|f + \Delta\psi_n^\delta\|_{L^\infty(\Omega(\alpha))} \,. \tag{2.17}$$

PROOF: This follows by (2.15) and by the properties of γ_ε. □

REMARK 2.8. Obviously $u \ge 0$ as the solution of (2.12). Therefore, on Ω_0, we have the pointwise estimate

$$0 \le u \le \psi_n^\delta + \|f + \Delta\psi_n^\delta\|_{L^\infty(\Omega)} \cdot \varepsilon \le \delta + \varepsilon\|f + \Delta\psi_n^\delta\|_{L^\infty(\Omega)} \,.$$

□

REMARK 2.9. An argument similar to Theorem 1.1 shows that the problem (2.11) (2.14) has at least one optimal pair, which we denote $\{u(\alpha_{n,\varepsilon}^\delta), \alpha_{n,\varepsilon}^\delta\}$. Moreover, since $\gamma_\varepsilon(y) = 0$ for $y \le 0$, then any admissible pair for (1.1), (1.3), (2.8) is also admissible for (2.11)-(2.14) with the same cost. It yields

$$J(\alpha_{n,\varepsilon}^\delta) + \frac{1}{\varepsilon}\|\gamma_\varepsilon(u(\alpha_{n,\varepsilon}^\delta) - \psi_n^\delta)\|^2_{L^2(\Omega(\alpha_{n,\varepsilon}^\delta))} \le J(\alpha^*) \,, \tag{2.18}$$

for n sufficiently large. □

By (2.17), (2.18) we see that the pair $\{u(\alpha_{n,\varepsilon}^\delta), \alpha_{n,\varepsilon}^\delta\}$ has nice properties. However, it doesn't satisfy the state equation (1.1) and this may cause troubles. In order to check this, we denote shortly by u^ε the solution of (1.1) corresponding to $\alpha_{n,\varepsilon}^\delta$.

Assume now that the cost functional (2.11) contains the term

$$\frac{1}{\varepsilon}\|\gamma_\varepsilon(u - \psi_n^\delta)\|^2_{L^q(\Omega(\alpha))}$$

with some $q > 2$. All the above results remain true and, by (2.18), we have

$$\|\gamma_\varepsilon(u(\alpha_{n,\varepsilon}^\delta) - \psi_n^\delta)\|_{L^q(\Omega(\alpha_{n,\varepsilon}^\delta))} \le C\varepsilon^{\frac{1}{2}}$$

with C independent of n, δ, ε.

It is possible to apply a result on the Lipschitzian dependence of the solution of variational inequalities with respect to the right-hand side, due to Brezis [2]. As we work in space dimension 2, we obtain

$$\|u(\alpha_{n,\varepsilon}^\delta) - u^\varepsilon\|_{L^\infty(\Omega(\alpha_{n,\varepsilon}^\delta))} \le C\|\gamma_\varepsilon(u(\alpha_{n,\varepsilon}^\delta) - \psi_n^\delta)\|_{L^q(\Omega(\alpha_{n,\varepsilon}^\delta))} \,.$$

The following result is proved

COROLLARY 2.10. *The pair $\{\alpha_{n,\varepsilon}^\delta, u^\varepsilon\}$ is suboptimal for the problem (1.1), (1.3), (1.4) in the following sense:*

i) *u^ε satisfies the state problem (1.1),*
ii) $J(\alpha_{n,\varepsilon}^\delta) \le J(\alpha^*)$,
iii) $\alpha_{n,\varepsilon}^\delta \in U_{ad}$,
iv) $0 \le u^\varepsilon|_{\Omega_0} \le \delta + \varepsilon\|f + \Delta\psi_n^\delta\|_{L^\infty(\Omega(\alpha_{n,\varepsilon}^\delta))} + C\varepsilon^{\frac{1}{2}}$.

The state problem (2.12)–(2.13) is approximated using the finite element method. It leads to the quadratic programming problem

$$\underset{x\in K_n^\delta}{\text{Minimize}}\ \frac{1}{2}x(\alpha)^T A(\alpha)x(\alpha) - F(\alpha)^T x(\alpha), \tag{3.1}$$

where K_n^δ is the closed, convex set

$$K_n^\delta = \{x \in \mathbf{R}^{n(h)} \mid \varphi_i \leq x_i \leq \psi_{n,i}^\delta, \quad i = 1, \dots, n(h)\}.$$

Here $x(\alpha)_i$, φ_i and $\psi_{n,i}^\delta$ denote the nodal values of u_h (the FE-approximation of u), φ and ψ_n^δ, respectively. $A(\alpha)$ and $F(\alpha)$ are the finite element stiffness matrix and the force vector.

Using the classical Kuhn-Tucker optimality conditions the problem (3.1) can be written as an equality

$$A(\alpha)x(\alpha) = F(\alpha) + \lambda(\alpha) + \mu(\alpha), \tag{3.2}$$

where $\lambda(\alpha)$ ($\mu(\alpha)$) is the vector of non-negative (non-positive) Lagrange multipliers corresponding to the constraints $\varphi_i \leq x(\alpha)_i$ ($x(\alpha)_i \leq \psi_{n,i}^\delta$).

We find that the penalty term in (2.11) contains essentially the norm of the generalized Lagrange multiplier corresponding to the state constraint which is added to the state problem.

Now the finite dimensional analogue for the problem (2.11) reads as follows:

$$\underset{\alpha\in\mathcal{U}}{\text{Minimize}}\ \{J(\alpha) + \frac{1}{\varepsilon}\|\mu(\alpha)\|^2_{\mathbf{R}^{n(h)}}\} \tag{3.3}$$

subject to

$$A(\alpha)x(\alpha) = F(\alpha) + \lambda(\alpha) + \mu(\alpha)\ .$$

REMARK 3.1. The mapping $\alpha \mapsto \mu(\alpha)$ is not necessarily differentiable (as the mapping $\alpha \mapsto x(\alpha)$ is only directionally differentiable). Therefore we are dealing with a non-smooth programming problem. □

REMARK 3.2. As the original state problem (1.1) was also a variational inequality, the computational work in solving the modified state inequality (3.1) is not increased. □

4. NUMERICAL EXAMPLES

We present two numerical examples. In minimizing the cost functional of (3.3) C. Lemarechal's implementation of the bundle method for non-smooth programming was used. The quadratic programming problem (3.1) was solved using a variant of the Gauss-Seidel method.

EXAMPLE 4.1. Let $\Omega_0 =]0.25, 0.5[\times]0.25, 0.75[$, $a = 0.6$, $b = 1.0$, $c = 2.0$, $f \equiv -1.0$, $\varphi(x_1, x_2) = -0.05(x_1^2 + (x_2 - 0.25)^2)$ and $\varepsilon = 10^{-5}$, $\delta = 10^{-6}$. The domain $\Omega(\alpha)$ was discretized using a finite element mesh with 128 triangles and 81 nodes. Therefore we have 9 degrees of freedom in the minimization problem (3.3).

As initial quess we choose $\alpha_i = 0.9$, $i = 0, \dots, 8$. With this choice the original (discrete) state problem and state constraint are satisfied. After 22 iterations the value of the cost functional was reduced from the initial value 0.900 to 0.777. The final triangulation and a contour plot of the corresponding state is shown in Figures 2a–2b. Ω_0 is indicated by dotted lines.

EXAMPLE 4.2. Let $\varphi(x_1, x_2) = -0.03\sin(\pi x_1)\sin(\pi x_2)$, $\varepsilon = 10^{-6}$, $\delta = 10^{-6}$ and let the rest of the data be the same as in the previous example. Now we use a slightly finer mesh with 256 triangles and 153 nodes and therefore we have 17 degrees of freedom in the minimization.

As initial quess we choose $\alpha_i = 0.7$, $i = 0, \dots, 16$. With this choice the original state problem is not satisfied. However, after 14 iterations a minimum was found with the final value of the cost functional being 0.688 (the initial value = 13.2). The final triangulation and a contour plot of the corresponding state is shown in Figures 3a–3b.

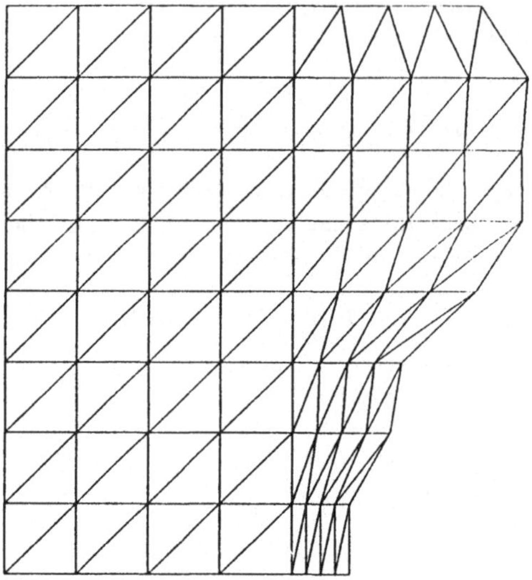

Figure 2a

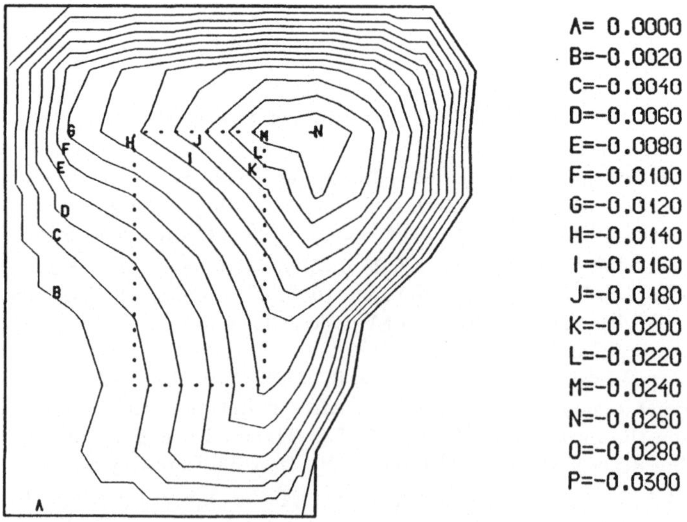

Figure 2b

Figure 3a

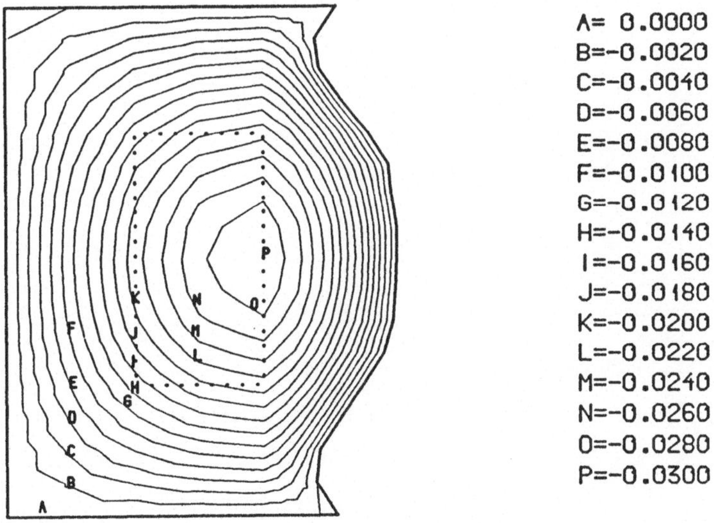

Figure 3b

REFERENCES

1. B. Benedict, J. Sokolowski and J.P. Zolesio, *Shape optimization for the contact problems*, in "Proc. of 11th IFIP Conference on System Modelling and Optimization," (P. Topft-Christiansen ed.) Lecture Notes in Control and Inform. Sciences 59, Springer-Verlag, New York, Berlin, Heidelberg, 1984, pp. 789–799.
2. H. Brezis, *Problèmes unilatéraux*, J. Math. Pures et Appl. **51** (1972).
3. J. Haslinger and P. Neittaanmäki, *On the design of the optimal covering of an obstacle*, in "Proc. of IFIP Conference of Boundary Control and Boundary Variations," (J.P. Zolésio ed.) Lecture Notes in Control and Information Sciences, Springer-Verlag, 1987.
4. J. Haslinger, P. Neittaanmäki and D. Tiba, *On state constrained optimal shape design problems*, in "Proc. Optimal Control of Partial Differential Equations II," (K.H. Hoffmann and W. Krabs ed.), Birkhauser, 1987, pp. 109–122.
5. P. Neittaanmäki, *On the optimal shape design problems*, in "Proc. of Int. Colloq. on Free Boundary Problems: Theory and Applications," Lecture Notes in Mathematics, Pitman, 1988 (to appear).
6. P. Neittaanmäki and D. Tiba, *A variational inequality approach to constrained control problems for parabolic systems*, Appl. Math. Optimiz. (to appear).
7. D. Tiba, *Une approche par inéqualitions variationelles pour les problèmes de contrôle avec contraintes*, C.R.A.S. Paris t. 302, Serie I **1** (1986).

SOME RESULTS ON OPTIMAL CONTROL FOR UNILATERAL PROBLEMS

Jean-Pierre Puel
Université d'Orléans et Laboratoire d'Analyse Numérique,
T.55-65, 5 ème étage, Université Paris VI,
4 Place Jussieu, 75252 PARIS CEDEX 05, France.

We shall consider some (a priori simple) examples of optimal control problems governed by variational inequalities corresponding to an obstacle constraint (or unilateral problems). Our aim is to show the difficulties encountered in obtaining "good" optimality conditions of first order, and to give several approaches and the corresponding results.

1 STATEMENT OF THE PROBLEM

1.1 Elliptic case

Let Ω be a bounded open subset of $\Re^n$ and let K be the closed convex subset of $H_0^1(\Omega)$ defined by

$$K = \{z \in H_0^1(\Omega), \quad z \geq 0 \quad \text{a.e. in } \Omega \}. \tag{1.1}$$

If U_{ad} denotes a nonempty closed convex subset of $L^2(\Omega)$ (set of admissible controls) and $< , >$ denotes the pairing between $H^{-1}(\Omega)$ and $H_0^1(\Omega)$, for $f \in L^2(\Omega)$ and $v \in U_{ad}$, we consider the "state equation" :

$$< -\Delta y(v), z - y(v) > \geq \int_\Omega (f+v)(z - y(v))dx, \quad \forall z \in K, \quad y(v) \in K. \tag{1.2}$$

It is standard that (1.2) defines a mapping $v \rightarrow y(v)$ from U_{ad} to $H_0^1(\Omega)$ which is Lipschitz continuous but not differentiable in general.

We now define the cost function

$$J(v) = \frac{1}{2}\int_\Omega |y(v) - z_d|^2 dx + \frac{N}{2}\int_\Omega |v|^2 dx, \tag{1.3}$$

where $\quad z_d \in L^2(\Omega) \quad$ and $\quad N > 0$.

The optimal control problem is then

$$\text{Find} \quad u \in U_{ad} \quad \text{such that} \quad J(u) = \min_{v \in U_{ad}} J(v). \tag{1.4}$$

This problem is a non-convex and non-differentiable optimization problem.
Existence of an optimal control for (1.4) can be obtained in a standard way (c.f. for example [M.P.1]) and, in general, there is no uniqueness.

The interesting question is : how to find "good" necessary conditions of first order which are satisfied by the optimal control, or in other words, how to find a "good" optimality system.

In order to point out the real difficulty we will give an analogous problem in $\Re$, with no partial differential operator.

Example 1.1

Let

$$K_0 = \{z \in \Re, \quad z \geq 0\}, \quad f = -1, \quad v \in \Re,$$

and consider the state equation

$$(y(v), z - y(v)) \geq (-1 + v, z - y(v)), \quad \forall z \in K_0, \quad y(v) \in K_0, \tag{1.5}$$

which is equivalent to

$$y(v) = (-1 + v)^+. \tag{1.6}$$

We now take

$$z_d = 1, \quad N = 1 \quad \text{and} \quad J_0(v) = (y(v) - 1)^2 + v^2.$$

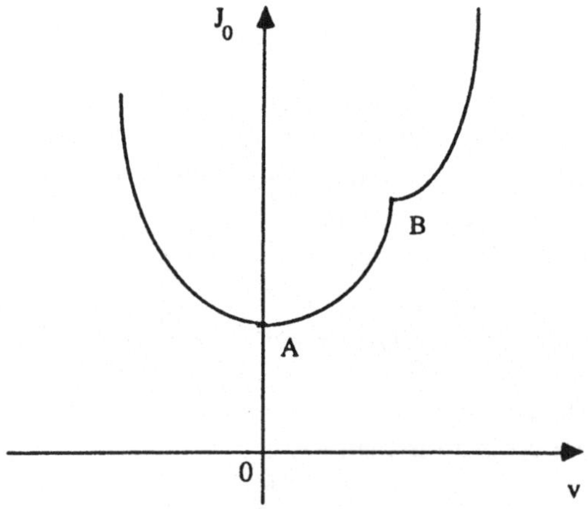

figure 1

We want to find an optimality system for which $A = (0,1)$ is admissible but not $B = (1,2)$!

1.2 Parabolic case

If T is a given positive number, let us write

$$Q = \Omega \times]0, T[,$$

and let U_{ad} be a non-empty closed convex subset of $L^2(Q)$.

For $f \in L^2(Q)$, $y_0 \in L^2(\Omega)$ and $v \in U_{ad}$, we define $y(v)$ as the solution of the following parabolic variational inequality

$$\left\{ \begin{array}{l} y(v) \in L^2(0,T;H_0^1(\Omega)), \quad \dfrac{\partial y(v)}{\partial t} \in L^2(0,T;H^{-1}(\Omega)), \\ y(v)(t) \in K \quad \text{for} \quad t \in]0,T[, \quad y(v)(0) = y_0, \quad \forall z \in K, \\ < \dfrac{\partial y(v)(t)}{\partial t} - \Delta y(v)(t), z - y(v)(t) > \geq \displaystyle\int_\Omega (f(t) + v(t))(z - y(v)(t))dx. \end{array} \right. \tag{1.7}$$

Here again it is well known that (1.7) defines a mapping $v \rightarrow y(v)$ from U_{ad} to $L^2(0,T;H_0^1(\Omega))$ which is Lipschitz continuous but not differentiable in general.
If z_d is an element of $L^2(Q)$ we define a new cost function

$$J(v) = \frac{1}{2}\int_Q |y(v) - z_d|^2 dxdt + \frac{N}{2}\int_Q |v|^2 dxdt, \tag{1.8}$$

and the optimal control problem is again

$$\text{Find} \quad u \in U_{ad} \quad \text{such that} \quad J(u) = \min_{v \in U_{ad}} J(v). \tag{1.9}$$

The existence of an optimal control for (1.9) is also standard and the main question is again : what is a "good" optimality system ?

For sake of simplicity, we shall restrict ourselves to considering the elliptic case and we will make some comments about the applicability of the methods to the parabolic situation.

2 THE PENALTY METHOD

The first natural idea to treat problem (1.4) is to replace the variational inequality (1.2) by a regularized problem via the penalty method. For example, let us consider

$$\beta(r) = \left\{ \begin{array}{lll} 0 & \text{if} & r \geq 0 , \\ -\frac{1}{2}r^2 & \text{if} & 0 \geq r \geq -1 , \\ (r + \frac{1}{2}) & \text{if} & -1 \geq r, \end{array} \right. \tag{2.1}$$

and the solution $y(v)$ of

$$-\Delta y_\varepsilon(v) + \frac{1}{\varepsilon}\beta(y_\varepsilon(v)) = f + v; \quad y_\varepsilon(v) \in H_0^1(\Omega). \tag{2.2}$$

If u is a solution of (1.4), we define a new cost function, which contains an adaptive term introduced by Barbu (c.f. [Ba.1], [Ba.2]) when there is no uniqueness in (1.4),

$$J_\varepsilon(v) = \frac{1}{2}\int_\Omega |y_\varepsilon(v) - z_d|^2 dx + \frac{N}{2}\int_\Omega |v|^2 dx + \frac{1}{2}\int_\Omega |v-u|^2 dx, \tag{2.3}$$

and we look for u_ε such that

$$u_\varepsilon \in U_{ad}; \quad J_\varepsilon(u_\varepsilon) = \min_{v\in U_{ad}} J_\varepsilon(v). \tag{2.4}$$

For every $\varepsilon > 0$ this problem admits a solution u_ε and, as the mapping $v \to J_\varepsilon(v)$ is differentiable, we can derive an optimality system in a standard way.

Proposition 2.1

If u_ε is a solution of (2.4) and $y_\varepsilon = y_\varepsilon(u_\varepsilon)$, there exists an adjoint state $p_\varepsilon \in H_0^1(\Omega)$ such that the triple $(u_\varepsilon, y_\varepsilon, p_\varepsilon)$ satisfies the optimality system

$$\left\{\begin{aligned} &-\Delta y_\varepsilon + \frac{1}{\varepsilon}\beta(y_\varepsilon) = f + u_\varepsilon; \quad y_\varepsilon \in H_0^1(\Omega), \\ &-\Delta p_\varepsilon + \frac{1}{\varepsilon}\beta'(y_\varepsilon)p_\varepsilon = y_\varepsilon - z_d; \quad p_\varepsilon \in H_0^1(\Omega), \\ &\int_\Omega (p_\varepsilon + Nu_\varepsilon + (u_\varepsilon - u), v - u_\varepsilon)dx \geq 0, \quad \forall v \in U_{ad}. \end{aligned}\right. \tag{2.5}$$

If we now let ε tend to zero, it is easy to show that

$$u_\varepsilon \to u \quad \text{in} \quad L^2(\Omega); \quad y_\varepsilon \to y = y(u) \quad \text{in} \quad H_0^1(\Omega); \quad p_\varepsilon \quad \text{is bounded in} \quad H_0^1(\Omega),$$

and therefore for a subsequence

$$p_\varepsilon \to p \quad \text{in} \quad H_0^1(\Omega) \quad \text{weakly}.$$

It is possible to obtain some additional information on (u, y, p) (c.f. [M.P.1]) which lead to the following optimality system.

Proposition 2.2

If u is a solution of (1.4) and $y = y(u)$, there exist $p \in H_0^1(\Omega)$, $\xi \in H^{-1}(\Omega)$, $\theta \in H^{-1}(\Omega)$ such that

$$\left\{\begin{array}{l} -\Delta y = f + u + \xi, \\ y \geq 0, \quad \xi \geq 0, \quad <\xi, y>= 0, \\ -\Delta p + \theta = y - z_d, \\ <\theta, y>= 0, \quad <\xi, p>= 0, \quad <\theta, p>\geq 0, \\ \int_\Omega (p + Nu, v - u)dx \geq 0, \quad \forall v \in U_{ad}. \end{array}\right. \tag{2.6}$$

This system is not a "good" optimality system because, for instance in Example 1.1, it allows $v = 1$ (point B) which clearly cannot be accepted even at the first order. In this example, as 0 is the unique optimal control, one can drop out Barbu's adaptive term, and doing this, one can show that the penalty method gives for small positive ε two extrema A_ε and B_ε converging when $\varepsilon \to 0$ to A and B. So, we cannot avoid point B using the penalty method.

Nevertheless, when $U_{ad} = L^2(\Omega)$, the last inequality in (2.6) becomes

$$p + Nu = 0,$$

and gives us the surprising information (regularity property) which will be useful in the sequel

$$Nu \in H_0^1(\Omega). \tag{2.7}$$

3 DIRECT METHOD

We shall restrict ourselves here to the case $U_{ad} = L^2(\Omega)$, and we know from (2.7) that any optimal control u satisfies $Nu \in H_0^1(\Omega)$.

3.1 Use of conical derivatives

Here we rapidly describe and discuss the method and results of [M.P.1] .
The mapping $v \to y(v)$ is not differentiable, but has at each point v a conical derivative which can be explicitely characterized (c.f. [M]) . Then $v \to J(v)$ also has a conical derivative $w \to D_cJ(v)[w]$ (which is not linear !) and a natural "good" optimality condition for (1.4) is now

$$D_cJ(v)[v - u] \geq 0, \quad \forall w \in U_{ad}. \tag{3.1}$$

Notice that in Example 1.1 only $u = 0$ (point A) satisfies this condition.
In [M.P.1] we have given an optimality system which is strictly stronger than (2.6) and which is equivalent to (3.1) .

If $y(v)$ is solution of (1.2) , we set

$$\xi(v) = -\Delta y(v) - f - v,$$

and we define the sets

$$Z_{y(v)} = \{x \in \Omega, \quad y(v)(x) = 0\} \quad \text{(defined up to a set of zero capacity)}, \tag{3.2}$$

$$S_{y(v)} = \{\varphi \in H_0^1(\Omega), \quad \varphi \geq 0 \quad \text{on} \quad Z_{y(v)}, \quad <\xi(v), \varphi>= 0\}. \tag{3.3}$$

We then have

Theorem 3.1

If u is a solution of (1.4) , $y = y(u)$ and $\xi = \xi(u)$ there exist $p \in H_0^1(\Omega)$, $\theta \in H^{-1}(\Omega)$ such that

$$\begin{cases} -\Delta y = f + u + \xi, \\ y \geq 0, \quad \xi \geq 0, \quad <\xi, y>= 0, \\ -\Delta p + \theta = y - z_d, \\ p \in S_y, \quad <\theta, \psi> \geq 0 \quad \forall \psi \in S_y, \\ p + Nu = 0. \end{cases} \tag{3.4}$$

We can notice that (3.4) implies

$$<\theta, y>= 0, \quad <\xi, p>= 0, \quad <\theta, p> \geq 0,$$

and therefore that (3.4) is stronger than (2.6) (when $U_{ad} = L^2(\Omega)$).
This method presents (at least !) two major drawbacks :

i) It requires that U_{ad} be the whole space $L^2(\Omega)$.

ii) It cannot be applied to the parabolic situation because for example in (1.7) one does not know whether $v \rightarrow y(v)$ has a conical derivative.

3.2 Method of constrained increments

We give here the main ideas of a method developped in [M.P.2] for the parabolic problem and in [M.P.3] and based on an idea of Bermudez and Saguez ([Be.Sa]).

We rewrite the variational inequality (1.2) as

$$\begin{cases} -\Delta y(v) = f + v + \xi(v), \\ y(v) \in H_0^1(\Omega), \quad \xi(v) \in H^{-1}(\Omega), \quad v \in U_{ad}, \\ y(v) \geq 0, \quad \xi(v) \geq 0, \quad <\xi(v), y(v)>= 0, \end{cases} \tag{3.5}$$

and we consider (3.5) as a set of coupled constraints.

We now define the functional $\tilde{J}$ on triples of independent variables (z, v, η) by

$$\tilde{J}(z,v,\eta) = \frac{1}{2}\int_\Omega |z - z_d|^2 dx + \frac{N}{2}\int_\Omega |v|^2 dx. \tag{3.6}$$

The problem is now

$$\begin{cases} \text{Find} \quad (y,u,\xi) \quad \text{satisfying (3.5) such that} \\ \tilde{J}(y,u,\xi) = \min\limits_{(z,v,\eta) \text{ satisfying(3.5)}} \tilde{J}(z,v,\eta). \end{cases} \tag{3.7}$$

Remarks

1) Of course if (z,u,ξ) is solution of (3.7), u is solution of (1.4) , $y = y(u)$ and $\xi = \xi(u)$, and vice versa.

2) We could try to penalize constraints (3.5) but this does not give any result.

3) We could also try to decouple constraints (3.5) via a Kuhn and Tucker argument. This works formally but requires unrealistic conditions.

The method is very simple and consists in considering increments (z,v,η) such that

$$\exists t_0 > 0, \quad \forall t, \quad 0 < t \leq t_0, \quad (y+tz, u+tv, \xi+t\eta) \quad \text{satisfy (3.5)} , \tag{3.8}$$

and in using the inequality

$$\tilde{J}(y,u,\xi) \leq \tilde{J}(y+tz, u+tv, \xi+t\eta), \quad \forall (z,v,\eta) \quad \text{satisfying (3.8)} . \tag{3.9}$$

Let us define the sets

$$C_y = \{z \in H_0^1(\Omega), \quad \exists t_0 > 0, \quad y + t_0 z \geq 0\},$$

$$C_\xi = \{\eta \in H^{-1}(\Omega), \quad \exists t_0 > 0, \quad \xi + t_0\eta \geq 0\},$$

$$C_u = \{v \in L^2(\Omega), \quad \exists t_0 > 0, \quad u + t_0 v \in U_{ad}\}.$$

These sets are convex cones which are not closed.

Lemma 3.1

The triple (z,v,η) satisfies (3.8) if and only if it satisfies

$$\begin{cases} (a) \quad z \in C_y, \quad v \in C_u, \quad \eta \in C_\xi, \\ (b) \quad -\Delta z - \eta - v = 0, \\ (c) \quad <\xi, z> = <\eta, y> = <\eta, z> = 0. \end{cases} \tag{3.10}$$

Proof : If (z,v,η) satisfies (3.8), it is clear that we have (3.10)(a) and (3.10)(b). We also have

$$<\xi + t\eta, y + tz> = 0, \quad \forall t, \quad 0 < t \leq t_0,$$

which means that

$$t[<\xi,z>+<\eta,y>]+t^2[<\eta,z>=0, \quad \forall t, \quad 0<t\leq t_0.$$

Then

$$<\xi,z>+<\eta,y>=0 \quad \text{and} \quad <\eta,z>=0.$$

But, as $\xi \geq 0$, $y+tz\geq 0$ and $<\xi,y>=0$, we have $<\xi,z>\geq 0$.
On the other hand, $\xi+t\eta\geq 0$, $y\geq 0$ and $<\xi,y>=0$; then $<\eta,y>\geq 0$, and we therefore obtain (3.10(c).

Conversely, if (z,v,η) satisfy (3.10) , as C_y , C_ξ and C_u are cones, there exists $t_0>0$ such that

$$\forall t, \quad 0<t\leq t_0, \quad y+tz\geq 0, \quad u+tv\in U_{ad}, \quad \xi+t\eta\geq 0.$$

The other relations in (3.5) are straightforward.

Using (3.9), we immediately obtain the following

Proposition 3.1

If (y,u,ξ) is an optimal control, then for every (z,v,η) satisfying (3.10) we have

$$\int_\Omega (y-z_d)zdx+\int_\Omega Nuvdx\geq 0. \tag{3.11}$$

This result is valid even if U_{ad} is any non-empty closed convex subset of $L^2(\Omega)$, but we can only give an interpretation of it if $U_{ad}=L^2(\Omega)$.

Theorem 3.2

If $U_{ad} = L^2(\Omega)$ and if (y,u,ξ) is an optimal control, there exist $p\in H_0^1(\Omega)$ and $\theta\in H^{-1}(\Omega)$ such that the following optimality system is satisfied

$$\begin{cases} (a) & -\Delta y=f+u+\xi, \\ (b) & y\geq 0, \quad \xi\geq 0, \quad <\xi,y>=0, \\ (c) & -\Delta p+\theta=y-z_d, \\ (d) & <p,\eta>\geq 0 \quad \forall\eta\in C_\xi, \quad <\eta,y>=0, \\ (e) & <\theta,z>\geq 0 \quad \forall z\in C_y, \quad <\xi,z>=0, \\ (f) & p+Nu=0. \end{cases} \tag{3.12}$$

Proof : As $Nu\in H_0^1(\Omega)$, we can take in (3.11) $v\in H^{-1}(\Omega)$ and (3.10) says that

$$v=-\Delta z-\eta \quad \text{(only constraint on } v).$$

We then obtain

$$\int_\Omega (y - z_d) z dx - < \Delta z + \eta, Nu > \geq 0, \quad \forall (z,\eta) \in C_y \times C_\xi, \quad <\xi, z> = <\eta, y> = <\eta, z> = 0. \tag{3.13}$$

Define p and θ by

$$p + Nu = 0 \quad ; \quad -\Delta p + \theta = y - z_d,$$

We can write (3.13) as

$$<\theta, z> + <\eta, p> \geq 0, \forall (z,\eta) \in C_y \times C_\xi, <\xi, z> = <\eta, y> = <\eta, z> = 0. \tag{3.14}$$

Taking $z = 0$, we obtain (3.12)(d) , and taking $y = 0$ we obtain (3.12)(e) , and this proves Theorem 3.2.

In fact, one can prove that (3.12) is equivalent to (3.4) , and in particular

Lemma 3.2

If S_y is defined by (3.3) , we have

$$S_y = \{q \in H_0^1(\Omega), \quad <q, \eta> \geq 0, \quad \forall \eta \in C_\xi, \quad <\eta, y> = 0\}.$$

Remarks

1) The results of Theorem 3.2 makes no use of the conical derivative. Using exactly the same type of argument , one can obtain an analogous result for the parabolic situation (c.f. [M.P.2])

2) Here again, we strongly use the hypothesis $U_{ad} = L^2(\Omega)$ in order to obtain an interpretation of (3.11).

4 DISCUSSION OF THE CASE $U_{ad} \neq L^2(\Omega)$

The problem is open in general. As already mentionned, the result of Proposition 3.1 is still valid, which means that we have

$$\int_\Omega (y - z_d) z dx + \int_\Omega Nuv dx \geq 0, \quad \forall (z, v, \eta) \quad \text{satisfying} \tag{3.11}$$

$$\begin{cases} (a) \quad z \in C_y, \quad v \in C_u, \quad \eta \in C_\xi, \\ (b) \quad -\Delta z - \eta - v = 0, \\ (c) \quad <\xi, z> = <\eta, y> = <\eta, z> = 0. \end{cases} \tag{3.10}$$

In order to obtain an optimality system expressed in an "usual" way with an adjoint state, we need the existence of a Lagrange multiplier to relax condition (3.10)(b).

Suppose there exists a Lagrange multiplier $p \in H_0^1(\Omega)$ such that (3.11) implies

$$\begin{aligned}&\int_\Omega (y - z_d)z dx + \int_\Omega N u v dx + < \Delta z + \eta + v, p > \geq 0,\\ &\forall (z, v, \eta) \text{ satisfying (3.10)(a) and (c)}.\end{aligned} \tag{4.1}$$

Then, one can easily obtain the following optimality system

$$\left\{\begin{aligned}&-\Delta y = f + u + \xi,\\ &y \geq 0, \quad \xi \geq 0, \quad <\xi, y> = 0,\\ &-\Delta p + \theta = y - z_d,\\ &<p, \eta> \ \geq \ 0 \quad \forall \eta \in C_\xi, \quad <\eta, y> = 0,\\ &<\theta, z> \ \geq \ 0 \quad \forall z \in C_y, \quad <\xi, z> = 0,\\ &\int_\Omega (p + Nu) v dx \ \geq \ 0 \ , \quad \forall v \in C_u.\end{aligned}\right. \tag{4.2}$$

This optimality system would the one expected.

Conversely, if there exist p satisfying (4.2), then (4.1) is satisfied and p is a Lagrange multiplier which relaxes condition (3.10)(b).

On the simple Example 1.1 , one can compute everything and one can try to obtain the existence of p which satisfies the adapted optimality system analogous to (4.2) when

$$U_{ad} = \{v \in \Re, \quad \alpha \leq v \leq \beta\}.$$

One can show that this implies conditions on α and β such as $\alpha \leq 0$ and $\beta > 0$.

Therefore, the existence of such a Lagrange multiplier is not always true and the above example suggests conditions on U_{ad} such that there might exist a solution. An answer to this open question would be an interesting result.

REFERENCES

[Ba.1] V.BARBU . Necessary conditions for nonconvex distributed control problems governed by elliptic variational inequalities. J. Math. Anal. Appl. 80 (1981), p : 566-597.

[BA.2] V.BARBU . Necessary conditions for control problems governed by parabolic variational inequalities. SIAM Journal on Control and Optimization, Vol. 19, 1, 1981, p : 64-86.

[Be.Sa] A.BERMUDEZ ; C.SAGUEZ . Optimal control of a Signorini problem. SIAM Journal on Control and Optimization, Vol. 25, 3, 1987, p : 576-582.

[M] F.MIGNOT : Contrôle dans les inéquations variationnelles elliptiques. J. Functional Analysis, 22 (1976), p : 130-185.

[M.P.1] F.MIGNOT ; J.P.PUEL : Optimal control in some variational inequalities. SIAM Journal on Control and Optimization, Vol. 22, 1984, p : 466-476.

[M.P.2] F.MIGNOT ; J.P.PUEL : Contrôle optimal d'un système gouverné par une inéquation variationnelle parabolique. Note C.R.A.S. Paris, t.298, Série I, 12, 1984.

[M.P.3] F.MIGNOT ; J.P.PUEL : Optimal control problems for elliptic and parabolic unilateral problems , to appear.

SHAPE SENSITIVITY ANALYSIS OF STATE CONSTRAINED OPTIMAL CONTROL PROBLEMS FOR DISTRIBUTED PARAMETER SYSTEMS

Murali Rao

University of Florida
Department of Mathematics
201 Walker Hall
Gainesville, Florida 32611
USA

and

Jan Sokołowski

Systems Research Institute
Polish Academy of Sciences
ul. Newelska 6
01-447 Warszawa
POLAND

1. Introduction

In the present paper we combine the previous result [8] on the differential stability of solutions of unilateral problems in $H^2(\Omega)$ with the material derivative method [19] for the shape sensitivity analysis of state constrained optimal control problem for elliptic equations. We use the same method of shape sensitivity analysis as in the case of unilateral problems [9-17]. For the related problems we refer the reader to the contributions of F. Abergel and R. Temam, E. Casas, M. Delfour, F. Mignot, J.-P. Puel, J.-P. Zolesio in the present Proceedings. State constrained problems were studied among others by E. Casas [2] and J.L. Lions [3].

The differential stability of solutions to variational inequalities is considered by F. Mignot [5]. For the related results on the optimal control of variational inequalities we refer the reader e.g. to A. Bermudez and C. Saguez [1] and F. Mignot and J.-P. Puel [6]. We derive our results in the case of a simple model problem however the method is general and can be applied in more general situations. The outline of the paper is following. In section 2 we derive the form of the tangent cone at an arbitrary point of the convex set

$$K = \{\phi \in H^2(\Omega) \cap H^1_0(\Omega) \mid \phi(x) \geq a(x) \text{ in } \Omega\} \qquad (1.1)$$

Using this result we are able to show that the set K is polyhedric in the sense of F. Mignot [5] and therefore the metric projection in $H^2(\Omega)$ onto the set $K \subset H^2(\Omega)$ is conically differentiable.

In section 3 we apply our result combined with the material derivative method [9-17] to the shape sensitivity analysis of state constrained optimal control problem. We use standard notation throughout the paper.

2. Closure of Tangent Cone

Let

$$K = \{\phi \in H^2(\Omega) \cap H_0^1(\Omega) \mid \phi \geq a \quad \text{q.e. in } \Omega\} \tag{2.1}$$

where a is given function in $H^2 \cap H_0^1$. We will define q.e. presently. K is closed and convex. Given $y \in K$ the tangent cone at y is by definition

$$C_y(K) = \{\phi \in H^2 \cap H_0^1 \mid \exists\, t>0\, ,\ y+t\phi \geq a \text{ q.e}\} \tag{2.2}$$

Theorem 1 below characterises the closure of the tangent cone $C_y(K)$ under one assumption. It actually is valid generally.

Theorem 1. Let

$$\Xi = \{y = a\}$$

and assume Ξ is compact. Then

$$\overline{C_y(K)} = \{\phi \in H^2 \cap H_0^1 \mid \phi \geq 0 \quad \text{q.e on } \Xi\} \tag{2.3}$$

A proof very similar to that of Theorem 1 gives also the following.

Corollary 1

$$\overline{C_y(K)} \cap [f - y]^{\perp} = \overline{C_y(K) \cap [f - y]^{\perp}} \tag{2.4}$$

here we denote by $y = P_k f$ the metric projection of $f \in H^2 \cap H_0^1$ onto set (2.1) and $[g]^{\perp}$ is the subspace of $H^2 \cap H_0^1$ orthogonal to an element $g \in H^2 \cap H_0^1$. Therefore the set (2.1) is polyhedric in the sense of Mignot [5].

Before we proceed let us establish the framework. It is not difficult to see that

$$H^2 \cap H_0^1 = \{G\, f \mid f \in L^2(\Omega)\} \tag{2.5}$$

where G is the Green function of Ω i.e. $G = (-\Delta)^{-1}$. We define the inner product in $H^2 \cap H_0^1$ by

$$(Gf,\ Gg) = \int_\Omega f(x)\ g(x)\ dx \tag{2.6}$$

We note that the corresponding topology is that inherited from H^2.

For purposes of this paper we define the C_2 - capacity $C(F)$ of a compact set $F \subset \Omega$ by

$$C_2(F) = \inf \{ \|f\|^2_{L^2} \mid Gf \geq 1 \text{ on } F \}$$

We extend this definition to all Borel sets by:

$$C_2(B) = \sup \{ C_2(F) \mid \text{compact } F \subset B \}$$

A statement holds q.e when it holds except for a G-Polar set i.e a set of C_2 - capacity zero. Observe that convergence of a sequence in $H^2 \cap H^1_o$ implies pointwise convergence (for a subsequence) off a G-polar set. For more on Capacities see [4].

Proof of Theorem 1. We start off by observing that $C_y(K)$ (and in particular also its closure) has the following properties:

1. It contains all non-negative elements of $H^2 \cap H^1_o$.
2. If $\phi_i \in C_y(K)$, $a_i \geq 0$ then
   $\Sigma a_i \phi_i \in C_y(K)$.
3. $\phi \in C_y(K)$, $0 \leq \psi \in C^\infty$ then
   $\psi\phi \in C_y(K)$
4. $\phi = 0$ in a neighbourhood of Ξ then $\phi \in \overline{C_y(K)}$

These properties are simple consequences of the definition of the tangent come. Property 4 above is immediate if ϕ is bounded and for general ϕ is by taking limits.

Since convergence in $H^2 \cap H^1_o$ implies q.e. convergence for a subsequence, it is clear that the left side of (2.3) is a subset of the right side.

Let $V \in H^2 \cap H^1_o$ and suppose $V \geq 0$ q.e. on $\Xi = \{y=a\}$. Our object is to show that $V \in \overline{C_y(K)}$. To this end let ϕ_o denote the unique element of $\overline{C_y(K)}$ such that

$$\| V-\phi_o \| = \inf \{ \|V-\phi\| \mid \phi \in \overline{C_y(K)} \} \tag{2.7}$$

Using simple arguments we see that (2.7) implies:

$$(\phi_o - V, \phi) \geq 0, \quad \phi \in \overline{C_y(K)} \tag{2.8}$$

For simplicity let us define the linear map

$$L\phi = (\phi_o - V, \phi), \quad \phi \in H^2 \cap H^1_o \tag{2.9}$$

Let $f_o \in L^2$ be such that

$$\phi_o - V = Gf_o \tag{2.10}$$

If $g \geq 0$, then $\phi = Gg \geq 0$ and hence belongs to $C_y(K)$. Using (2.9) we see that $\int f_o g \geq 0$. This says that $f_o \geq 0$ a.e. If $0 \leq \phi \in C_o^\infty$ then again using (2.9) we see

$$\int f_o \Delta\phi \leq 0, \quad 0 \leq \phi \in C_o^\infty$$

ie. that f_o is superharmonic. By Riesz decomposition we may write

$$f_o = G\mu + h_o \tag{2.11}$$

where μ is a positive Radon-measure and h_o is positive harmonic in Ω. For clarity we break up the proof into small steps.

<u>Step 1</u>. For all $\phi \in H^2 \cap H_o^1$

$$\int |\phi| d\mu \leq \|L\| \, \|\phi\|_{H^2 \cap H_o^1} \tag{2.12}$$

Indeed let $0 \leq f \in L^2$. There is a sequence of non-negative elements of C_o^∞ which <u>increases</u> pointwise to Gf.

$$Gf = \lim \phi_n, \quad 0 \leq \phi_n \in C_o^\infty$$

From (2.11) and (2.8)

$$L(Gf) \geq L(\phi_n) = \int \phi_n d\mu$$

By monotone convergence we get

$$\int (Gf) d\mu = \lim \int \phi_n d\mu \leq L(Gf);$$

Now if $\phi = Gf$ then

$$\int |\phi| d\mu \leq \int (G|f| d\mu \leq L(G|f|) \leq \|L\| \|f\|_{L^2} = \|L\| \|\phi\|_{H^2 \cap H_o^1} \tag{2.13}$$

(2.13) in particular tels us that if ϕ_n converges to ϕ in $H^2 \cap H_o^1$, it also converges in $L^1(\mu)$.

<u>Step 2</u>. If $\phi \in H^2 \cap H_o^1$ has compact support then

$$\int \phi d\mu = L\phi \tag{2.14}$$

Indeed for such ϕ, there is a sequence $\phi_n \in C_o^\infty$ converging to ϕ in $H^2 \cap H_o^1$. Then from Step 1, ϕ_n also converges in $L^1(\mu)$ to ϕ and, L agrees with μ on C_o^∞. Thus (2.14) is valid.

<u>Step 3</u>. If $\phi \in \overline{C_y(K)}$ then

$$0 \leq \int \phi d\mu \leq L\phi \tag{2.15}$$

Indeed let $0 \leq \psi \leq 1$, $\psi \in C_o^\infty$. Then $\psi\phi \in \overline{C_y(K)}$ <u>and also</u> $(1-\psi)\phi \in \overline{C_y(K)}$.

Thus

$$\int \psi\phi d\mu = L(\psi\phi) \leq L\phi$$

because $L\phi = L(\psi\phi) + L(1-\psi)\phi$
and the last term is non-negative.

Now we let ψ iscrease to 1 on Ω.

Step 4. μ is concentrated on Ξ. Indeed since $y-a\geq 0$, if $0\leq\phi\leq 1$, $\phi\in C_o^\infty$ and $-1\leq t\leq 1$, we have

$$y-a+t\phi(y-a) \geq 0$$

In other words $t\phi(y-a)\in C_y(K)$. Using

$$\int t\phi(y-a)d\mu \geq 0 \tag{2.15}$$

$-1\leq t\leq 1$ or that $\int\phi(y-a)\ d\mu = 0$

Since $y > a$ off Ξ, this can only be true if μ is concentrated on Ξ.

Step 5. $\mu=0$. To show this note first that

$$\int\phi_o d\mu = 0 \tag{2.16}$$

Indeed we know $L\phi_o = 0$. So since $\phi_o \geq 0$ on Ξ, from (2.15), (2.16) is seen to be valid. Now $\phi_o - V = Gf_o$ and we knew that $f_o\geq 0$.

So $\phi_o - V$ is non-negative superharmonic and so either identically zero ar strictly positive everywhere. Since $\int(\phi_o-V)d\mu=0$, we must have $\mu=0$.

Step 6. We claim $h_o=0$. For this we use Property 4 of $\overline{C_y(K)}$. Let D be a relatively compact open set containing Ξ. Using Theorem g and Theorem 16 of [4] we see that there is a, $0\leq f\in L^2$ such that $Gf\geq 1$ on D. Let $\phi\in C_o^\infty$ s.t. $\phi\equiv 1$ on D. Then ϕ-Gf vanishes on D and hence $\phi-Gf\in C_y(K)$.

Hence $L(\phi-Gf)=0$

But $L(\phi-Gf)=\int f_o\Delta(\phi-Gf) = \int h_o[\Delta\phi+f] = \int h_o f$

because h_o is harmonic. Since $f\geq 0$ we get h_o 0. Thus L=0 or that $V\in\overline{C_y(K)}$

3. Shape Sensitivity Analysis.

Let $\{\Omega_\varepsilon\}\subset R^n$, n=1,2,3, $\varepsilon\in[0,\delta)$, be a given family of domains. We will consider the differential stability with respect to the parameter ε 0, at $\varepsilon=0^+$ of solutionsof the following optimal control problem

Problem (P_ε)

Find an element $u_\varepsilon\in L^2(\Omega_\varepsilon)$ which minimizes the cost functional

$$I_\varepsilon(y,u)=\tfrac{1}{2}\int_{\Omega_\varepsilon}(y-z_d)^2dx + \tfrac{\alpha}{2}\int_{\Omega_\varepsilon}(u)^2dx \tag{3.1}$$

over the space $L^2(\Omega_\varepsilon)$, subject to the state constrainsts

$$y \in K(\Omega_\varepsilon) = \{y \in H^2(\Omega_\varepsilon) \cap H^1_0(\Omega_\varepsilon) \mid |y(x)| \leq 1 \quad \text{in } \Omega_\varepsilon\} \qquad (3.2)$$

Here $z_d \in H^1(R^n)$ is a given element. For a given control $u \in L^2(\Omega_\varepsilon)$, the state $y \in H^2(\Omega_\varepsilon) \cap H^1_0(\Omega_\varepsilon)$ is given by the unique solution of the following state equation

$$-\Delta y = u, \quad \text{in } \Omega_\varepsilon \qquad (3.3)$$

$$y = 0, \quad \text{on } \partial\Omega_\varepsilon \qquad (3.4)$$

Let us observe that, in view of (3.3), the cost functional (3.1) takes the form

$$I_\varepsilon(y) = \tfrac{1}{2}\int_{\Omega_\varepsilon} (y-z_d)^2 dx + \tfrac{\alpha}{2}\int_{\Omega_\varepsilon} (\Delta y)^2 dx \qquad (3.5)$$

In order to solve problem (P_ε), $\varepsilon \in [0,\delta)$, it is sufficient to find the unique minimizer of functional (3.5) over the set (3.2) or equivalently the unique solution of the following variational inequality

$$y_\varepsilon \in K(\Omega_\varepsilon)$$

$$\alpha\int_{\Omega_\varepsilon} \Delta y_\varepsilon \Delta(\phi - y_\varepsilon)dx + \int_{\Omega_\varepsilon} (y_\varepsilon - z_d)(\phi - y_\varepsilon)dx \geq 0 \qquad \phi \in K(\Omega_\varepsilon) \qquad (3.6)$$

The unique optimal control $u_\varepsilon \in L^2(\Omega_\varepsilon)$ can be determined from the state equation (3.3), i.e.

$$u_\varepsilon = -\Delta y_\varepsilon, \quad \text{in } \Omega_\varepsilon \qquad (3.7)$$

In order to state our result on the shape differential stability of solutions to problem (P_ε) we define the family $\{\Omega_\varepsilon\}$ using the standard method (see [18] for details).

<u>3.1. Family $\{\Omega_\varepsilon\}$</u>

Let $V(.,.) \in C([0,\delta); C^1(R^n;R^n))$ be a given vector field. We denote by

$$T_\varepsilon(V) : R^n \to R^n, \quad \varepsilon \in (0,\delta) \qquad (3.8)$$

the mapping defined as follows

$$T_\varepsilon(V)(X) = x(\varepsilon), \quad \varepsilon \in (0,\delta) \qquad (3.9)$$

where x(0) is given by the unique solution of the following system of ordinary differential equations

$$\frac{d}{dt}x(t) = V(t,x(t)), \qquad t \in (0,\delta) \tag{3.10}$$

$$x(0) = X \tag{3.11}$$

We denote

$$\Omega_\varepsilon = T_\varepsilon(V)(\Omega) = \{x \in R^n \mid \exists X \in \Omega \text{ such that } x(0)=X \text{ and } x(\varepsilon)=x\} \tag{3.12}$$

In particular we have for $\varepsilon=0$

$$\Omega = T_o(V)(\Omega) \tag{3.13}$$

We will denote by $DT_\varepsilon(X)$ the Jacobian of the mapping (3.9) evaluated at $X \in R^n$, by $DT_\varepsilon^{-1}(X)$ inverse of matrix $DT_\varepsilon(X)$ and by $*DT_\varepsilon^{-1}(X)$ the transpose of $DT_\varepsilon^{-1}(X)$.

3.2. Shape Derivative of Optimal Control.

Let us recall [9,10] that the shape derivative u' of an optimal control u_o in the direction of a vector field V(.,.) is defined as follows

$$u' = u - \nabla u_o \cdot V(0) \tag{3.14}$$

where

$$u = \lim_{\varepsilon \downarrow 0} (u_\varepsilon \circ T_\varepsilon - u_o)/\varepsilon \tag{3.15}$$

$$V(0) = V(0,.) \tag{3.16}$$

We refer the reader to [9,10] for related results on the shape sensitivity analysis of control constrained optimal control problems for distributed systems and to [13-17] for the shape sensitivity analysis of variational inequalities.

We assume that the boundary $\Gamma=\partial\Omega$ of the domain Ω is smooth and we denote

$$v(x) = \langle V(0,x), n(x) \rangle_{R^n}, \quad x \in \partial\Omega \tag{3.17}$$

the normal component on Γ of the vector field V(0).

Using the method of shape sensitivity analysis introduced in [9,10] we obtain the following result.

Theorem 2

The shape derivative u' of an optimal control $u_o \in L^2(\Omega)$ in the direction of a vector field $V(.,.)$ is given by the unique solution of the following state constrained optimal control problem.

Problem (Q)

Find an element $u' \in S_V(\Omega)$ which minimizes the cost functional

$$I(u) = \tfrac{1}{2}\int_\Omega (z)^2 dx + \frac{\alpha}{2}\int_\Omega (u)^2 dx + \int_{\partial\Omega} v g(z) d\Gamma + \int_{\partial\Omega} \frac{\partial v}{\partial n} h(z) d\Gamma \tag{3.18}$$

over the cone $S_V(\Omega)$.

The cone $S_V(\Omega)$ is defined as follows

$$S_V(\Omega) = \{\phi \in H^2(\Omega) \mid \phi = -v\frac{\partial y_o}{\partial n} \text{ on } \Gamma,$$
$$\phi = 0 \text{ q.e. on } \Xi^o, \tag{3.19}$$
$$\phi \leq 0 \text{ q.e. on } \Xi^-,$$
$$\phi \geq 0 \text{ q.e. on } \Xi^+\}$$

where

$$\Xi^o = \text{supp } \mu_o \tag{3.20}$$
$$\Xi^+ = \Xi_1 \setminus \text{supp } \mu_o \tag{3.21}$$
$$\Xi^- = \Xi_2 \setminus \text{supp } \mu_o \tag{3.22}$$
$$\Xi_1 = \{x \in \Omega \mid y_o(x) = -1\} \tag{3.23}$$
$$\Xi_2 = \{x \in \Omega \mid y_o(x) = 1\} \tag{3.24}$$

and μ_o is the Radon measure defined by the equation

$$\int_\Omega \phi d\mu_o = \int_\Omega \{y_o\phi + \alpha\Delta y_o \Delta\phi\}dx - \int_\Omega z_d \phi dx \tag{3.25}$$

$$\phi \in H^2(\Omega) \cap H_o^1(\Omega) \subset C_o(\bar{\Omega})$$

For a given control $u \in L^2(\Omega)$, the state $z = z(u;.) \in H^1(\Omega)$ for the problem (Q) is given by the unique solution of the following elliptic equation

$$-\Delta z = u, \quad \text{in } \Omega \tag{3.26}$$

$$z = -v\frac{\partial y_o}{\partial n}, \quad \text{on } \Gamma \tag{3.27}$$

For ϕ sufficiently smooth, the distributions $g(\phi)$, $h(\phi)$ on Γ have the following representations

$$g(\phi) = \alpha[-\Delta y_o \Delta\phi + 2H(\Delta y_o \frac{\partial\phi}{\partial n} + \Delta\phi \frac{\partial y_o}{\partial n})$$
$$-\nabla(\Delta\phi) . \nabla y_o - 2 \frac{\partial}{\partial n}(\Delta\phi) \frac{\partial y_o}{\partial n} \tag{3.28}$$
$$-\nabla(\Delta y_o) . \nabla\phi - 2 \frac{\partial}{\partial n}(\Delta y_o) \frac{\partial\phi}{\partial n})]$$

$$h(\phi) = -2(\Delta y_o \frac{\partial\phi}{\partial n} + \Delta\phi \frac{\partial y_o}{\partial n}) \tag{3.29}$$

The proof of Theorem 2 is omitted here.

References.

[1] Bermudez A. and Saguez C.: Optimal control of a Signorini problem. SIAM J. Control and Optimization (to appear).

[2] Casas E.: Control of an elliptic problem with pointwise state constraints. SIAM J. Control and Optimization 24(6) (1986) pp.1309-1318.

[3] Lions J.L.: Controle optimal de systemes gouvernes par des equations aux derivées partielles.

[4] Meyers N.G.: A Theory of Capacities. Math.Scand. 26 (1970).

[5] Mignot F.: Controle dans les inequations variationelles elliptiques. J. Functional Analysis 22 (1976) pp. 130-185.

[6] Mignot F. and Puel J.-P.: Controle optimal d'un systeme gouverné par une inéquation variationnelle parabolique. C.R.A.S. t.298, Série I, No. 12 (1984), pp. 277-280.

[7] Murat F. and Simon J.: Sur le controle par un domaine géométrique Publication de l'Université Paris 6 No. 76015, 1977.

[8] Rao M. nad Sokołowski J.: Sensitivity analysis of unilateral problems in $H^2(\Omega)$ and applications. (to appear).

[9] Sokołowski J.: Sensitivity analysis of control constrained optimal control problems for distributed parameter systems. SIAM J. Control and Optimization 25(6), November 1987.

[10] Sokołowski J.: Shape sensitivity analysis of boundary optimal control problems for parabolic systems. SIAM J. Control and Optimization (to appear).

[11] Sokołowski J.: Sensitivity analysis of contact problems with adhesive friction. Applied Mathematics and Optimization (to appear).

[12] Sokołowski J.: Sensitivity analysis of optimal control problems for parabolic systems (in: Control Problems for Systems Described as Partial Differential Equations and Applications , eds. I. Lasiecka and R. Triggiani, Springer Verlag (to appear).

[13] Sokołowski J.: Shape sensitivity analysis of nonsmooth variational problems. in: Boundary Control and Shape Optimization, ed. J.-P. Zolesio, Springer Verlag (to appear).

[14] Sokołowski J. and Zolesio J.-P.: Shape sensitivity analysis of unilateral problems. SIAM J. Mathematical Analysis, 18(5) September 1987.

[15] Sokołowski J. and Zolesio J.-P.: Shape design sensitivity analysis of plates and plane elastic solids under unilateral constraints. Journal of Optimization Theory and Applications, August 1987, pp. 361-382.

[16] Sokołowski J. and Zolesio J.-P.: Shape sensitivity analysis of an elastic-plastic torsion problem. Bulletin of the Polish Academy of Sciences, Technical Sciences, Vol. 33, No 11-12 (1985) pp. 579-586.

[17] Sokołowski J. and Zolesio J.-P.: Dérivée par rapport an domaine de la solution d'un probleme unilatéral, C.R. Acad. Sc. Paris 3C1 (1985), pp. 103-106.

[18] Zolesio J.-P.: Identification de domaines par déformations. Thése d'Etat, l'Université de Nice, 1979.

[19] Zolesio J.-P.: The material derivative (or speed) method for shape optimization. in: Optimization of Distributed Parameter Structures Vol. 2, eds. E.J. Haug and J. Cea, Sijthoff and Noordhoff, Alphen aan den Rijn, The Netherlands, 1981.

[20] Rao M. and Sokolowski J., Sensitivity analysis of obstacle problem for Kirchhoff plate, INRIA, Rapport de Recherche (1937).

APPROXIMATION OF DISCRETE-TIME LQR PROBLEMS FOR BOUNDARY CONTROL SYSTEMS WITH CONTROL DELAYS

I.G. Rosen
Department of Mathematics
University of Southern California
Los Angeles, California, 90089-1113, USA

In this short note we consider the extension and application of the approximation theory for discrete-time linear-quadratic regulator problems with either bounded or unbounded inputs we developed earlier (see, for example, [2], [3], [8]) to boundary control systems with control delays. More precisely, we synthesize our earlier, existing results for distributed systems with boundary controls and for systems with control delays into a theory which is applicable to systems that simultaneously exhibit both forms of unbounded input. Our primary intent here is to briefly outline the formulation of the problem, to simply describe the approximation theory, and to present some preliminary numerical results. Consequently, our treatment will be largely formal with a more detailed and precise discussion of our results to follow in a forthcoming paper.

Let H and W be Hilbert spaces with W densely and continuously embedded in H. We consider abstract boundary control systems of the form

$$\dot{w}(t) = \Delta w(t), \quad t > 0 \tag{1}$$

$$\Gamma w(t) = \beta v(t) + \Lambda v_t, \quad t \geq 0 \tag{2}$$

$$w(0) = w^0, \quad v_0 = v^0 \tag{3}$$

where $\Delta \varepsilon \mathcal{L}(W, H)$, $\Gamma \varepsilon \mathcal{L}(W, R^{\ell})$, $\beta \varepsilon \mathcal{L}(R^m, R^{\ell})$ and $v \varepsilon L_2(-\rho, t_1; R^m)$ for some $\rho > 0$ and every $t_1 > 0$. Let $Y = L_2(-\rho, 0; R^m)$, $V = H^1(-\rho, 0; R^m)$ and for $u \varepsilon L_2(-\rho, t_1; R^m)$ and $t \varepsilon [0, t_1]$, let u_t denote that element in Y given by $u_t(\theta) = u(t+\theta)$, $-\rho \leq \theta \leq 0$. We assume that $w^0 \varepsilon H$, $v^0 \varepsilon Y$ and that $\Lambda \varepsilon \mathcal{L}(V, R^{\ell})$ is of the form

$$\Lambda\psi = \sum_{i=1}^{\nu} L_i \psi(-\rho_i) + \int_{-\rho}^{0} L(\theta)\, \psi(\theta)\, d\theta, \qquad \psi \varepsilon V$$

with $L_i \varepsilon \mathcal{L}(R^m, R^{\ell})$, $i = 1, 2, \ldots, \nu$, $L \varepsilon L_2(-\rho, 0; \mathcal{L}(R^m, R^{\ell}))$ and $0 < \rho_1 < \rho_2 < \ldots < \rho_\nu \leq \rho$. We assume further that Γ is surjective, that its null space, $\mathcal{N}(\Gamma) = \{\phi \varepsilon W: \Gamma\phi = 0\}$ is dense in H and that the operator A_0: $\mathrm{Dom}(A_0) \subset H \to H$ defined to be the restriction of the operator Δ to $\mathcal{N}(\Gamma) = \mathrm{Dom}(A_0)$ is the infinitesimal generator of a C_0-semigroup $\{T_0(t): t \geq 0\}$ of bounded linear operators on H.

We define the Hilbert space H_1 to be the dual of the space $Dom(A_0^*)$ endowed with the graph Hilbert space norm associated with the operator A_0^*. Then H is densely and continuously embedded in H_1 and the semigroup $\{T_0(t): t \geq 0\}$ can be uniquely extended to a C_0-semigroup on H_1. Its generator is the extension of A_0 to the operator $\hat{A}_0$ in $\mathcal{L}(H,H_1)$ given by $(\hat{A}_0\phi)(\tilde{\phi}) = <\phi, A_0^* \tilde{\phi}>_H$ for $\phi \varepsilon H$ and $\tilde{\phi} \varepsilon Dom(A_0^*)$. We shall require the assumption that for each $t_1 > 0$ and $u \varepsilon L_2(0, t_1; W)$ we have $\int_0^{t_1} T_0(t_1 - s)(\Delta - \hat{A}_0) u(s)\, ds \varepsilon H$ and that there exists a positive constant μ which may depend upon t_1 such that $\left| \int_0^{t_1} T_0(t_1 - s)(\Delta - \hat{A}_0) u(s)\, ds \right|_H \leq \mu |u|_{L_2(0, t_1; W)}$ where the integrals in the above expressions are understood to be integrals in H_1.

Recalling that Γ was assumed to be surjective, let $\Gamma^+ \varepsilon \mathcal{L}(R^\ell, W)$ denote any right inverse of Γ and define the operators $B_0 \varepsilon \mathcal{L}(R^m, H_1)$ and $L_0 \varepsilon \mathcal{L}(V, H_1)$ by $B_0 u = (\Delta - \hat{A}_0)\Gamma^+ \beta u$ and $L_0 \psi = (\Delta - \hat{A}_0)\Gamma^+ \Lambda \psi$ respectively for $u \varepsilon R^m$ and $\psi \varepsilon V$. The operators B_0 and L_0 are well defined (i.e. are independent of the particular choice of $\Gamma^+ \varepsilon \mathcal{L}(R^\ell, W)$) since if Γ_1^+ and Γ_2^+ are two distinct right inverses of Γ then $R(\Gamma_1^+ - \Gamma_2^+) \subset \mathcal{N}(\Gamma)$ and consequently $(\Delta - \hat{A}_0)(\Gamma_1^+ - \Gamma_2^+) = (\Delta - A_0)(\Gamma_1^+ - \Gamma_2^+) = 0$.

For each $t_1 > 0$ the function $w \varepsilon C([0, t_1]; H)$ given by

$$w(t) = T_0(t)w^0 + \int_0^t T_0(t-s)(B_0 v(s) + L_0 v_s) ds$$

is referred to as the weak solution to the abstract boundary control system (1) - (3) and as the mild solution to the initial value problem

$$w(t) = A_0 w(t) + B_0 v(t) + L_0 v_t, \qquad \text{in } H_1, \text{ a.e.}$$
$$w(0) = w^0, \quad v_0 = v^0.$$

Following the treatment of abstract systems with control delays given by Ichikawa in [4] we set $X = H \times Y$ and let $\{U_0(t) : t \geq 0\}$ denote the semigroup of left translation on Y. Its generator is the operator D_0: $Dom(D_0) \subset Y \to Y$ given by $D_0\psi = D\psi = \psi'$ for $\psi \varepsilon Dom(D_0) = \{v \varepsilon V: v(0) = 0\}$. For each $t > 0$ define the operator $\mathcal{T}(t) \varepsilon \mathcal{L}(X)$ by

$$\mathcal{T}(t)x = \mathcal{T}(t)(\phi, \psi) = (T_0(t)\phi + \int_0^t T_0(t-s)L_0 U_0(s)\psi ds, U_0(t)\psi).$$

The one parameter family $\{\mathcal{T}(t): t \geq 0\}$ forms a C_0-semigroup of bounded linear operators on X with infinitesimal generator $\mathcal{A}$: $Dom(\mathcal{A}) \subset X \to X$ given by $\mathcal{A}(\phi,\psi) = A_0(\phi - \Gamma^+ \Lambda \psi) + \Delta\Gamma^+\Lambda\psi, D_0\psi)$ for all $(\phi,\psi) \varepsilon Dom(\mathcal{A}) = \{(w,v) \varepsilon X: w \varepsilon W, v \varepsilon Dom(D_0), \Gamma w - \Lambda v = 0\}$.

Define the operators $B_1 \varepsilon \mathcal{L}(R^m, V')$ and $\gamma \varepsilon \mathcal{L}(V, R^m)$ by $(B_1 u)(\psi) = u^T \psi(0)$ and $\gamma\psi = \psi(0)$ respectively. The operator D_0 admits a unique extension to an operator $D_0 \varepsilon \mathcal{L}(V,V')$ given by

$\hat{\mathcal{D}}_0(\psi)(\tilde{\psi}) = \int_{-\rho}^{0} \tilde{\psi} D\psi - \psi(0)^T \tilde{\psi}(0)$ for $\psi, \tilde{\psi} \in V$. If we let $\gamma^+ \in \mathcal{L}(R^m, V)$ denote the right inverse of γ given by $(\gamma^+ u)(\theta) = u, -\rho \le \theta \le 0, u \in R^m$ then $B_1 u = (D - \hat{\mathcal{D}}_0)\gamma^+ u$.

Let Z be the Hilbert space $H_1 \times V'$ and define the operator $\mathcal{B} \in \mathcal{L}(R^m, Z)$ by $\mathcal{B}u = (B_0 u, B_1 u)$. Then (see [4]) $x(t) = (w(t), v_t)$ is given by

$$x(t) = \mathcal{T}(t)x^0 + \int_0^t \mathcal{T}(t - s)\mathcal{B}v(s)ds, \qquad t \ge 0 \tag{4}$$

where $x^0 = (w^0, v^0)$ and the integral is understood to be an integral in Z. The function x given by (4) is an element in $C([0,t_1]; X)$ and is referred to as a weak solution to the initial value problem

$$\dot{x}(t) = \mathcal{A}x(t) + \mathcal{B}v(t), \qquad \text{in Z, a.e.}$$
$$x(0) = x^0.$$

The corresponding discrete-time system in the state space X is found by letting τ denote the length of the sampling interval and considering piecewise constant controls of the form $v(t) = u_k, t \in [k,\tau, (k + 1)\tau]$, $k = 0,1,2,\ldots$. Then, setting $x_k = x(k\tau)$, $k = 0,1,2,\ldots$, $T = \mathcal{T}(\tau) \in \mathcal{L}(X)$ and

$B = \int_0^\tau \mathcal{T}(s)\mathcal{B}\,ds \in \mathcal{L}(R^m, X)$, we obtain

$$x_{k+1} = Tx_k + Bu_k, \quad k = 0,1,2,\ldots \tag{5}$$
$$x_0 = x^0. \tag{6}$$

It is not difficult to show that with

$$T(\phi, \psi) = T_0(\tau)\phi + \int_0^\tau T_0(\tau - s) L_0 U_0(s)\psi ds, U_0(\tau)\psi)$$

$$\equiv (T_0\phi + S_0\psi, U_0\psi)$$

for $(\phi,\psi) \in X$, the discrete-time input operator B is given by

$$Bu = (\int_0^\tau T_0(s)(B_0 + L_0\gamma^+)u ds - S_0\gamma^+ u, (I - U_0)\gamma^+ u).$$

If, in addition, Γ^+ is chosen so that $\mathcal{R}(\Gamma^+) \subset \mathcal{N}(\Delta)$ then

$$Bu = ((I - T_0)\Gamma^+(\beta + \Lambda\gamma^+)u - S_0\gamma^+ u, (I - U_0)\gamma^+ u).$$

The linear-quadratic regulator (LQR) problem involves the determination of a control input sequence $\bar{u} = \{\bar{u}_k\}_{k=0}^\infty \in \ell_2(0, \infty; R^m)$ which minimizes the quadratic performance index

$$J(u) = \sum_{k=0}^{\infty} \langle Qx_k, x_k\rangle_X + u_k^T R u_k$$

where $u = \{u_k\}_{k=0}^\infty \in \ell_2(0,\infty; R^m)$, $x = \{x_k\}_{k=0}^\infty$ is given by (5), (6), $Q \in \mathcal{L}(X)$ is nonnegative self-adjoint and R is an m x m positive definte symmetric matrix. The operator Q is typically of the form $Qx = Q(\phi,\psi) = (Q_0\phi, 0)$ where $Q_0 \in \mathcal{L}(H)$ is nonnegative, self-adjoint and $x = (\phi,\psi) \in X$.

We now summarize the results in [2] (primarily due to Zabczyk, [9]) concerning the closed-loop solution of the discrete-time LQR problem. An input sequence $\bar{u} = \{\bar{u}_k\}_{k=0}^{\infty} \in \ell_2(0, \infty, R^m)$ is called admissible for the initial conditions $x^0 \in X$ if $J(u) < \infty$. If there exists an admissible control for each $x^0 \in X$ and if u admissible for x^0 implies $\lim_{k\to\infty} |x_k|_X = 0$, then there exists a unique nonnegative, self-adjoint solution $\Pi \in \mathcal{L}(X)$ to the operator algebraic Riccati equation

$$\Pi = T^*(\Pi - \Pi B(R + B^*\Pi B)^{-1}B^*\Pi)T + Q. \tag{7}$$

The optimal control sequence is given in linear state feedback form by

$$\bar{u}_k = -F\bar{x}_k, \quad k = 0, 1, 2, \ldots$$

where $F = (R + B^*\Pi B)^{-1} B^*\Pi T$, the optimal state trajectory $\bar{x} = \{\bar{x}_k\}_{k=0}^{\infty}$ satisfies the recurrence

$$\bar{x}_{k+1} = S\bar{x}_k, \qquad k = 0,1,2,\ldots$$

with $x_0 = x^0$ and $S = T - BF \in \mathcal{L}(X)$ and $J(\bar{u}) = \min J(u) = \langle \Pi x^0, x^0\rangle_X$. If Q is coercive then S will have spectral radius less than one and it will be uniformly exponentially stable. Since $F \in \mathcal{L}(X,R^m)$, there exists a vector $f = (f_1, f_2, \ldots f_m)^T$ with $f_j = (f_j, g_j) \in X$ ($f_j \in H$, $g_j \in Y$) such that $\bar{u}_k = -\langle f, \bar{x}_k\rangle_X$, $k = 0,1,2,\ldots$or $[\bar{u}_k]_j = -\langle f_j, \bar{w}_k\rangle_H - \langle g_j, \bar{v}_k\rangle_Y$, $k = 0,1,2,\ldots$, $j = 1,2,\ldots, m$ where $\bar{x}_k = (\bar{w}_k, \bar{v}_k)$. The vector f is referred to as the optimal functional feedback control gain.

Our approximation theory can be outlined as follows. For each $N = 1,2,\ldots$ and $M = 1,2,\ldots$ Let H_N and Y^M be finite dimnensional subspaces of H and Y respectively and let $P_N: H \to H_N$ and $P^M: Y \to Y^M$ denote the corresponding orthogonal projections. Let $X_N^M = H_N \times Y^M$ and define $P_N^M: X \to X_N^M$ by $P_N^M(\phi,\psi) = (P_N\phi, P^M\psi)$. Let $T_N^M, Q_N^M \in \mathcal{L}(X_N^M)$, $B_N^M \in \mathcal{L}(R^m, X_N^M)$ and consider the finite dimensional operator algebraic Riccati equation in X_N^M given by

$$\Pi_N^M = (T_N^M)^*(\Pi_N^M - \Pi_N^M B_N^M(R + (B_N^M)^*\Pi_N^M B_N^M)^{-1}(B_N^M)^*\Pi_N^M)T_N^M + Q_N^M. \tag{8}$$

Under the usual stabilizability and detectability conditions from the finite dimensional linear-quadratic theory, the equation (8) admits a unique nonnegative self-adjoint solution $\Pi_N^M \in \mathcal{L}(X_N^M)$.

We define approximating optimal control input sequences $\bar{u}_N^M = \{\bar{u}_{N,k}^M\}_{k=0}^{\infty}$ by

$$\bar{u}_{N,k}^M = -F_N^M P_N^M \bar{x}_{N,k}^M, \quad k = 0,1,2,\ldots$$

where $F_N^M \in \mathcal{L}(X_N^M, R^m)$ is given by $F_N^M = (R + (B_N^M)^*\Pi_N^M B_N^M)^{-1}(B_N^M)^*\Pi_N^M T_N^M$ and $\bar{x}_N^M = \{\bar{x}_{N,k}^M\}_{k=0}^{\infty}$ satisfies the recurrence

$$\bar{x}_{N,k+1}^M = S_N^M \bar{x}_{N,k}^M, \quad k = 0,1,2,\ldots$$

with $\bar{x}^M_{N,0} = x^0$ and $\mathcal{S}^M_N = T - BF^M_N P^M_N \ \varepsilon\ \mathcal{L}(X)$. Once again there exists a vector

$f^M_N = (f^M_{N,1} \ldots f^M_{N,m})^T$ with $f^M_{N,j} = (f^M_{N,j}, g^M_{N,j}) \ \varepsilon\ X^M_N$, $j = 1,2,\ldots,m$ for which $\bar{u}^M_{N,k} = -< f^M_N, \bar{x}^M_{N,k} >_X$

$k = 0, 1,2,\ldots$ or $[u^M_{N,k}]_j = -< f^M_{N,j}, \bar{w}^M_{N,k} >_H - < g^M_{N,j}, v^M_{N,k} >_Y$, $k = 0,1,2,\ldots$, $j = 1,2,\ldots,m$ where

$\bar{x}^M_{N,k} = (\bar{w}^M_{N,k}, \bar{v}^M_{N,k})$. The vector f^M_N is referred to as the approximating optimal functional feedback control gain.

A formal convergence theory based upon the treatment in [2] can be outlined as follows. Assume that $P^M_N \to I$, $T^M_N P^M_N \to T$, $(T^M_N)^* P^M_N \to T^*$ and $Q^M_N P^M_N \to Q$ strongly on X and that $B^M_N \to B$ (and therefore that $(B^M_N)^* P^M_N \to B^*$) in norm as $N,M \to \infty$. Let $S^M_N = T^M_N - B^M_N F^M_N \ \varepsilon\ \mathcal{L}(X^M_N)$. If nonnegative self-adjoint solutions Π^M_N to the algebraic Riccati equation (8) exist and are bounded, uniformly in N and M then $\Pi^M_N P^M_N$ converges weakly to a nonnegative self-adjoint solution Π to (7). If, in addition, the S^M_N are uniformly exponentially stable, uniformly in N and M, then $\Pi^M_N P^M_N \to \Pi$ strongly on X as $N,M \to \infty$. If $\Pi^M_N P^M_N \to \Pi$ weakly (strongly) then $F^M_N P^M_N \to F$ strongly (in norm), $S^M_N P^M_N \to S$ strongly (strongly) and $f^M_{N,j} \to f_j$ weakly (strongly or in norm) in X, $j = 1,2 \ldots$, m, as $N,M \to \infty$.

To construct the operators T^M_N, B^M_N, andQ^M_N we assume that we have approximations to the operators A_0 and D_0, $A_{0,N}$ and D^M_0 respectively, for which $T_{0,N} P_N \equiv T_{0,N}(\tau) P_N \equiv \exp(A_{0,N}\tau) P_N \to T_0(\tau) = T_0$ strongly on H and $U^M_0 P^M \equiv U^M_0(\tau)P^M \equiv \exp(D^M_0\tau)P^M \to U_0(\tau) = U_0$ strongly on Y as $N,M \to \infty$. We assume that $L^M_{0,N} \ \varepsilon\ \mathcal{L}(Y^M, H_N)$ is an approximation to L_0 for which

$S^M_{0,N} P^M \equiv \int_0^\tau T_{0,N}(t-s) L^M_{0,N} U^M_0(s) P^M ds \to S_0$ as $N,M \to \infty$. We set

$$\mathcal{A}^M_N = \begin{bmatrix} A_{0,N} & L^M_{0,N} \\ 0 & D^M_0 \end{bmatrix}$$

and then

$$T^M_N \equiv \mathcal{T}^M_N(\tau) \equiv \exp(\mathcal{A}^M_N \tau) = \begin{bmatrix} T_{0,N} & S^M_{0,N} \\ 0 & U^M_0 \end{bmatrix}.$$

We also require that the $A_{0,N}$, D^M_0 and $L^M_{0,N}$ be constructed so that $(T^M_N)^* P^M_N \to (T^M_N)^*$ as $N,M \to \infty$. Assuming that $\mathcal{R}(\Gamma^+) \subset \mathcal{N}(\Delta)$ (which is frequently the case) we define $B^M_N \ \varepsilon\ \mathcal{L}(R^m, X^M_N)$ by

$$B_N^M = \begin{bmatrix} (I - T_{0,N}) P_N \Gamma^+ (\beta + \Lambda\gamma^+) - S_{0,N}^M P^M \gamma^+ \\ (I - U_0^M) P^M \gamma^+ \end{bmatrix}.$$

When $\mathcal{R}(\Gamma^+) \subset H_N$ and $\mathcal{R}(\gamma^+) \subset Y^M$ for all N and M, the projections P_N and P^M appearing in the definition of B_N^M above may be omitted. Finally we set $Q_N^M = P_N^M Q$. When $Qx = Q(\phi,\psi) = (Q_0\phi, 0)$, we have $Q_N^M x_N^M = Q_N^M (\phi_N, \psi^M) = (P_N Q_0 \phi_N, 0)$.

We present an example which serves to illustrate the application of the general theory outlined above. We consider a one dimensional heat equation with Neumann boundary control with delay. The boundary control system is given by

$$\begin{aligned} w_t(t,\eta) &= a\, w_{xx}(t,\eta), \quad 0<\eta<1, \ t>0 \\ w(t,0) &= 0, \qquad t>0 \\ aw_x(t,1) &= bv(t) + cv(t-\rho), \quad t>0 \\ w(0,\eta) &= w^0(\eta), \quad 0 \le \eta \le 1 \\ v(\theta) &= v^0(\theta), \quad -\rho \le \theta \le 0 \end{aligned}$$

where $a,b,c,\rho \in R$, $a, \rho \ge 0$, $w^0 \in L_2(0,1)$ and $v^0 \in L_2(-\rho, 0)$. In this example $\ell = m = 1$. We take $H = L_2(0,1)$ and $Y = L_2(-\rho,0)$ each endowed with the usual L_2 inner product and $W = H^2(0,1) \cap H_L^1(0,1)$ endowed with the usual H_2 inner product where $H_L^1(0,1) = \{\phi \in H^1(0,1) : \phi(0) = 0\}$. The operators Δ,Γ,β and Λ are given by $\Delta\phi = aD^2\phi$, $\Gamma\phi = aD\phi(1)$, $\beta u = bu$ and $\Lambda\psi = c\psi(-\rho)$ for $\phi \in W$, $u \in R$ and $\psi \in V = H^1(-\rho,0)$.

We have W densely and continuously embedded in H, $\mathcal{N}(\Gamma) = \{\phi \in H^2(0,1) : \phi(0) = D\phi(1) = 0\}$ dense in H and A_0: $\mathrm{Dom}(A_0) \subset H \to H$ given by $A_0\phi = aD^2\phi$ for $\phi \in \mathrm{Dom}(A_0) = \mathcal{N}(\Gamma)$. The operator A_0 is negative definite, and self-adjoint and it is the infinitesimal generator of the uniformly exponentially stable analytic semigroup $\{T_0(t) : t \ge 0\}$ of bounded self-adjoint linear operators on H. The operator $\hat{A}_0$ is given by $(\hat{A}_0\phi)(\tilde{\phi}) = \langle \phi, A_0\tilde{\phi} \rangle_{L_2}$ for $\phi \in L_2(0,1)$, $\tilde{\phi} \in \mathcal{N}(\Gamma)$. It can be shown (see [1], [7]) that for $u \in L_2(0, t_1; W)$ we have $\int_0^{t_1} T_0(t_1 - s)(\Delta - \hat{A}_0)u(s)\,ds \in H$ and $\left| \int_0^{t_1} T_0(t_1 - s)(\Delta - \hat{A}_0)u(s)\,ds \right|_H \le \mu |u|_{L_2(0,t_1;W)}$. We choose $\Gamma^+ \in \mathcal{L}(R,W)$ as $(\Gamma^+u)(\eta) = (\eta/a)u$, $0 \le \eta \le 1$.

Note that we have $\mathcal{R}(\Gamma^+) \subset \mathcal{N}(\Delta)$. The nonnegative self-adjoint operator $Q_0 \in \mathcal{L}(H)$ is assumed to be of the form $Q_0\phi = q\phi$, $\phi \in L_2(0,1)$ with $q \in R$, $q \ge 0$ and the 1×1 positive definite symmetric matrix R is assumed to be given by $R = r$ with $r \in R$ and $r > 0$.

For each N = 1,2,... let $\{\phi_N^j\}_{j=0}^N$ denote the usual linear B- splines on [0,1] defined with respect to the uniform mesh {0, 1/N, 2/N, ... , 1}. We take $H_N = \mathrm{span}\{\phi_N^j\}_{j=1}^N \subset \mathrm{Dom}((-A_0)^{1/2})$ and let P_N: $H \to H_N$ denote the corresponding orthogonal projections. Let $\mathcal{P}_N$ be the orthogonal projection of $\mathrm{Dom}((-A_0)^{1/2})$ endowed with the usual inner product onto H_N and define $A_{0,N} \in \mathcal{L}(H_N)$

via $A_{0,N}^{-1} = \mathcal{P}_N A_0^{-1}$ restricted to H_N. It is not difficult to show (see [2]) that the $A_{0,N}$ are well defined, that this is a standard linear spline based Ritz-Galerkin scheme, and via the Trotter-Kato theorem from the approximation theory for linear semigroups (see [6]) that $T_{0,N}(t) P_N \equiv \exp(A_{0,N}t) P_N \to T_0(t)$ strongly on H and uniformly in t for t in compact intervals as $N \to \infty$.

For the hereditary component of the system we employ a recent scheme due to Ito and Kappel [5]. For each $M = 1,2,\ldots$, and $j = 1,2,\ldots, M$ let $\chi_j^M \in L_2(-\rho,0)$ denote the characteristic function for the interval $[-j\rho/M, -(j-1)\rho/M)$. Set $Y^M = \text{span}\{\chi_j^M\}_{j=1}^M$ and let P^M denote the corresponding orthogonal projection of Y onto Y^M. If we define Y_1^M to be the span of $\{\psi_j^M\}_{j=1}^M$ where $\{\psi_j^M\}_{j=0}^M$ are the usual linear B-splines defined on the interval $[-\rho,0]$ with respect to the uniform mesh $\{-\rho,\ldots, -\rho/M, 0\}$, then the restriction of P^M to Y_1^M is a bijection onto Y^M. Also, the restriction of the operator D_0 to Y_1^M has range in Y^M. We define $D_0^M \in \mathcal{L}(Y^M)$ by $D_0^M = D_0(P^M)^{-1}$ restricted to Y^M and $U_0^M(t) = \exp(D_0^M t)$, $t \geq 0$. As an approximation to L_0 we take $L_{0,N}^M \in \mathcal{L}(Y^M, H_N)$ given by $L_{0,N}^M = \hat{P}_N L_0 (P^M)^{-1}$ where $\hat{P}_N$ denotes the unique bounded linear extension of P_N to H_1.

The development of a complete convergence theory for the scheme outlined above based upon the results in [5] is currently in progress with our findings to be reported in a forthcoming paper. We have, however, carried out some preliminary numerical studies. In the heat equation example outlined above, we took $a = b = c = q = r = 1$, $\rho = .5$, $\tau = .01$ and $N = M = 2, 4, 6,$ and 8. We obtained the approximating optimal functional feedback control gains plotted in Figures 1 and 2 below. Note that $\mathcal{f}_N^M = (f_N^M, g_N^M) \in X_N^M = H_N \times Y^M$ (i.e. $f_N^M \in \text{span}\{\phi_N^j\}_{j=1}^N$, $g_N^M \in \text{span}\{\chi_j^M\}_{j=1}^M$) and

$$\bar{u}_{N,k}^M = -\int_0^1 f_N^M(\eta) w(k\tau,\eta) d\eta - \int_{-.5}^0 g_N^M(\theta) v(k\tau + \theta) d\theta, \qquad k = 0,1,2,\ldots$$

The matrix equations corresponding to the finite dimensimal approximating algebraic Riccati equations (8) were solved for every N and M via eigenvalue - eigenvector decomposition of the associated discrete-time Hamiltonian matrix (i.e. the Potter method). Matrix Riccati equations of order as high as 16 were solved without difficulty on an IBM PC AT microcomputer.

Matrix exponentials necessary to form the matirx representations for the operators T_N^M and B_N^M were also computed via eigenvalue - eigenvector decomposition. In carrying this out, it was not surprising to discover that the fact that $\sigma(D_0) = \emptyset$ causes some difficulties. Indeed, it turns out that $\sigma(D_0^M) = \{\frac{-2M}{\rho}\}$ for every $M = 1,2,\ldots$, with the corresponding eigenspace being of dimension 1. Consequently the application of standard QR software to compute $\exp(\mathcal{A}_N^M \tau)$ will in general not work. However, by taking advantage of the relatively simple form of the matrix representations for the operators D_0^M and by exploiting the block triangular structure of the matrix representations for the $\mathcal{A}_N^M$, it is not difficult to determine the similarity transform which puts the matrix representation of $\mathcal{A}_N^M$ in Jordan canonical form. It is then of course a rather simple matter to compute the required matrix exponential.

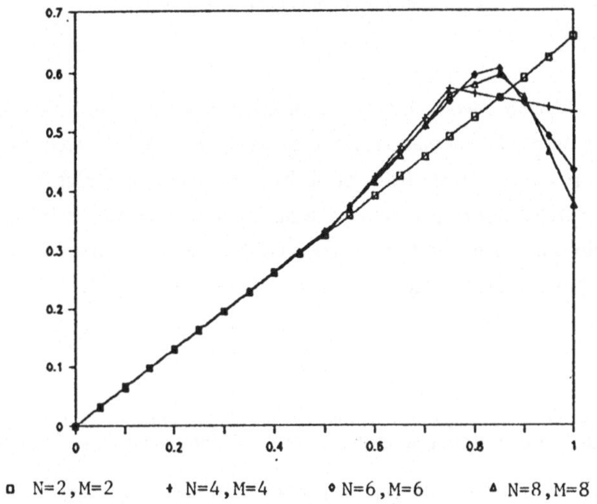

Figure 1. $f_N^M(\eta)$, $0 \leq \eta \leq 1$, N = M = 2,4,6,8.

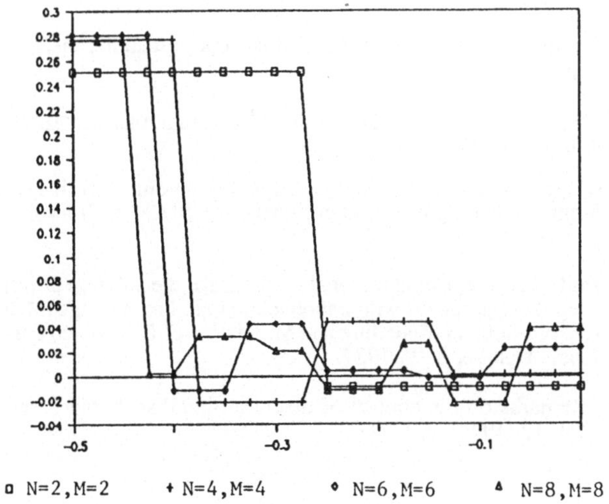

Figure 2. $g_N^M(\theta)$, $-.5 \leq \theta \leq 0$, N = M = 2,4,6,8.

As with our theoretical investigations, our numerical efforts are continuing. The extensive testing of our general approach and its application to a variety of examples in a variety of computing environments is currently in progress and will be reported on elsewhere.

Acknowledgment: The author would like to gratefully acknowledge Mr. Milton Lie for his assisitance in carrying out the numerical computations reported on above. This research was supported in part by the United States Air Force Office of Scientific Research under contract No. AFOSR - 84 - 0393. This research was carried out in part while the author was a visiting scientist at the Institute for Computer Applications in Science and Engineering (ICASE) at the NASA Langley Research Center in Hampton, VA which is operated under NASA contract No. NAS1-18107.

References

[1] Curtain, R.F. and D. Salamon, Finite dimensional compensators for infinite dimensional systems with unbounded input operators, SIAM J. Control and Opt., 24 (1986), 797 - 816.

[2] Gibson, J.S. and I.G. Rosen, Numerical approximation for the infinite-dimensional discrete-time optimal linear-quadratic regulator problem, SIAM J. Control and Opt., to appear.

[3] Gibson, J.S. and I.G. Rosen, Approximation of discrete-time LQG compensators for distributed systems with boundary input and unbounded measurement, Automatica, to appear.

[4] Ichikawa, A., Quadratic control of evolution equations with delays in control, SIAM J. Control and Opt, 20 (1982), 645 - 668.

[5] Ito, K. and F. Kappel, A uniformly differentiable approximation scheme for delay systems using splines, preprint.

[6] Pazy, A., Semigroups of Linear Operators and Applications to Partial Differential Equations, Springer-Verlag, New York, 1983.

[7] Pritchard, A.J. and D. Salamon, The linear quadratic control problem for infinite dimensional systems with unbounded input and output operators, SIAM J. Control and Opt., 25 (1987), 121 - 144.

[8] Rosen, I.G. amd M.A. Lie, Computational methods for the solution of infinite dimensional discrete-time regulator problems with unbounded input, Proceedings IMACS/IFAC International Symposium on Modelling and Simulation of Distributed Parameter Systems, Hiroshima, Japan, October 6 - 9, 1987.

[9] Zabczyk, J., Remarks on the control of discrete-time distributed parameter systems, SIAM J. Control and Opt. 12 (1974), 721 - 735.

SHAPE OPTIMIZATION OF STRUCTURES WITH POINTWISE STATE CONSTRAINTS

B. Rousselet
Département de Mathématiques
Université de Nice
06034 NICE Cedex
and
INRIA, B.P. 105 78153 Le Chesnay Cedex
FRANCE

1. INTRODUCTION

In this lecture we describe some Lipschitz continuity and differentiability results wich provide a justification of design sensitivity of mechanical structures; with these differentiability results and with the Hahn-Banach theorem, we obtain necessary optimality conditions for some non convex optimization problems with pointwise state constraints. In the same conference several authors (including BONNANS-CASAS; NEITTAANMAKI; TEMAM-ABERGEL) have presented results on this last topic but for different equations and using different techniques. The present proof stems from the treatment of duality in convex optimization in [6].

More precisely, the problem to be adressed is the minimization of a functional $J_0(x,u)$ subject to constraints $J_i(x,u) \leq 0$, where the *state* u is solution of a partial differential equation which depends on the design variable $x \in X$ some Banach space; the equation may be time dependent but x is not; we are not adressing control but *design* problems. Relevant *applications* include :
(i) conventional optimization of structures : beams and plates with variable thickness $e(\xi)$, being design variable;
(ii) *shape* optimization of plates, plane or three dimensional elasticity, heat equation; the design variable is here the shape of the boundary;
(iii) heat equation on a surface; arches and shells; here the design variable may be the thermal conductivity or the thickness or/and the *shape* of the (middle) *surface*.

A typical optimization project is then to minimize the weight of the structure with constraints on displacement, stress, strain energy. The situation (i) has been intensively studied; many examples may be found in [20] and [27]. As well as examples of the situation (ii) which is now intensively studied; for case (i) some differentiability results for repeated eigenvalues may be found in [41].

Situation (ii) traces back to [19] but a starting point for computer oriented research seems to be [8]. Numerous theoretical as well as application oriented papers have adressed the topic including, [7,12,16,17,26,28,29,31,32,33,35,36,42]. INRIA Lectures have been organized by Pironneau in 1982 and Céa-Rousselet in 1983 [30,10].

Situation (iii) has received less attention; some references are [1,2,3,13,15, 21].

Mechanical structures considered here may be loaded *statically* or *dynamically*; the last case is governed by equations well posed in the sense of Petrowsky and is naturally more technical; the case of eigenvalues has been considered in [31]; application to midline sensitivity for linear arch buckling is in [14]. The related problem of sensitivity of solutions of optimal control problems has been adressed by several authors including [25].

To illustrate the results of the following paragraph two basic examples will be considered which fall in the standard variational frame, find $u \in V$

$$\forall v \in V \qquad a(x;u,v) = \ell(v) \tag{1.1}$$

where a is a continuous, coercive bilinear form on V and $\ell \in V'$, $x \in X$ a Banach space of design variables.

(i) stationary *heat equation;* $H_0^1(\Omega) \subset V \subset H^1(\Omega)$

$$a(c;u,v) = \int_\Omega c(\xi) \;\; \text{grad } u.\text{grad } v \; d\xi$$

$$\ell(v) = \int_\Omega f \; v \; d\xi + \int_{\Gamma_0} g \; v \; d\sigma$$

the thermal conductivity c may depend on ξ; it may be a design variable as in [9]; the shape of the boundary may be a design variable.

(ii) *plane stress;* $(H_0^1(\Omega))^2 \subset V \subset (H^1(\Omega))^2$; the linear strain is

$$\gamma_{\alpha\beta} = \frac{1}{2}\,(u_{\alpha,\beta} + u_{\beta,\alpha}) \qquad 1 \leq \alpha,\beta \leq 2$$

and for homogeneous, isotropic elastic material we have the stress-strain relation :

$$n_{\alpha\beta} = \frac{Ee(\xi)}{1-\nu^2}\,|\,(1-\nu)\gamma_{\alpha\beta} + \nu\,\delta_{\alpha\beta}\,\gamma_{xx}\,|$$

with convention of repeated indices, E Young's modules, ν Poisson's ratio and e thickness of the slab; it may depend on ξ; the strain energy is

$$a(e;u,u) = \int_{\Omega} n_{\alpha\beta}(e;u)\gamma_{\alpha\beta}(u)d\xi$$

and the external virtual work

$$\ell(v) = \int_{\Omega} f_{\alpha} v_{\alpha} d\xi + \int_{\Gamma_0} g_{\alpha} v_{\alpha} d\sigma$$

the variable thickness may be a design variable as well as the shape of the boundary.

2. LIPCHITZ CONTINUITY AND DIFFERENTIABILITY RESULTS

To compute the derivative of a functional $J(x,u)$ it is quite natural to use the chain rule; set $j(x) = J(x,u_x)$ then

$$\frac{dj}{dx} y = \frac{\partial J}{\partial x} y + \frac{\partial J}{\partial u} \frac{\partial u}{\partial x} y \tag{2.1}$$

where $\partial u/\partial x\ y$ is the solution of the partial differential equation differentiated with respect to x (in a sense to be defined). The justification of this process is not trivial. Consider the case of an abstract variational problem (1.1) where the bilinear form a and the linear form ℓ depend on a design variable x.

In the most standard case, the functional J is well defined for $u \in V$; the justification of the chain rule then relies on the differentiability of $x \to u_x$ from X to V. Several authors have proved this last result by using the classical implicit function theorem [17,28,42] in the frame of shape optimization and [11] in the frame of identification. The basic hypothesis of this theorem when applied in our context means that the equation (1.1) defines an isomorphism from the space V' of right hand sides to the space V of solutions; this is well known. But for second order time dependent equations well posed in the sense of Petrowsky this is wrong (as indicated in [24], chapter 5). We have developped a method which starts *from scratch* and provides results for these equations.

In the frame of shape optimization the following results were proved in [35]; a detailed version of the present paragraph is in [40] with mechanical examples and analysis of implementation. For the abstract variational problem (1.1), the following lemma is easily proved :

Lemma 2.1 : If the bilinear form a is coercive uniformly in x and if the linear form is bounded uniformly in x, then

$$||u_x|| \leq c \text{ where } c \text{ does not depend on } x. \tag{2.2}$$

With the following notations :

$$\delta a(u,v) = a(x+y;u,v) - a(x;u,v)$$

$$\delta \ell(v) = \ell(x+y;v) - \ell(x;v) \tag{2.3}$$

$$\delta u = u_{x+y} - u_x$$

we are going to obtain a result of Lipschitz continuity of the state with respect to design variable.

<u>Proposition 2.2</u> : Under hypothesis of lemma 2.1, if moreover

$$|\delta a(u,v)| \leq c_a |y|_X \, ||u||_V \, ||v||_V \tag{2.4}$$

$$| \delta \ell(u,v)| \leq c_\ell \, |y|_X \, ||u||_V \, ||v||_V \tag{2.5}$$

then

$$||\delta u||_V \leq c_u \, |y|_X \, ||u||_V \tag{2.6}$$

Proof : Consider (1.1) for the value (x+y) of the design variable and expand it with notations (2.3) :

$$a(x;u_x,v) + \delta a(u_{x+y},v) + a(x;\delta u,v) = \ell(x;v) + \delta \ell(v)$$

simplify using (1.1), set $v = \delta u$, use coercitivity of a, and assumptions (2.4), (2.5) ; we obtain :

$$\gamma \, ||\delta u||_V^2 \leq |y|_X \, (c_a \, ||u_{x+y}|| + c_\ell) \, ||\delta u||_V$$

simplification and use of lemme 2.1 yield the desired result. ||

Now set

$$\delta^2 a(u,v) = a(x+y;u,v) - a(x;u,v) - a'(x,y;u,v)$$

$$\delta^2 \ell(v) = \ell(x+y;v) - \ell(x;v) - \ell'(x,y;v) \tag{2.7}$$

$$\delta^2 u = u_{x+y} - u_x - u'_{x,y} \, .$$

With a similar method one can prove (see |35|,|40|) the following theorem for *differentiable dependence in the functional space* V.

Theorem 2.3 : Under hypothesis of lemma 2.1, if moreover there exists $\frac{\partial a}{\partial x}(x,y;u,v)$ which satisfies

$$|\frac{\partial a}{\partial x}(x,y;u,v)| \leq c\ |y|_X\ ||u||_V\ ||v||_V \quad \text{and} \tag{2.8}$$

$$|\delta^2 a(u,v)| \leq c\ |y|_X\ \varepsilon(y)\ ||u||_V\ ||v||_V \quad \text{with } \varepsilon(y) \to 0 \text{ when } y \to 0 \text{ in } x \tag{2.9}$$

and a *similar* hypothesis for the linear form ℓ then

$$||\delta^2 u||_V \leq c\ |y|_X\ \varepsilon(y) \quad \text{(Frechet differentiability)} \tag{2.10}$$

with $u'_{x,y}$ being solution of the differentiated equation

$$\forall\ v \in V \quad a(x;u'_{x,y},v) = -\frac{\partial a}{\partial x}(x,y;u,v) - \frac{\partial \ell}{\partial x}(x,y;v) \tag{2.11}$$

Corollary 2.4 : Under hypothesis of Theorem 2.3, any functional $j(x) = J(x,u_x)$ defined for $u \in V$ is Frechet differentiable.

This is an obvious application of the chain rule in Banach space (see e.g. |18|).

Examples

(i) $j(x) = a(x;u_x,u_x)$

(ii) for the heat equation, with $\Omega_0 \subset \Omega$ we can consider :

$$j(x) = \int_{\Omega_0} |\nabla u_x|^2 \ ; \ j(x) = \int_{\Gamma_0} |u_x - u_0|^2\ d\xi$$

(iii) for plane elasticity we can also consider $j(x) = \int_{\Omega_0} F(n_x)d\xi$ where F is some function of the stress tensor (e.g. Von Mises function).

Counterexamples could be for heat equation

$$j(x) = \int_{\Gamma_0} |\nabla u_x|^2 \quad \text{or similarly} \quad j(x) = \int_{\Gamma_0} F(n_x)d\xi \quad \text{for plane elasticity;}$$

pointwise values of gradients of u would be other counterexamples. Indeed previous counterexamples will be *examples* of the next result of *differentiable dependence of strong solutions*.

Theorem 2.5 : Under hypothesis of Theorem 2.3, let $\mathcal{A}$ be an elliptic differential operator of order $2m$ associated to the bilinear form a; assume an a priori inequality holds :

$$||u||_{H^{2m}(\Omega)} \leq c_1 \, ||f||_{L^2(\Omega)} + c_2 \, ||u||_{H^m(\Omega)} \tag{2.12}$$

and assume (with obvious notations)

$$|(\frac{\partial \mathcal{A}}{\partial x} y)u|_{L^2(\Omega)} \leq c\mathcal{A}|y|_X \, ||u||_{H^{2m}(\Omega)} \tag{2.13}$$

$$|\delta^2 \mathcal{A} u|_{L^2(\Omega)} \leq \varepsilon(y) \, |y|_X \, ||u||_{H^{2m}(\Omega)} \quad \text{with } \varepsilon(y) \to 0 \text{ when } y \to 0 \text{ in } X \tag{2.14}$$

then

$$||\delta^2 u||_{H^{2m}(\Omega)} \leq \varepsilon(y) \, |y|_X \tag{2.15}$$

in other terms $x \to u_x$ is Frechet differentiable from X to $H^{2m}(\Omega)$.

This result may be proved with a similar method using a priori inequality (2.12). With smoother data we could prove differentiability in a space of more regular functions.

In the case of an equation well posed in the sense of Petrowsky, obviously we need more regularity on the data to get differentiability in $L^2(0,T;V)$ than the related result of the static case; see [35].

3. NECESSARY OPTIMALITY CONDITIONS WITH POINTWISE STATE CONSTRAINTS

We consider u_x the solution of equation (1.1); second order time dependent equations are considered in [37] and in [35] in the frame of shape optimal design; in this reference a *detailled treatment* of the results of this paragraph is to be found. We assume that the data depend smoothly enough on the design variable so that theorem 2.5 with $H^{2m}(\Omega) \subset \mathcal{C}(\overline{\Omega})$; hitherto the functional

$$j(x) = \operatorname*{Sup}_{\xi \in \Omega_0} |u_x(\xi)| \tag{3.1}$$

with $\Omega_0 \subset \Omega$ can be shown to be directionaly differentiable and regularly localy convex (with terminology of [22]); its directional differential is

$$j'(x,y) = \operatorname*{Sup}_{\xi \in E(x)} u'_{x,y}(\xi) \frac{u_x(\xi)}{|u_x(\xi)|} \quad \text{with } E(x) \text{ being the set of } \xi \in \Omega_0 \text{ where } u_x(\xi) = j(x) . \tag{3.2}$$

Then we definethe subdifferential of such a functional to be :

$$\partial j(x) = \{g \in X' \mid \forall \, y \in X \ \langle g,y\rangle \leq j'(x,y)\} \tag{3.3}$$

Next theorem provides a caracterisation of this subdifferential :

Theorem 3.1 : If the dual of X' with weak topology is isomorphic to X and with previous hypothesis we have :

(i) $\partial j(x)$ is convex, weakly closed in X';

(ii) $g \in \partial j(x)$ if and only if

$$\langle g,y \rangle = \int_{\overline{\Omega}} u'_{x,y}\ s(u_x)\ d\mu(\xi) \tag{3.4}$$

with

$$\begin{cases} \mu \in \mathcal{C}(\Omega)' \quad \mu \geq 0 \\ \int d\mu = 1 \qquad \text{and} \quad s(u) = \dfrac{u}{|u|} \\ \operatorname{Supp} \mu = E(x) \ . \end{cases} \tag{3.5}$$

We *sketch* only the *proof* of the second point. Set

$$\mathcal{M} = \{ g \in X' \mid \langle g,y \rangle = \int_{\overline{\Omega}} u'_{x,y}\ s(u_x) d\mu \ \text{ with } \ \mu \ \text{ satisfying (3.5)}. \}$$

It is quite simple to notice that $\mathcal{M}$ is a subset of $\partial j(x)$. To get the other inclusion, we prove successively that $\mathcal{M}$ is weakly closed (using compactness of the set of measures satisfying (3.5) [18]; convexity is obvious. Then we argue by contradiction; let $g_0 \notin \mathcal{M}$, a corollary of the Hahn-Banach Theorem (see e.g. [7]) yields the existence of a linear form on X' with weak topology i.e. by hypothesis $y \in X$, such that for any measure μ satisfying (3.5) :

$$\int_{\overline{\Omega}} u'_{x,y}\ s(u_x) d\mu \leq \alpha < \langle g_0,y \rangle$$

from which we get $g_0 \notin \partial j(x)$; the end of the contradiction argument.

By using also the Hahn-Banach Theorem we can prove :

Theorem 3.2 : With hypothesis of Theorem 3.1, the minimum of $j(x) = \sup_{\xi \in \Omega_0} |u_x(\xi)|$ under the constraint $\int_\Omega x(\xi) d\xi \leq c$ satisfies the necessary optimality condition : there exists $\lambda \geq 0$ and $\mu \in \mathcal{C}(\overline{\Omega})'$ satisfying (3.5) such that :

$$\int_\Omega u'_{x,y}\ s(u_x)\ d\mu(\xi) + \lambda \int_\Omega y\ d\xi = 0 \tag{3.6}$$

$$\lambda \left(\int_\Omega x\ d\xi - c \right) = 0 \ . \tag{3.7}$$

The measure μ appears as a generalized Lagrange multiplier. As usual to obtain a usable necessary condition, we introduce an adjoint state which here should be a weak solution with a right hand side involving the measure μ : for any $w \in H^{2m}(\Omega)$ (with some boundary condition)

$$- \int_{\Omega} p \mathcal{A} w \, d\xi = \int_{\Omega} s(u_x) w \, d\mu \ . \tag{3.8}$$

Using this, (3.6) may be written :

$$\int_{\Omega} p \mathcal{A}'_{x,y} u \, d\xi + \lambda \int_{\Omega} y \, d\xi = 0 \ .$$

However it should be emphasized that p is not uniquely defined by (3.8) as μ is not known. The situation is quite similar in control with pointwise state constraints (see [5]).

4. CONCLUSION

We have presented a *systematic approach* to prove differentiability with respect to design variables; it starts from scratch and has the following features
(i) it is quite elementary to prove differentiability of $x \to u_x$ from X to V,
(ii) the extension to get differentiability into $H^{2m}(\Omega)$ is quite natural,
(iii) if the bilinear form depends Lipschitz continuously on the data we prove easily that the state u_x depends in the same way on the data,
(iv) it yields similar results for an equation well posed in the sense of Petrowsky,
(v) it may be extended to nonlinear situations; obviously this should be studied case by case; see [35] for a nonlinear hyperbolic equation with a nonlinearity of type $|u|^p u$ (studied in [23]).

Then we have derived necessary optimality conditions for optimization problems with state constraints; the proof relies only on the well known theorem of *Hahn-Banach*.

Application to sensitivity with respect to midline of arches is in [14]; extension to shell is adressed in [13] and [2]. Surface sensitivity for the heat equation is described in [40].

An analysis of the implementation is in [39]. Implementation in Modulef library is under checking for basic situations and in progress for shells (partly sponsored by a contract between INRIA and AEROSPATIALE).

REFERENCES

1 N. BANICHUK, Determining the Optimal Form of Curved Elastic Bars, Makanika Tverdogo Tela, vol. 10, N° 6 (1975), 124-133.

2 BERNADOU, M.; PALMA, F.; ROUSSELET, B., Optimisation de forme d'une coque mince élastique sous différents critères, Rapport de Recherche INRIA (1987).

3 J.M. BOISSERIE, R. GLOWINSKI, Optimization of the thickness law for thin axisymmetric shells, Comp. Struct., 331-343 (1978).

4 B. BUDIANSKY, J.L. FRAUENTHAL, J.W. HUTCHINSON, On optimal Arches, J. of Appl. Mech., vol. 36, (1969), 880-882.

5 CASAS, Control of an elliptic problem with pointwise state constraints; SIAM J. Control and Optimization, vol. 24, (1986), 1309-1318.

6 J. CEA, Optimisation : théorie et algorithmes. Paris, Dunod, (1971).

7 J. CEA, Conception optimale ou identification de formes, calcul rapide de la dérivée directionnelle de la fonction coût, M2AN, vol. 20 (1986), 371-402.

8 J. CEA, A. GIOAN, J. MICHEL, Adaptation de la méthode du gradient à un problème d'identification de domaine. Computing methods in applied science and engineering, Springer Verlag, Berlin (1974).

9 J. CEA, C. MALANOWSKI, An example of a maximum problem in partial differential equations, Siam J. Control, 8 (1970), 305-316.

10 J. CEA, B. ROUSSELET, Conception optimale de forme, Ecole INRIA (26-30 Septembre 1983).

11 G. CHAVENT, Analyse fonctionnelle et identification de coefficients répartis dans les équations aux dérivées partielles, Université de Paris, (1971), Thèse d'Etat.

12 D. CHENAIS, Sur une famille de variétés à bord lipchitziennes : application à un problème d'identification de domaines, Annales de l'Institut Fourier, 27, (1977), 201-231.

13 D. CHENAIS, Optimal design of midsurface of shells : differentiability proof and sensitivity computation, (1987).

14 D. CHENAIS, B. ROUSSELET, Différentiation du champ de déplacements dans une arche par rapport à la forme de la surface moyenne en élasticité linéaire, C.R.Acad. Sci. Paris, 298, (1984), 533-536.

15 D. CHENAIS, B. ROUSSELET, Dépendance de la charge critique de flambement d'une arche par rapport à l'épaisseur et à la surface moyenne, J. Appl. Mathematics and Optimization (1987, to appear).

16 K. DEMS, Z. MROZ, Variational approach by means of adjoint systems to structural optimization and sensitivity analysis, part II, Structure shape variation, Int. J. Solids Struct., 20 (1984), 527-552.

17 A. DERVIEUX, B. PALMERIO, Une formule de Hadamard dans les problèmes d'identification de domaines, Comptes Rendus de l'Académie des Sciences, (1975), 280, 1697-1700, et 280, 1761-1764.

18 J.A. DIEUDONNE, Eléments d'analyse. Paris, Gauthier-Villars, (1969), tomes I et II.

19 J. HADAMARD, Mémoire sur le problème d'analyse relatif à l'équilibre des plaques élastiques encastrées, Mémoire des savants étrangers, (1908), 33.

20 E.J. HAUG, J. CEA, Optimization of distributed parameter structures, Sijthoff & Noordhoff, Alphen aan den Rijn, Netherlands, (1981).

21 I. HLAVACEK, Optimization of the shape of axisymmetric shells" Aplihace Matewatki, 269-294 (1983).

22 A.D. IOFFE, V.M. TIHOMIROV, Theory of extremal problems, Amsterdam-New York-Oxford, Norh Holland, (1979).

23 J.L. LIONS, Quelques méthodes de résolution des problèmes aux limites non linéaires, Paris, Dunod, (1969).

24 J.L. LIONS, E. MAGENES, Problèmes aux limites non homogènes et applications. Paris, Dunod, (1968), Vol. I et II.

25 K. MALANOWSKI, J. SOKOLOWSKI, Sensitivity of solutions to convex, control constrainted optimal control problems for distributed parameter Systems, J. of Math. Analysis and appli., 120, (1986), 240-263.

26 M. MASMOUDI, Outils pour la conception optimale de formes, thèse d'Etat, Université de Nice, 1987.

27 MOTA SOARES, Computer aided optimal design : structural and mechanical systems, Springer (1987).

28 F. MURAT, J. SIMON, Sur le contrôle par un domaine géométrique. Publication du L.A. 189, Université de Paris VI, (1976).

29 O. PIRONNEAU, Optimal shape design for elliptic systems, Springer Verlag, New York,(1984).

30 O. PIRONNEAU, Optimisation de forme dans les systèmes à paramètres distribués résolution numérique et applications, Ecole INRIA, Rocquencourt, France, (8-10 novembre 1982).

31 B. ROUSSELET, Etude de la régularité des valeurs propres par rapport à des déformations Lipschitziennes du domaine, C.R. Acad. des Sci., Série I, 283, (1976), 507.

32 B. ROUSSELET, Identification de domaines et valeurs propres, Université de Nice, thèse, France (1977).

33 B. ROUSSELET, Communications at NATO Institute:Optimization of distributed parameter structures, (see. Ref. 20) (1981).

34 B. ROUSSELET, Note on Design Differentiability of the Static Response of Elastic Structures, Struct. Mech., 10 (3), (1982), 353-358.

35 B. ROUSSELET, Quelques résultats en optimisation de domaines, Thèse d'Etat, Université de Nice (1982).

36 B. ROUSSELET, Shape design sensitivity of a membrane, J.O.T.A., 40, (1983), 595-623.

37 B. ROUSSELET, Static and Dynamic loads, pointwise constraint in structural optimization, C.R. du colloque "Optimization : theory and algorithms", Confolans, France (1981), M. Dekker (1983).

38 B. ROUSSELET, Principes d'analyse de sensitivité ; utilisation pour la conception optimale, INRIA (1986).

39 B. ROUSSELET, Shape design sensitivity, from partial differential equation to implementation, Enf. Opt, 11 (1987), 151-171.

40 B. ROUSSELET, Lectures Notes at 3è curso de Mecanica teorica e aplicada, projeto otimo, fundamentos e aplicaciones, L.N.C.C., Rio de Janeiro (1987).

41 B. ROUSSELET, D. CHENAIS, Continuité et différentiabilité d'éléments propres, application à l'optimisation de structures, J.A.M.O., to appear.

42 J.P. ZOLERIO, Identification de Domaines par déformations, Université de Nice, France, (1979), thèse d'Etat.

SOME CONTROL-THEORETIC QUESTIONS FOR A FREE BOUNDARY PROBLEM[1]

Thomas I. Seidman
Department of Mathematics and Statistics,
University of Maryland Baltimore County,
Baltimore, MD 21228, USA.
(BITNET:seidman@umbc; arpanet:seidman@umbc3).

ABSTRACT: We consider a number of control-theoretic questions related to a free boundary problem modeling growth and dissolution of a crystal grain surrounded by a dilute solution of the same substance.

KEY WORDS: *Free boundary problem, nonlinear, parabolic, partial differential equation, boundary control, optimal control.*

1. Introduction

Consider a solution of some substance surrounding a 'crystal grain' of the pure material. We assume radial symmetry[2] in $\mathbb{R}^d$ and consider the concentration C of the solution for $t > 0$ and $R(t) < r = |x| < 1$ where $L = 1$ is the fixed outer radius and $R(t) > 0$ is the radius of the grain at time t.

The underlying model involves diffusion/reaction in the solution with the concentration C satisfying

$$\dot{C} = \Delta C + F(C) = r^{1-d}(r^{d-1}C_r)_r + F(C) \qquad \text{in } \mathcal{Q} \tag{1.1}$$

where F is some nonlinear reaction rate, for simplicity, taken nonincreasing in C and $\mathcal{Q} := \{(t,r) \in \mathbb{R}_+ \times \mathbb{R}^2 : R(t) < r < 1\}$. Conservation of mass gives the boundary condition:

$$-C_r = -\dot{R}[1 - C] \qquad \text{at } r = R(t), \tag{1.2}$$

[1]This research was partially supported under grants #AFOSR-87-0190 and #AFOSR-87-0350.

[2]Physically, of course, we would have $d = 2$ or 3. It would certainly be of interest to consider less symmetrical settings but these would seem to be far more complicated — e.g., the free boundary would then be a *time-varying surface* and the function H appearing in the analogue of (1.4), governing the 'flow' of the surface along its normal, would become a function of the concentration C at the boundary and the local radius of curvature R.

while at the outer boundary we impose the Dirichlet condition:

$$C = \gamma \qquad \text{at } r = 1. \tag{1.3}$$

We also, of course, have initial conditions.

In comparison with [3], note that (with no loss of generality) we are here scaling t, x, C so that the (constant) diffusion coefficient is $D \equiv 1$, the 'outer radius' is $L = 1$, the 'pure' concentration in the crystal is $\Gamma = 1$.

The rate of growth or dissolution of the grain, i.e., the motion of the free boundary surface $\{r = R(t)\}$, is not specified *a priori* but is governed by an equation of the form:

$$\dot{R} = H(R, \omega) \qquad [\omega(t) := C(t, R(t))]. \tag{1.4}$$

Thus we have a ***free boundary problem***, here called the ***direct problem***, defined by coupling (1.1) and (1.4) with the associated conditions.

A principal ***control-theoretic*** concern will be to view the data γ appearing in (1.3) as a possible ***boundary control*** and then to consider such questions as the existence of an optimal control for standard optimality criteria. An interesting alternative is to take the functions $F(\cdot)$ and $H(\cdot)$ appearing in (1.1) and (1.4) to depend also on the temperature ϑ and assume the availability of a 'heater' with radially symmetric effect β and variable intensity — so $\vartheta(t, r) = \alpha(t)\beta(r)$. Such a setting, in which the intensity $\alpha(\cdot)$ would be taken as under our control (within some specified range) is physically plausible, perhaps more so than the boundary control setting. Apart from optimality, we also formulate (without much in the way of results already obtained) some questions of feedback stabilization and identification.

Most of the technical work of this paper is in the next section, discussing the direct problem from the perspective of the control-theoretic applications. These can then be presented comparatively briefly in the final section.

2. The direct problem

The paper [3] presents a detailed argument for the existence and uniqueness of solutions to the direct problem. Our present considerations are somewhat different since we are not as concerned with uniqueness but are extremely concerned to know that we obtain limiting solutions by a weak limit of controls. Further, the treatment in [3] does not consider as general boundary data γ as we would like. We therefore sketch the relevant theory for the direct problem.

Our strategy is first to obtain existence for a modified problem and then to show, using a maximum principle argument, that the modification is nugatory — the solution stays within the range for which the two problems coincide — so one has, indeed, a solution of the original problem. Initially we restrict our consideration to a fixed time interval $[0,T]$ with T depending only on R_0.

We begin by assuming:

(H-1): $F : I\!R \to I\!R$ is continuous and nondecreasing; $F(0) \geq 0 \geq F(1)$.

(H-2): $H : (0,1) \times I\!R \to I\!R$ is continuous with $H(\cdot,0) \leq 0$.

(H-3): $0 < R_0 < 1$ and $C_0 \in L^\infty(R_0,1)$ with $0 \leq C_0 \leq 1$.

(H-4): γ is in $\mathcal{G}_T := \{\gamma \in L^2(0,T) : 0 \leq \gamma \leq 1\}$.

as our basic set of hypotheses.

Let us restate the problem: we wish to find a pair $[R,C]$ of functions such that R satisfies on the interval $[0,T]$ the ordinary differential equation

$$\dot{R} = H(R,\omega) \qquad \text{with } R(0) = R_0, \tag{2.1}$$

$$\omega(t) := C(t, R(t)) \tag{2.2}$$

and C satisfies on $\mathcal{Q} = \mathcal{Q}_T$ the parabolic moving boundary problem

$$(1.1) \text{ with } C = C_0 \text{ at } t = 0 \text{ and } (1.2),\ (1.3) \tag{2.3}$$

For the *modified* problem, define functions $\hat{H}$, $\hat{\varphi}$: $(0,1) \times I\!R \to I\!R$ by:

$$\hat{H}(R,\omega) := \begin{cases} H(R,0) & \text{if } \omega \leq 0 \\ H(R,1) & \text{if } \omega \geq 1 \\ H(R,\omega) & \text{for } 0 \leq \omega \leq 1; \end{cases} \tag{2.4}$$

$$\hat{\varphi}(R,\omega) := \begin{cases} -\hat{H}(R,\omega)[1-\omega] & \text{if } \omega \leq 1 \\ -\hat{H}(R,\omega)[1-\omega]/2 & \text{if } \omega \geq 1 \end{cases} \tag{2.5}$$

and then replace $-H[1-\omega] =: \varphi$ by $\hat{\varphi}$ in (1.2), obtaining

$$C_t = r^{1-d}(r^{d-1}C_r)_r + F(C) \qquad \text{with } C = C_0 \text{ at } t = 0 \tag{2.6}$$

$$\text{and} \begin{cases} \text{(i)} & -C_r = \hat{\varphi}(R,\omega) & \text{at } r = R(t) \\ \text{(ii)} & C = \gamma & \text{at } r = 1, \end{cases}$$

and H by $\hat{H}$ in (2.1), obtaining

$$\dot{R} = \hat{H}(R,\omega) \qquad \text{with } R(0) = R_0. \tag{2.7}$$

We refer to this modified problem [(2.6), (2.7), (2.2)] as **(MP)** = $\mathbf{(MP)}_T$ and to the original problem as **(OP)**.

Suppose, now, we fix $0 < R_- < R_0 < R_+ < 1$ and find upper and lower bounds h_-, h_+ for H on the compact set $[R_-, R_+] \times [0,1]$ so $h_- \leq \hat{H}(R,\omega) \leq h_+$ on $[R_-, R_+] \times \mathbb{R}$. Choose $T > 0$ so

$$R_- \leq R_0 + h_- T \leq R_0 + h_+ T \leq R_+. \tag{2.8}$$

This is then the T for consideration of $\mathbf{(MP)}_T$.

We will use a 'fixpoint approach' to existence. Given a function $\omega(\cdot)$, we solve (2.7) to obtain $R(\cdot)$ and set

$$h(\cdot) := \hat{H}(R,\omega) = \dot{R} \quad \text{ and } \rho(\cdot) := 1/[1 - R(\cdot)]. \tag{2.9}$$

One easily sees that for every possible solution $R(\cdot)$ of (2.7) we always have the bounds $R_- \leq R \leq R_+$ on $[0,T]$ by the definition of T and $h = \hat{H}$ must satisfy $h_- \leq h \leq h_+$. Also, we note that $\rho(\cdot)$ is in a fixed compact subset of $C[0,T]$ with $1 < 1/[1 - R_+] \leq \rho \leq 1/[1 - R_-]$.

To treat the moving boundary problem, we make the change of variables $r \mapsto y := \rho[1-r] = \frac{1-r}{1-R}$ and set $u(t,y) = C(t,r)$. The parabolic problem (2.3) then becomes

$$u_t = \rho^2 u_{yy} - \psi u_y + F(u) \text{ on } \tilde{Q}_T \qquad \text{with } u = u_0 \text{ at } t = 0 \tag{2.10}$$

$$\text{and } \begin{cases} \text{(i)} & \rho^2 u_y = \tilde{\varphi} \quad \text{at } y = 1 \\ \text{(ii)} & u = \gamma \quad \text{at } y = 0. \end{cases}$$

Here $\tilde{Q}_T := (0,T) \times (0,1)$ and, with R, ρ, h as above, we have set

$$\psi(t,y) = \psi(t,y;\omega) \;:=\; \rho\left[\frac{(d-1)\rho(t)}{\rho(t)-1} + yh(t)\right]; \tag{2.11}$$

$$\tilde{\varphi}(t) = \tilde{\varphi}(t;\omega) \;:=\; \rho(t)\hat{\varphi}(R(t),\omega(t)). \tag{2.12}$$

Note that ψ, $\tilde{\varphi}$, and $\psi_y = \rho h$ are uniformly bounded, independently of ω.

Provided we show that the solution u of (2.10) will have enough regularity at $y = 1$ to justify taking the trace (as a suitable function on $[0,T]$), we can now define a map $\mathbf{T} = \mathbf{T}_\gamma : \omega \mapsto \omega_{new}$ by setting

$$[\mathbf{T}_\gamma(\omega)] = \omega_{new} := u(\cdot, 1) \tag{2.13}$$

with u obtained as above. We will next verify, for $\mathbf{T}$, the hypotheses of the Schauder Fixpoint Theorem.

Introduce the solution $z = z(\cdot;\gamma,\rho)$ of the linear partial differential equation

$$z_t = \rho^2 z_{yy} \text{ on } \tilde{Q}_T \tag{2.14}$$

with $z = u_0$ at $t = 0$, $\rho^2 z_y = 0$ at $y = 1$, and $z = \gamma$ at $y = 0$. In view of **(H-3)**, **(H-4)**, we note that one also has $0 \le z \le 1$ by the maximum principle. Now set $v := u - z$ so

$$v_t = \rho^2 v_{yy} - \psi v_y + F(v+z) - \psi z_y \quad \text{on } \tilde{Q}_T \tag{2.15}$$

with $v = 0$ at $t = 0$, $\rho^2 v_y = \tilde{\varphi}$ at $y = 1$, and $v = 0$ at $y = 0$ (without γ appearing explicitly). Multiplying this by v and integrating over $\tilde{Q}_t$ gives the identity

$$\begin{aligned} \tfrac{1}{2}\|v(t)\|^2 &+ \int_0^t \rho^2\|v_y\|^2 \\ &= \int_0^t [\tilde{\varphi} - \psi z]v|_{y=1} + \int_0^t\int_0^1 (z-v)\psi v_y + \int_0^t\int_0^1 [F(z) + z\psi_y]v \\ &\qquad + \int_0^t\int_0^1 [F(v+z) - F(z)][(v+z) - z]. \end{aligned}$$

As usual,[3] this provides uniform bounds (independent of ρ,γ) on the solution v in $L^\infty((0,T) \to L^2(0,1)) \cap L^2((0,T) \to H^1(0,1))$ and so, using the equation, on v_t in $L^2((0,T) \to H^{-1}(0,1))$. By the Aubin Compactness Theorem [1], this shows that v is in a compact, convex subset of $\mathcal{V} := L^2((0,T) \to H^s(0,1))$ (with, e.g., $s < 1$) independent of ρ,γ. A similar argument shows that v is in a compact subset of $\mathcal{U} := C([0,T] \to L^2(0,1))$. By trace theory (taking $s > /2$) $v(\cdot,1)$ is in a fixed compact subset of $L^2(0,T)$.

Returning to (2.14), we consider the further change of variables

$$\tau = \vartheta(t) = \vartheta(t;\rho) := \int_0^t \rho^2, \qquad t = \Theta(\tau) := \int_0^\tau \rho^{-2}, \tag{2.16}$$

so $\Theta = \vartheta^{-1}$ for given $\rho(\cdot)$. Setting $\varsigma(\tau,y) = \varsigma(\tau,y;\gamma,\rho) = z(t,y;\gamma,\rho)$, we then have

$$\varsigma_t = \varsigma_{yy} \quad \text{on } \tilde{Q}_T \tag{2.17}$$

[3]Note that $|v(t,1)| \le C\|v(t,\cdot)\|^2_{H^1}$ and the last integral is non-positive by **(H-1)**. Recall that $\rho > 1$ and we have uniform bounds for $\rho, \tilde{\varphi}, \psi, z, \psi_y$. One can then estimate the terms on the right so as to apply the Gronwall Inequality.

with $\varsigma = 0$ at $t = 0$, $\varsigma_y = 0$ at $y = 1$, $\varsigma = \hat{\gamma}$ at $y = 0$, and

$$\hat{\gamma}(\tau) = \hat{\gamma}(\tau;\rho) = [\mathbf{L}_\rho \gamma](\tau) := \gamma(\Theta(\tau;\rho)) = \gamma(t). \tag{2.18}$$

Note that the linear maps $\mathbf{L}_\rho$ are continuous from L^2 to L^2 (uniformly in ρ for the functions ρ arising here — already noted to be in a compact subset of $C[0,T]$). Note also that $[\rho \mapsto \mathbf{L}_\rho^*]$ is uniformly continuous from (the relevant compact subset of) $C[0,T]$ to the strong (pointwise) operator topology for $\mathbf{L}_\rho^*$; from this it follows that the map: $[\gamma,\rho] \mapsto \hat{\gamma}$ is continuous (topologizing γ and $\hat{\gamma}$ in L^2_{weak} and ρ in $C[0,T]$). It follows from linearity, closedness, and well-known interior regularity results that the solution map: $\hat{\gamma} \mapsto \varsigma$ for (2.17) is continuous and compact to $\mathcal{V}_\varepsilon := L^2(\to H^s(\varepsilon,1))$ $(s < 1)$ for each $\varepsilon > 0$ from L^2_{weak}. Inverting $\mathbf{L}_\rho$ to obtain z again, this means that the map: $[\gamma,\rho] \mapsto z(\cdot;\gamma,\rho)$ defined by (2.14) is also continuous and compact to $L^2(\tilde{\mathcal{Q}}_T)$ — indeed, to $\mathcal{U}$, using the bound $0 \leq z \leq 1$ and obtaining continuity for each $z : [0,T] \to L^2(0,1)$ by a density argument — and also to each $\mathcal{V}_\varepsilon$. It follows that the map: $[\gamma,\rho] \mapsto z(\cdot,1;\gamma,\rho)$ to $L^2(0,T)$ is also continuous and compact.

Combining this with our compactness result for $v(\cdot,1)$, we see that

$$\mathbf{T}_\gamma(\omega) = z(\cdot,1) + v(\cdot,1)$$

lies in a *compact* subset $\mathcal{W}$ of $L^2(0,T)$. Re-interpreting $\mathbf{T}_\gamma$ as its restriction to $\mathcal{W}$, we have shown that $\mathbf{T}_\gamma$ is a self-map of a compact set (which we may assume is also convex). Note that $\mathcal{W}$ is here independent of $\gamma \in \mathcal{G}_T$.

Returning to (2.15), by a standard estimate[4] the map: $[\omega, z] \mapsto v$ defined by (2.15), (2.11), (2.12), etc., is continuous to $\mathcal{V}$ (taking $\omega \in \mathcal{W}$ and $z \in L^2(\mathcal{Q}_T)$ with $0 \leq z \leq 1$). Combining this with our earlier continuity result for z (and using the continuity of the map: $\omega \mapsto \rho : \mathcal{W} \to C[0,T]$), this gives the continuity of $\mathbf{T}_\gamma$. The Schauder Fixpoint Theorem is now applicable to give existence (but, of course, not uniqueness) of a solution pair $[R,u]$ for (2.7), (2.10) with $\omega = u(\cdot,1)$.

This solves $(\mathbf{MP})_T$ (for fixed $\gamma \in \mathcal{G}_T$) since we may obviously invert the '$r \mapsto y$-substitution' to obtain a solution C — with the same regularity as u — for (2.6). Next we wish to observe a property of continuous dependence on $\gamma \in L^2_{weak}(0,T)$. The nature of our analysis makes it convenient to work with

[4]For solutions v_1, v_2 corresponding to two pairs of functions $[\omega_1, z_1]$, $[\omega_2, z_2]$ set $v = v_1 - v_2$. Much as in the estimate above (giving the *a priori* bound), we multiply the resulting equation by v, integrate over $\tilde{\mathcal{Q}}_t$, estimate, and apply the Gronwall Inequality.

u rather than with C but, in any reasonable sense, the notions of convergence will correspond.

Suppose $\gamma_k \rightharpoonup g_*$ (weak convergence in $L^2(0,T)$, subject to **(H-4)**) and let $[R_k, u_k]$ be corresponding solutions of $(\mathbf{MP})_T$. We wish to know, at least for a subsequence, that $[R_k, u_k] \to [R_*, u_*]$ in some suitable sense, where $[R_*, u_*]$ is a solution corresponding to γ_*. We use, subscripted by k, the notation of the analysis above (at the fixpoint of $\mathbf{T}_{\gamma_k}$, of course). Thus, e.g., $\{\omega_k\}$ is in $\mathcal{W}$ (which we have already noted is independent of γ and compact). Extracting a subsequence if necessary, we may thus assume $\omega_k \to \omega_*$ in the sense of $L^2(0,T)$ for *some* function $\omega_* \in \mathcal{W}$ and have $[R_k, \rho_k] \to [R_*, \rho_*]$ in $C[0,T]$ and $h_k \to h_*$ from (2.7), (2.9). Note that this gives suitable convergence of $\psi_k, (\psi_k)_y$ from (2.11) and $\tilde{\varphi}_k$ from (2.12) to the corresponding $\psi_*, (\psi_*)_y$ and $\tilde{\varphi}_*$. It also gives $\hat{\gamma}_k \rightharpoonup \hat{\gamma}_*$ in L^2 and so $z_k = z(\cdot; \gamma_k, \rho_k) \to z_* = z(\cdot; \gamma_*, \rho_*)$ in each $\mathcal{V}_\varepsilon$, as noted above. It follows that $z_k(\cdot, 1) \to z_*(\cdot, 1)$ in $L^2(0,T)$ and also that $z_k \to z_*$ in $\mathcal{U}$. We now have, again as noted above, $v_k \to v_*$ in $\mathcal{V}$ so $v_k(\cdot, 1) \to v_*(\cdot, 1)$ in $L^2(0,T)$. Thus, $u_k := z_k + v_k \to z_* + v_* =: u_*$ in, say, the senses of $\mathcal{U}$ and of $\mathcal{V}_\varepsilon$ for each $\varepsilon > 0$ so u_* satisfies (2.10). Finally, we have convergence of the traces: $u_k(\cdot, 1) \to u_*(\cdot, 1) =: \mathbf{T}_{\gamma_*}(\omega_*)$. Since $\omega_k = \mathbf{T}_{\gamma_k}(\omega_k) := u_k(\cdot, 1)$ and $\omega_k \to \omega_*$ by construction, we have $\mathbf{T}_{\gamma_*}(\omega_*) = \omega_*$ so the limit $[R_*, u_*]$ is indeed a solution of the coupled system $(\mathbf{MP})_T$ using γ_*.

Finally, we use a maximum principle argument to show that each solution $[R, C]$ of (2.6), (2.7) is actually already a solution of (2.3), (2.1). Clearly, this will be the case if $0 \le C \le 1$ ae on $\tilde{Q}_T$ so $\hat{H}, \hat{\varphi}$ coincide with the original functions H, φ. To this end, multiply (2.6) by w, taken as either $w_0 := \min\{C, 0\}$ or $w_1 := \max\{C-1, 0\}$. In either case we have

$$w = 0 \text{ at } r = 1,\ t = 0,\ wF(C) \le 0,\ wC_t = (\tfrac{1}{2}w^2)_t,\ w_r C_r = w_r{}^2$$

and

$$\frac{d}{dt}\left(\frac{1}{2}\int_{R(t)}^1 r^{d-1} w^2\right) = -\tfrac{1}{2} R^{d-1} w^2 \dot{R} + \int_{R(t)}^1 r^{d-1} w C_t.$$

A standard computation then gives

$$\frac{1}{2}\int_{R(t)}^1 r^{d-1} w^2 + \int_0^t \int_{R(t)}^1 r^{d-1} w_r{}^2 \le \int_0^t R^{d-1} w[\hat{\varphi} - \tfrac{1}{2} w \dot{R}]|_{r=R(\cdot)}. \tag{2.19}$$

CASE 1: $\boxed{w = w_1}$ When $w_1|_{r=R} \ne 0$ we have $\omega = C > 1$ so (2.5) gives $\hat{\varphi} := \frac{1}{2}(C-1)\dot{R} = \frac{1}{2} w_1 \dot{R}$. Thus, the integrand of the right hand side of (2.19) vanishes identically and one concludes that $w_1 \equiv 0$, i.e., $C \le 1$.

CASE 2: $\boxed{w = w_0}$ When $w_0|_{r=R} \neq 0$ we have $w = \omega < 0$ and $\hat{\varphi} = (\omega - 1)\dot{R}$ so $[\hat{\varphi} - \frac{1}{2}\dot{R}] = [\frac{1}{2}w]\dot{R}$ and also $\dot{R} = \hat{H}(R,0) \leq 0$. Thus the integrand of the right hand side of (2.19) must be non-positive ae and one concludes, as above, that $w_0 \equiv 0$, i.e., $C \geq 0$. □

Summarizing the arguments above, we see that we have shown:

Direct Theorem 1: *Consider the system* (2.3), (2.1), (2.2) *with initial data satisfying* **(H-3)** *and with* T *given as in* (2.8)*; assume* **(H-1,2)**. *Then for each* $\gamma \in \mathcal{G}_T$ *there is always at least one solution* $[R,C]_\gamma$ *on the interval* $[0,T]$. *This solution lies in a fixed compact set, independent of* $\gamma \in \mathcal{G}_T$, *in* $H^1(0,T) \times \mathcal{U}$. *Further, if* H *is Lipschitzian and one additionally has* $\gamma \in H^1(0,T)$, *then*[5] $u \in \mathcal{V}$ *and* R *are unique.* □

Direct Theorem 2: *Consider the system as in Direct Theorem 1 with fixed initial data as in* **(H-3)**. *Then the graph* $\{[\gamma, R, u]\}$ *is compact in* $\mathcal{G}_T \times H^1(0,T) \times [\mathcal{U} \cap \mathcal{V}_\epsilon]$ *where* $\mathcal{G}_T$ *is taken with the* L^2_{weak}*-topology.* □

So far we have determined the value of T by (2.8). Before proceeding further, a word is in order about the *global existence* of solutions: we note that it is always possible to restart at $t = T$, taking $R(T), C(T)$ as new initial data but the new T obtained for the restarted problem may be smaller and proceeding recursively need not yield an extended solution defined for all t — in general, the interval of definition can be expected to vary with the solution.

Direct Theorem 3: *Let* $[R_k, u_k]$ *be solutions — defined on time intervals* $[0,\tau_k)$ *— corresponding to* γ_k $(0 \leq \gamma_k \leq 1)$ *with fixed initial data satisfying* **(H-3)**. *Suppose* $\gamma_k \rightharpoonup \gamma_*$ *in* $\mathcal{G}_{\tau_*}$ *with* $\tau_* := \lim \tau_k$. *Then there is a solution* $[R_*, u_*]$ *— defined on* $[0,\tau_*)$ *— corresponding to* γ_* *and such that, possibly for a subsequence, one has suitable convergence* $[R_k, u_k] \to [R_*, u_*]$ *on compact subintervals of* $[0,\tau_*)$.

PROOF In view of Direct Theorem 2, we see (possibly extracting a subsequence) that $R_k \to R_*$ uniformly on the fixed initial interval $[0,T]$ and that $u_k \to u_*$ suitably where $[R_*, u_*]$ is a solution corresponding to γ_*. In particular, $R_k(T) \to R_*(T)$ so $T_k \to T_*$ for the 'guaranteed existence' times. This gives $[R_*, u_*]$ defined up to $\tau = T + \lim T_k$ and $[R_k, u_k] \to [R_*, u_*]$. We apply this argument

[5] See [3] for the uniqueness argument. The hypothesis that $\gamma \in H^1(0,T)$ is certainly stronger than needed but makes it easy to show the map: $[\gamma, \rho] \mapsto z$ is Lipschitzian to $\mathcal{V}$. One certainly needs $\gamma \in H^{1/2}(0,T)$ to have $z \in \mathcal{V}$, rather than $\mathcal{V}_\epsilon$, as here; compare [4].

recursively (using a diagonal process to get the final subsequence) to obtain the result. □

We will also want to know similar results in preparation for the alternate mode of control under consideration.

Direct Theorem 4: *Let $F = F(\cdot,\vartheta,C)$ be bounded, continuous in ϑ, C, nonincreasing in C, with $F(\cdots,0) \geq 0 \geq F(\cdots,1)$. Let $H = H(\vartheta,R,\omega)$ be continuous and bounded for R in $[R_-,R_+] \subset (0,1)$ and $0 \leq \omega \leq 1$. Fix $\gamma = \gamma_0 = const.$ $(0 \leq \gamma_0 \leq 1)$ in (2.3.ii), $0 < R < 1$ in (2.1), and $C_0 = u_0$ in (2.3) with $0 \leq u_0 \leq 1$. Let $\beta(\cdot)$ be a fixed L^2 function of $r \in [R_-,R_+]$. Then the map:*

$$\alpha \longmapsto \vartheta = \alpha(t)\beta(r) \longmapsto [R,u] \quad (\leftrightarrow [R,C]) \tag{2.20}$$

is continuous from $L^2_{weak}(0,T)$ to $H^1(0,T)\times[\mathcal{U}\cap\mathcal{V}]$. If one has solutions $[R_k,C_k]$ — defined on maximal intervals $[0,\tau_k)$ — corresponding to $\alpha_k \rightharpoonup \alpha_*$, then one has suitable convergence to a limit solution — defined on $[0,\tau_*)$ with $\tau_* = \lim \tau_k$ — corresponding to α_*.

PROOF These hypotheses give, for each fixed α, the hypotheses of Direct Theorem 1 so the map: $\alpha \mapsto [R,u] \leftrightarrow [R,C]$ is well-defined. Noting the uniqueness now available with constant γ, the proof of continuity is much along the same lines as the argument for Direct Theorem 2 showing closedness of the graph $\{[\gamma,R,u]\}$. Finally, the extension to maximally defined solutions is much as for Direct Theorem 3. □

3. Some control-theoretic questions

In this section we will be considering a number of control-theoretic problems for our system. We assume throughout that F,H satisfy **(H-1,2)** and that we always have $0 \leq \gamma \leq 1$. Note that most of the technical preparation has already been done in the previous section so our presentation here is quite brief.

First, consider the *minimum time problem*, taking γ as the control: we are given a *target set* $\mathcal{T}$ and seek a boundary control γ_* for which $[R_*,C_*]|_{t=\tau} \in \mathcal{T}$ with τ as small as possible. Actually, it is more convenient to think of $\mathcal{T}$ as a subset of $(0,1)\times L^2(0,1)$ and ask that $[R_*(\tau),u_*(\tau)] \in \mathcal{T}$.

Control Theorem 1: *Consider the system as in Direct Theorems 1,2,3. Let $\mathcal{T}$ be closed in $(0,1)\times L^2(0,1)$ and suppose the set*

$$\mathcal{S} := \{[\tau,\gamma,R,u] : \gamma \in \mathcal{G}_\tau,\ (2.10) \text{ on } \tilde{\mathcal{Q}}_\tau,\ (2.1),\ (2.2) \text{ on } [0,\tau],\ [R(\tau),u(\tau)] \in \mathcal{T}\}$$

is nonempty. Then τ attains its minimum: there exists some γ_* and a corresponding solution $[R_*, u_*]$ of (2.10), (2.1) for which $[R_*(\tau), u_*(\tau)] \in \mathcal{T}$ with $\tau = \tau_{min}$.

PROOF Let $\{[\tau_k, \gamma_k, R_k, u_k]\} \subset \mathcal{S}$ be a minimizing sequence for τ. Since, by assumption, we have $0 \leq \gamma_k \leq 1$ on $[0, \tau_{min} + \varepsilon]$ (say, extending each γ_k as 0 past τ_k), we may extract a subsequence weakly convergent in $L^2(0, \tau_{min} + \varepsilon)$ to some γ_*. Now apply Direct Theorem 3. Note that the closure assumption for $\mathcal{T}$ ensures that $R_k(\tau_k)$ must be bounded away from 0, 1 so the intervals in question are not *maximal* intervals of definition but can be prolonged by some uniform T. Hence the application of Direct Theorem 3 gives uniform convergence in $\mathbb{R} \times L^2(0,1)$ of $[R_k, u_k]$ on, say, $[0, \tau_{min} + \varepsilon]$. Thus we have $[R_k(\tau_k), u_k(\tau_k)] \to [R_*(\tau_{min}), u_*(\tau_{min})]$, whence $[R_*(\tau_{min}), u_*(\tau_{min})] \in \mathcal{T}$. □

One might reasonably conjecture a 'bang-bang principle': that the time-optimal control satisfies $\gamma_* \equiv 0, 1$ ae. The difficulty of *proving* this is indicated by the fact that the corresponding assertion is open for boundary control of the one-dimensional heat equation. We also note that essentially the same argument as above, but now using Direct Theorem 4, gives existence of a time-optimal control when we use temperature as our control mode (with some constraint giving coercivity).

Now consider an *optimal control problem* with an integral cost functional. As a typical example of the kind of result available here, we take the control to be α as in (2.20) and seek to minimize

$$J := \int_0^T \int_{R(t)}^1 |C - C^*| + \int_0^T |\alpha - \alpha^*|^2 \tag{3.1}$$

for some given $[T, C^*, \alpha^*]$.

Control Theorem 2: *Consider the system as in Direct Theorem 4 with fixed initial data and γ_0. Let $T > 0$ and assume the problem is feasible: there is some $\alpha \in L^2(0,T)$ for which a solution exists on the interval $[0,T]$. Let C^* and α^* be given in $L^1(\check{Q}_T)$ and $L^2(0,T)$, respectively. Then J attains its minimum: there is a control α_* for which there is a solution $[R_*, C_*]$ on $[0,T)$ and such that J attains its infimum.*

PROOF Take a minimizing sequence $\{\alpha_k\}$ for J and, noting the coercivity of J so this sequence is bounded in $L^2(0,T)$, extract a weakly convergent subsequence. Then apply Direct Theorem 4 to obtain α_* and $[R_*, u_*]$ on[6] $[0,T)$. Consider each C_k extended as 0 for $0 \le r < R_k(t)$ and similarly for C_*. On each $\tilde{Q}_\tau$ $(\tau < T)$, we have L^2 convergence $C_k \to C_*$ so, extracting a subsequence, we may assume $C_k \to C_*$ pointwise ae on $\tilde{Q}_T$. Since $0 \le C_k \le 1$ ae, the Dominated Convergence Theorem ensures suitable convergence for the first term of (3.1). The second term is $\|\alpha\|^2$ and so is weakly lower semicontinuous. Hence, $J(\alpha_*) \le \lim J(\alpha_k) := \inf J$. □

A similar result holds when one uses boundary control (except that coercivity is 'built in' by the assumption: $0 \le \gamma \le 1$ which we have needed.) Assuming differentiability of F, H with respect to ϑ, it is tedious but not difficult to compute (formally!) the derivative of the solution with respect to the control for either of the possibilities. This should be justifiable and lead to regularity rsults for the optimal controls (e.g., along the lines of Barbu's work) but this is conjectural.

Finally, we note some problems for which, as yet, no results are available. The particular model originally treated in [2] took H in (1.4) to have the special form:

$$H(R, \omega) = K[\omega - G(R)] \tag{3.2}$$

with K a positive constant and $G(\cdot)$ an empirically determined positive (smooth) function on $(0, L]$. Assuming this form, one could seek to identify the function $G(\cdot)$ (over an appropriate range) from observations of $R(\cdot)$. With exact observation (and assuming K and the data $C_0, \gamma(\cdot)$ are known) we could solve the uncoupled equation (2.3) for u to obtain $\omega = u(\cdot, 1)$ and then use the relation: $G(R) = \omega - \dot{R}/K$. The problem is to obtain a useable result if, say, K, γ are unknown — or to find a control $\gamma(\cdot)$ which makes the determination of $G(\cdot)$ most reliable over the largest range.

Another class of problems is related to stabilizing the solution near a steady state. Some consideration of stability is provided in [2] for (3.2) and $\gamma = const.$ but we envision, here, the construction of a stabilizing feedback in a more general context, say, by linearizing the problem around the steady state. It would be of particular interest to seek a feedback of the form:

$$\gamma := \{c_1 \text{ if } R > R_0; \quad c_2 \text{ if } R < R_0; \quad \text{in } [c_1, c_2] \text{ for } R = R_0\}$$

[6]It is conceivable that we have $R_* \to 0$ or $R_* \to 1$ as $t \to T$ but only in such a way might $[R_*, C_*]$ fail to be a solution on the closed interval $[0, T]$: one cannot have infinite oscillation of R_* since H has been assumed uniformly bounded for compact subintervals of $(0, 1)$. If one *does* have $R_* \to 0$, the equation (1.1) continues to hold in $(0, 1)$ at $t = T$.

where $0 \leq c_1 < c < c_2 < 1$. Here, the steady state solution considered has $\gamma \equiv c$ and $R \equiv R_0$. Still more interesting would be a feedback of thermostat type: γ switches from c_1 to c_2 when R moves 'left' of some prescribed $R_- < R_0$ and switches from c_2 to c_1 when R moves right of some $R_+ > R_0$; compare [5].

References

[1] J.P. Aubin, Un théorème de compacité, CRAS de Paris **265**, pp. 5042-5045 (1963).

[2] F. Conrad and M. Cournil, Free boundary problems in dissolution-growth processes, preprint (1987).

[3] F. Conrad, D. Hilhorst, and T.I. Seidman, Well-posedness of the free boundary problem for a dissolution-growth process, to appear.

[4] J.L. Lions and E. Magenes, Non-Homogeneous Boundary Value Problems and Applications, vol. II, Springer-Verlag, Berlin (1972).

[5] T.I. Seidman, Switching systems, I: general theory, to appear.

DIFFERENTIATION ON A LIPSCHITZ MANIFOLD

Jacques Simon
Analyse Numérique, Tour 55-65, 5e étage
Université Pierre et Marie Curie
4 Place Jussieu, 75252 PARIS CEDEX 05

INTRODUCTION

Let Γ be a manifold of $\mathbb{R}^N$ and let v be a vector field defined in all of $\mathbb{R}^N$ representing the variations of the manifold. The transported manifold is

$$\Gamma + v = \{x + v(x) : x \in \Gamma\}$$

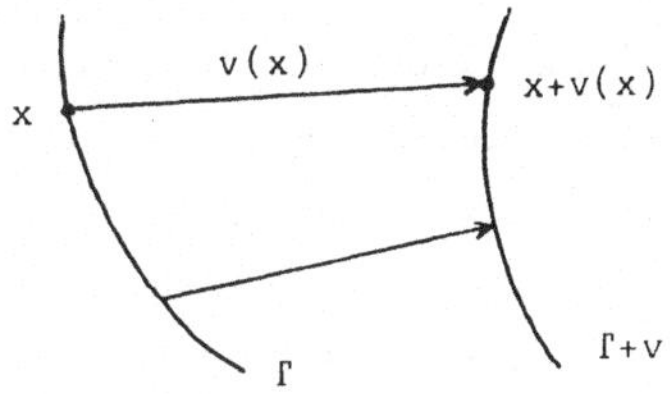

A function g being given in all of $\mathbb{R}^N$, we are interested in the variations of $\int_{\Gamma+v} g\, ds$ with respect to the variations of Γ.

If all the data are smooth enough the following expansion holds, see Simon J. (1980),

$$\int_{\Gamma+v} g\, ds = \int_\Gamma g\, ds + \int_\Gamma v_n\left(Hg + \frac{\partial g}{\partial n}\right) ds + O(v) \tag{1}$$

where v_n is the normal component of v,

H is the mean curvature of Γ,

$\partial g/\partial n$ is the normal derivative,

$O(v)/\|v\| \to 0$ as $\|v\| \to 0$ for some convenient norm.

If Γ is a C^2 manifold, H is continuous and this expansion is allowed for smooth v and g.

What happens if Γ has a "corner point" M ?

In dimension N = 2 the curvature is proportional to the derivative of the slope. At M the slope jumps, thus H(M) is a Dirac mass at M. Then the integral of $v_n Hg$ must be replaced by :

$$\int_\Gamma v_n\, Hg\, ds \qquad \int_{\Gamma - \{M\}} v_n\, Hg\, ds + \text{singularity at } M.$$

Such singularity calculus are done by Zolesio J.P. (1984).

We intend to prove that there is no singularity, and that

$$\int_{\Gamma+v} g\, ds = \int_\Gamma g\, ds + \int_\Gamma \left(\nabla_\Gamma \cdot (gv) + v_n \frac{\partial g}{\partial n}\right) ds + O(v) \tag{2}$$

where ∇_Γ is the "derivative along Γ".

The operator ∇_Γ is of common use in differential geometry for C^1 manifolds. Here ∇_Γ will be defined for a Lipschitz manifold Γ, and the expansion (2) will be proved for such a manifold. In a Lipschitz manifold a corner is no more a singular point, then it don't needs special calculus and it don't involves singularity.

Further developments

1. For a C^2 manifold a Stokes theorem on Γ imply that

$$\int_\Gamma \nabla_\Gamma \cdot (gv)ds = \int_\Gamma v_n \, Hg \, ds$$

so that one finds again the expansion (1).

2. Similar calculus of variations may be done for functionals which are defined via boundary value problems. For example the energy of an electric capacitor may be expanded with respect to the variations of a Lipschitz boundary, without singularity at the possible "corners", as in the smooth case (Simon J. 1980).

Outlines

The proofs requiring a full book, here are given only the precise definitions and statements.

1. Lipschitz manifolds.
2. Integration on Γ and normal vector.
3. Differentiation on Γ.
4. Divergence, curvature, Stokes formula and traces on Γ.
5. Variations of the integral on Γ.

1. LIPSCHITZ MANIFOLDS

For $x = (x_1,\dots,x_N) \in \mathbb{R}^N$ denote $x_* = (x_1,\dots,x_{N-1}) \in \mathbb{R}^{N-1}$.

Definition. A subset $\Gamma \subset \mathbb{R}^N$ is a Lipschitz manifold if there exists $a>0$, $b\geq 0$ such that :

$\forall z \in \Gamma$, there exists Euclidian coordinates $(x_1,\dots,x_N)$ and a function ψ defined on a subset $\mathcal{O}_* \subset \mathbb{R}^{N-1}$ such that

$$\mathcal{O}_* = \{x_* \in \mathbb{R}^{N-1} : |x_* - z_*| < a\}$$

$$|\psi(x_*) - \psi(y_*)| \leq b|x_* - y_*| \qquad \forall x_*, y_* \in \mathcal{O}_*$$

$$\begin{cases} x_N = \psi(x_*) \Rightarrow x \in \Gamma \\ x_N \neq \psi(x_*),\ |x_N - \psi(x_*)| < a \Rightarrow x \notin \Gamma . \ \blacksquare \end{cases}$$

Consider the subset $\Lambda \subset \Gamma$ defined by

$$\Lambda = \{x \in \mathbb{R}^N : x_* \in \mathcal{O}_*,\ x_N = \psi(x_*)\}.$$

Then a one to one map $\mathcal{T}$ from $\mathcal{O}_*$ onto Λ is defined by

$$\mathcal{T}(x_*) = (x_*, \psi(x_*))$$

In all the following Γ is a Lipschitz manifold. It may be represented as follows .

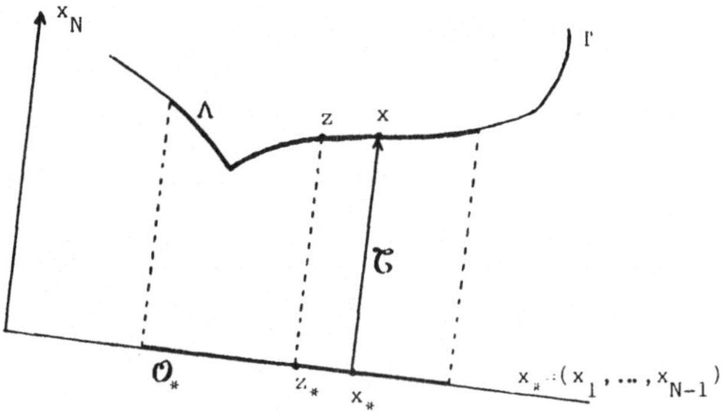

2. INTEGRATION ON Γ AND NORMAL VECTOR

Let f be a continuous function on Γ.

Then $f \circ \mathcal{T}$ is continuous on $\mathcal{O}_*$ and

$$\int_\Lambda f \, ds = \int_{\mathcal{O}_*} (f \circ \mathcal{T}) \lambda \, dx_*$$

where the area function $\lambda \in L^\infty(\mathcal{O}_*)$ is defined by

$$\lambda = (|\nabla_* \psi|^2 + 1)^{1/2} = \left(\left|\frac{\partial \psi}{\partial x_1}\right|^2 + \ldots + \left|\frac{\partial \psi}{\partial x_{N-1}}\right|^2 + 1 \right)^{1/2} .$$

Using a locally finite family of maps, and a corresponding partition of unity on Γ in function β_k, one defines

$$\int_\Gamma f \, ds = \sum_k \int_{\Lambda_k} \beta_k f \, ds .$$

Definition. $L^1(\Gamma)$ is the completion of the space $C(\Gamma)$ equipped with the norm $\int_\Gamma |f| ds$. ∎

By linear continuous extension, one defines

- the integral $\int_\Gamma f \, ds$ for $f \in L^1(\Gamma)$,
- the map $f \to f \circ \mathcal{T}$ which is one to one from $L^1(\Lambda)$ onto $L^1(\mathcal{O}_*)$.

Moreover $L^\infty(\Gamma)$ is defined as a subset of $L^1(\Gamma)$ if Γ is bounded (and of $L^1_{loc}(\Gamma)$ in the general case).

The normal vector n is defined by the

THEOREM. There exists a unique $n \in L^\infty(\Gamma;\mathbb{R}^N)$ such that, for every map,

$$n = \frac{\xi}{|\xi|} \quad \text{in } \Lambda,$$

where $\xi \in L^\infty(\Lambda;\mathbb{R}^N)$ is defined by

$$\xi = (\nabla_* \psi, -1) \circ \mathcal{T}^{-1} = \left(\frac{\partial \psi}{\partial x_1}, \ldots, \frac{\partial \psi}{\partial x_{N-1}}, -1 \right) \circ \mathcal{T}^{-1} . \; ∎$$

Let us recall that every Lipschitz function on an open set has distribution derivatives that belong to L^∞. Since ψ is Lipschitz in $\mathcal{O}_* \subset \mathbb{R}^{n-1}$, it has a gradient

$$\nabla_* \psi = \left(\frac{\partial\psi}{\partial x_1}, \ldots, \frac{\partial\psi}{\partial x_{N-1}}\right) \in L^\infty(\mathcal{O}_* ; \mathbb{R}^{N-1}).$$

3. DIFFERENTIATION ON Γ

Let $f \in L^1(\Gamma)$ be such that, for every map $\mathcal{T}$, $f \circ \mathcal{T} \in W^{1,1}(\mathcal{O}_*)$. By definition this is $f \in W^{1,1}(\Gamma)$.

Then $\nabla_*(f \circ \mathcal{T}) \in L^1(\mathcal{O}_* ; \mathbb{R}^{N-1})$ and $D_* f$ is defined on Λ by

$$D_* f = \left(\nabla_*(f \circ \mathcal{T}), 0\right) \circ \mathcal{T}^{-1} \in L^1(\Lambda; \mathbb{R}^N).$$

The derivative $\nabla_\Gamma f$ is defined on Λ by

$$\nabla_\Gamma f = D_* f - n(n \cdot D_* f)$$

where $\cdot$ is the Euclidian product.

THEOREM. There exists a unique $\nabla_\Gamma f \in L^1(\Gamma; \mathbb{R}^N)$ which satisfies the above definitions for every map $\mathcal{T}$. ∎

This operator ∇_Γ being defined on a Lipschitz manifold, it has no singularity at the possible corners.

Denote that $\nabla_\Gamma f$ has "tangential values". More precisely

THEOREM. $n \cdot \nabla_\Gamma f = 0$. ∎

Indeed on every Λ, $n \cdot \nabla_\Gamma f = n \cdot D_* f - (n \cdot n)(n \cdot D_* f)$.

Comparison with the usual derivative of differential geometry.

Let Γ be a C^1 manifold and $f \in C^1(\Gamma)$.

The present definition gives $\nabla_\Gamma f \in C(\Gamma; \mathbb{R}^N)$.

In differential geometry one defines the derivative at any point $z \in \Gamma$, $(Df)(z)$, which is a linear function on the tangent plane T_z to Γ at z, see Spivak M. (1974).

They are related by

$$(Df)(x)(\varphi) = (\nabla_\Gamma f)(z) \cdot \varphi \quad \forall \varphi \in T_z$$

which means that $(\nabla_\Gamma f)(z)$ is the vector in T_z which defines the derivative $(Df)(z)$.

The important point is the method used for the definition : In differential geometry $(Df)(z)$ is defined by mapping on the tangent plane T_z which varies with z. This is not easy to extend to Lipschitz case since then T_z is not defined for all z.

In the present method $(\nabla_\Gamma f)(z)$ is defined by mapping on an arbitrary plane $\mathcal{O}_*$ which is the same for all z. ∎

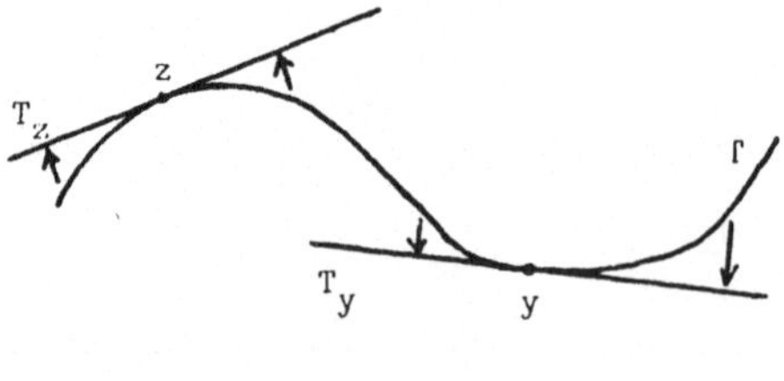

method of differential geometry
(for C^1 case)

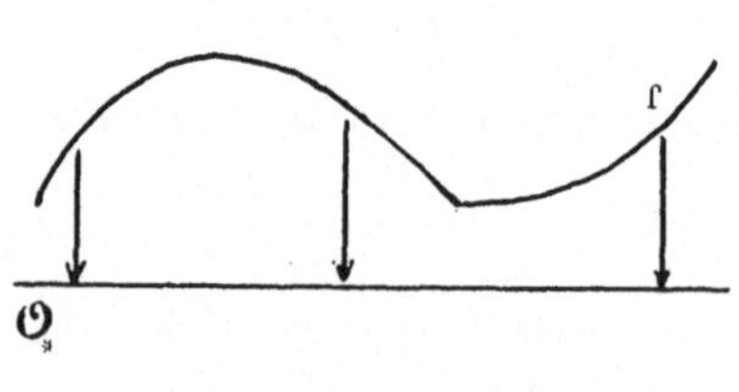

present method
(extends to Lipschitz case)

The operator ∇_Γ has the usual properties of differentiation operators ; for example

THEOREM. Let f be a Lipschitz function on Γ.
Then $f \in W^{1,1}(\Gamma)$, and $\nabla_\Gamma f \in L^\infty(\Gamma;\mathbb{R}^N)$. ■

4. DIVERGENCE, CURVATURE, STOKES FORMULA AND TRACES ON Γ.

Let us give some more definitions and properties, which are necessary for the application to the calculus of variations of an integral.

Divergence. For a vector valued function h on Γ, it is defined by

THEOREM. Let $h \in W^{1,1}(\Gamma;\mathbb{R}^N)$. Then there exists a unique $\nabla_\Gamma \cdot h \in L^1(\Gamma)$ such that, for every Euclidian coordinates

$$\nabla_\Gamma \cdot h = \sum_{i=1}^{N} (\nabla_\Gamma h_i)_i . \ ■$$

Curvature. A manifold Γ is said to be Lip^2 if it is a Lipschitz manifold and if the function ψ in the definition of Γ have Lipschitz derivatives with a uniform constant.

THEOREM. If Γ is Lip^2 then $n \in W^{1,\infty}(\Gamma;\mathbb{R}^N)$. It follows that $\nabla_\Gamma \cdot n \in L^1(\Gamma)$. ■

By definition, the mean (or Hicks) curvature is then

$$H = \nabla_\Gamma \cdot n$$

If $N = 2$ and Γ is C^2, then $H(z)$ is the usual curvature at a point $z \in \Gamma$.

Stokes formula. If h has "tangential values", then $\nabla_\Gamma \cdot h$ has a null integral :

THEOREM. Let Γ be Lip^2 and $h \in W^{1,1}(\Gamma;\mathbb{R}^N)$ be such that $h \cdot n = 0$. Then

$$\int_\Gamma \nabla_\Gamma \cdot h \, ds = 0. \ ■$$

Traces. Let now g be a function defined in all of $\mathbb{R}^N$ and Γ be a Lipschitz manifold. It's boundary value is given by :

THEOREM. There exists a unique linear map $g \to g|_\Gamma$ from $W^{1,1}(\mathbb{R}^N)$ in $L^1(\Gamma)$ which is continuous and such that $g|_\Gamma$ is the restriction of g to Γ when $g \in C(\Gamma)$. ■

If g is smooth its trace $g|_\Gamma$ has a derivative $\nabla_\Gamma g$ which is related to the trace of ∇g by the following formula .

THEOREM. Let $g \in W^{2,1}(\mathbb{R}^N)$. Then $g|_\Gamma \in W^{1,1}(\Gamma)$ and

$$\nabla_\Gamma g = \nabla g - n(n \cdot \nabla g) \quad \text{on } \Gamma. \ ■$$

5. VARIATIONS OF THE INTEGRAL ON Γ.

Now we are able to give a precise statement of the expansion of the integral $\int_{\Gamma+v} g\,ds$ with respect to v, in the case where Γ has some corners.

THEOREM. Let Γ be a Lipschitz manifold, v be a Lipschitz vector field from $\mathbb{R}^N$ into $\mathbb{R}^N$ with norm < 1, and $g \in W^{2,1}(\mathbb{R}^N)$.

Then $\Gamma+v$ is a Lipschitz manifold and

$$\int_{\Gamma+v} g\,ds = \int_\Gamma g\,ds + \int_\Gamma \left(\nabla_\Gamma \cdot (gv) + v_n \frac{\partial g}{\partial n}\right) ds + \mathcal{O}(v)$$

where $\frac{\mathcal{O}(v)}{\|v\|_{Lip}} \to 0$ as $\|v\|_{Lip} \to 0$. ∎

All the terms of the expansion are defined. Indeed

- $g|_\Gamma \in W^{1,1}(\Gamma)$, $v|_\Gamma \in W^{1,\infty}(\Gamma;\mathbb{R}^N)$ then $gv|_\Gamma \in W^{1,1}(\Gamma;\mathbb{R}^N)$ thus $\nabla_\Gamma \cdot (gv) \in L^1(\Gamma)$
- $v_n = v \cdot n \in L^\infty(\Gamma)$
- $\frac{\partial g}{\partial n} = n \cdot \nabla g|_\Gamma \in L^1(\Gamma)$ since $\nabla g \in W^{1,1}(\Gamma;\mathbb{R}^N)$

Remark. It is necessary to define the derivative along Γ for the term $\nabla_\Gamma \cdot (gv)$. Indeed this term cannot be defined to be the trace on Γ of $\nabla \cdot (gv)$ since this last term is in $L^\infty(\mathbb{R}^N)$ so that its trace is not defined.

In the case of a smooth boundary, the standard expansion follows by using Stokes formula:

COROLLARY. If moreover Γ is Lip^2, then

$$\int_\Gamma \nabla_\Gamma \cdot gv\,ds = \int_\Gamma H\,gv_n\,ds$$

thus

$$\int_{\Gamma+v} g\,ds = \int_\Gamma g\,ds + \int_\Gamma v_n \left(Hg + \frac{\partial g}{\partial n}\right) ds + \mathcal{O}(v). \ ∎$$

CONCLUSION

In a smooth manifold with corners, a corner is a singular point. In the more general family of Lipschitz manifolds, a corner is no more singular.

The differentiation operator ∇_Γ being defined for a Lipschitz manifold, it has no singularity at the possible corners of a manifold.

So, some calculus on smooth (or C^1) manifolds may be extended to the case of manifolds with corners.

This is done here for the expansion of an integral on Γ. Similar results may be done for the expansion of physical quantities defined via boundary value problems such as the drag of a body. More generally the differential calculus on manifolds with corners may be useful in a lot of problems.

It remains to give the full proves, which require a full book.

REFERENCES

MURAT F., SIMON J. (1976). Sur le contrôle par un domaine géométrique. Publ. of Paris 6 University.

SIMON J. (1980). Differentiation with respect to the domain in boundary value problems. Numer. Funct. Anal. and Optimiz. 2 (7 and 8), 649-687.

SPIVAK M. (1974). Differential geometry, Van Nostrand.

ZOLESIO J.P. (1984). Gradient des coûts gouvernés par des problèmes de Neumann posés sur des ouverts anguleux en optimisation de domaine. Publ. of Montréal University.

SHARP REGULARITY THEORY FOR SECOND ORDER HYPERBOLIC EQUATIONS OF NEUMANN TYPE

R. Triggiani
Department of Applied Mathematics, Thornton Hall, University of Virginia
Charlottesville, Virginia 22903, U.S.A.

1. Regularity problem, preliminaries, and statement of main results.

We report the results of joint work with I. Lasiecka to appear in full in the lengthy paper [L-T.5]. Let $x > 0$ be a scalar positive variable, t be a real variable, and $y = [y_1, \cdots, y_{n-1}]$ be an (n-1) -dimensional vector with real components. In symbols: $x \in R^1_{x+}$; $t \in R^1_t$; $y \in R^{n-1}_y$. Let

$$\Omega \equiv R^1_{x+} \times R^{n-1}_y, \quad \Gamma \equiv R^{n-1}_y = \Omega|_{x=0} \tag{1.1}$$

be, respectively, an n-dimensional half-space Ω with boundary Γ. On Ω we consider the second order differential operator

$$P(x,y; D_t,D_x,D_y) \equiv -aD_t^2 + \sum_{1,j=1}^{n-1} a_{ij} D_{y_i} D_{y_j} + 2 \sum_{j=1}^{n-1} a_{n_j} D_{y_j} D_x + D_x^2 \tag{1.2}$$

with space-dependent, but time-independent coefficients

$$a \equiv a(x,y), \; a_{ij} \equiv a_{ij}(x,y), \; i=1, \cdots, n; \; [x,y] \in \Omega; \quad j=1, \cdots, n-1 \tag{1.3}$$

satisfying the symmetricity condition $a_{ij} = a_{ji}$, $i,j = 1, \cdots, n-1$. Here and throughout we use the notation $D_t \equiv \frac{1}{\sqrt{-1}} \frac{\partial}{\partial t}$; $D_x \equiv \frac{1}{\sqrt{-1}} \frac{\partial}{\partial x}$, $D_{y_j} \equiv \frac{1}{\sqrt{-1}} \frac{\partial}{\partial y_j}$ etc.

On Γ, the boundary of the half-space Ω, we consider the first order operator

$$B(y;D_x,D_y) \equiv D_x + \sum_{j=1}^{n-1} b_j D_{y_j} \quad \text{on } x = 0 \tag{1.4}$$

with space-dependent, but time-independent coefficients

$$b_j \equiv b_j(y), \; y \in \Gamma \tag{1.5}$$

The present paper investigates regularity properties of the solution $u(t,x,y)$ of the following second order hyperbolic mixed problem with Neumann boundary conditions

$$\begin{cases} P(x,y;D_t,D_x,D_y)u = f(t,x,y) & \text{on } \Omega,\ t>0 \quad \text{(a)} \\ B(y;D_x,D_y)u = g(t,y) & \text{on } \Gamma,\ t>0 \quad (1.6)\ \text{(b)} \\ u|_{t=0} = u_0;\ D_t u|_{t=0} = u_1 & \text{on } \Omega,\ t=0 \quad \text{(c)} \end{cases}$$

at least for a few specific fundamental function spaces for f and g. Other classes of function spaces will be examined in a subsequent paper. Generally, we are interested in the continuity of the map from the data (u_0, u_1, f, g) in preassigned function spaces (possibly, subject to compatibility conditions) into the solution u, u_t, $\cdots$ and possibly its trace $u|_\Gamma$, $\cdots$ in suitable (optimal) function spaces. Throughout the paper, problem (1.6) will be subject to the following assumptions:

(i) the coefficients a, a_{ij}, a_{n_j} of P and b_j of B are assumed real, time independent, sufficiently smooth in the space variables, and constant outside a compact set $\mathcal{K}_{xy}$ of $R^1_{x+} \times R^{n-1}_y = \Omega$;

(ii) the boundary Γ (x=0) is non-characteristic for P and P is "regularly hyperbolic with respect to t", i.e. the characteristic polynomial of P,

$$p(x,y;\tau,\xi,\eta) \equiv -a\tau^2 + \sum_{i,j=1}^{n-1} a_{ij}\,\eta_i\,\eta_j + 2\xi \sum_{j=1}^{n-1} a_{n_j}\,\eta_j + \xi^2 \tag{1.7a}$$

$$\equiv -a\tau^2 + [\xi + \sum_{j=1}^{n-1} a_{n_j}\,\eta_j]^2 + \sum_{i,j=1}^{n-1} a_{ij}\,\eta_i\,\eta_j - \left[\sum_{j=1}^{n-1} a_{n_j}\,\eta_j\right]^2 \tag{1.7b}$$

has two real and distinct roots in τ, for $(x,y) \in \Omega$ and (ξ, η) on the unit sphere $\xi^2 + |\eta|^2 = 1$, where $|\eta|^2 = \sum_{j=1}^{n-1} \eta_j^2$. If we consider $\eta = 0$ and $\xi = 1$, this requirement yields the condition

$$\min a(x,y) > 0 \quad \text{in } \Omega; \tag{1.8}$$

moreover, if we consider the points of the unit sphere in (ξ, η) which lie also on the hyperplane $\xi + \sum_{j=1}^{n-1} a_{n_j}\,\eta_j = 0$, this requirement yields the necessary condition, which is plainly also sufficient, that the quadratic form in η

$$d(x,y;\eta) \equiv a^2(x,y)\left\{\sum_{i,j=1}^{n-1} a_{ij}(x,y)\eta_i\,\eta_j - \left[\sum_{j=1}^{n-1} a_{n_j}(x,y)\eta_j\right]^2\right\} \tag{1.9}$$

(independent of ξ) be positive definite

$$d(x,y;\eta) > 0 \quad (x,y) \in \Omega \quad |\eta|^2 \neq 0; \tag{1.10}$$

(iii) the first order operator $\tilde{D}_x$ defined by

$$\tilde{D}_x \equiv D_x + \sum_{j=1}^{n-1} a_{n_j}(x,y)\,D_{y_j} \tag{1.11}$$

restricted on the boundary Γ, coincides with B; i.e.

$$B \equiv \tilde{D}_x |_{x=0}; \text{ i.e. } b_j(y) \equiv a_{n_j}(0,y), \quad j = 1, \cdots n-1 \tag{1.12}$$

The following results are known and provide the a-priori regularity needed in the subsequent development.

Lemma 1.1 Let $u_0 = u_1 = 0$ in (1.6c) and let $0 < T < \infty$.

a) Let $g \equiv 0$ and $f \in L_1(0,T; L_2(\Omega))$ in (1.6). Then

$u \in C([0,T]; H^1(\Omega))$

$u_t \in C([0,T]; L_2(\Omega))$ (a fortiori $u \in H^1([0,T] \times \Omega)$) continuously.

b) Let $f \equiv 0$ and $g \in L_2(0,T; L_2(\Gamma))$ in (1.6). Then

$u \in C([0,T]; H^{1/2}(\Omega))$

$u_t \in C([0,T]; H^{-1/2}(\Omega))$ (a fortiori $u \in H^{1/2}([0,T] \times \Omega)$)

(Lions-Magenes vol. II, p. 120 provide only $L_2(0,T; \cdot)$; but this can be improved to $C([0,T]; \cdot)$ with the same space regularity, as e.g. in [L-T.2], [L-T.4].) □

Trace theory applied to Lemma 1.1a) then gives

$$\left.\begin{array}{l} f \in L_1(0,T; L_2(\Omega)) \\ g = 0 \\ u_0 = u_1 = 0 \end{array}\right\} \rightarrow u|_\Sigma \in C([0,T]; H^{1/2}(\Gamma)) \text{ continuously.} \tag{1.13}$$

A main goal of the present paper is to show the following

Main Theorem 1.2 Let $g = 0$ and $u_0 = u_1 = 0$ and let $f \in L_2(Q_+)$, $Q_+ = R^1_{t+} \times \Omega$. Then,

a) if $\Sigma_+ = R^1_{t+} \times \Gamma$, the trace $u|_\Sigma$ of the solution to (1.6) satisfies $u|_\Sigma \in H^{3/5}(\Sigma_+)$ continuously: there is a constant $C > 0$ independent of f such that

$$\| u|_\Sigma \|_{H^{3/5}(\Sigma_+)} < C \| f \|_{L_2(Q_+)}. \tag{1.14}$$

b) In the special cases where the coefficients a_{ij}, $i, j = 1, \cdots n-1$; a_{n_j}, $j=1, \cdots n-1$ either do not depend on x, or else do not depend on y, then $u|_\Sigma \in H^{2/3}(\Sigma_+)$ continuously: there is a constant $C > 0$ independent of f such that

$$\| u|_\Sigma \|_{H^{2/3}(\Sigma_+)} < C \| f \|_{L_2(Q_+)} \tag{1.15}$$

Remarks 1.1

(i) The general case (1.14) represents an improvement by "1/10" ($\frac{1}{2} + \frac{1}{10} = \frac{3}{5}$) in the space regularity of the trace over (1.13).

(ii) Addition of a first order differential operator to P does not affect the results. □

A second main result of this paper is the following

Main Theorem 1.3 Let $f = 0$, $u_0 = u_1 = 0$, and $g \in L_2(\Sigma_+)$. Then, continuously for any $\epsilon > 0$:

a)
$$\begin{cases} u \in H^{3/5-\varepsilon}(Q_+) \text{ (improvement by } \frac{1}{10} - \varepsilon \text{ over Lemma 1.1b)} & (1.16) \\ \text{AND} \\ u|_\Sigma \in H^{1/5-\varepsilon}(\Sigma_+) & (1.17) \end{cases}$$

b) In the special cases where the coefficients a_{ij}, a_{n_j} $i,j=1, \cdots n-1$ either do not depend on x, or else do not depend on y, then

$$\begin{cases} u \in H^{2/3}(Q_+) & (1.18) \\ \text{AND} \\ u|_\Sigma \in H^{1/3}(\Sigma_+) \quad \square & (1.19) \end{cases}$$

Remarks 1.2

(i) For dim $\Omega > 2$ and the Laplacian case one can show that $u \notin H^{3/4+\varepsilon}(Q)$, $\forall \varepsilon > 0$ [L-T.3] -

(ii) Result (1.17) is a regularity result. Trace theory applied to interior regularity (1.16) gives only $H^{3/5-\varepsilon-1/2=1/10-\varepsilon}(\Sigma)$, a result worse than (1.17) by $\frac{\text{"1"}}{10}$. Similarly, trace theory applied to (1.18) gives $H^{2/3-1/2=1/6}(\Sigma)$, a result worse than (1.19) by $\frac{\text{"1"}}{6}$.

(iii) The regularity in (1.18)-(1.19) coincides with that proved directly, by eigenfunction expansions, for the Laplacian Δ on a sphere $= \Omega$ [L-T.1]

(iv) Direct computations, by eigenfunction expansions, with the Laplacian on a parallelepiped Ω produced $u \in H^{3/4-\varepsilon}(Q)$, $\varepsilon > 0$ [L-T.1].

The proofs of Theorems 1.2 and 1.3 are very lengthy and technical and are to be found in [L-T.5].

Literature

[L.1] J. L. Lions, private communication, May 1984.

[L-M.1] J. L. Lions and E. Magenes, Nonhomogeneous boundary value problems and applications, Vols. I, II, Springer-Verlag, Berlin-Heidelberg, New York, 1972.

[L-T.1] I. Lasiecka and R. Triggiani, "A cosine operator approach to modelling $L_2(0,T; L_2(\Gamma))$ - boundary input hyperbolic equations," Applied Mathem. & Optimiz., 7, 35-93 (1981).

[L-T.2] I. Lasiecka and R. Triggiani, "Regularity of hyperbolic equations under $L_2(0,T; L_2(\Gamma))$ - boundary terms," Applied Mathem. & Optimiz., 10, 275-286 (1983).

[L-T.3] I. Lasiecka and R. Triggiani, "Trace regularity of the solutions of the wave equations with homogeneous Neumann boundary conditions." J.M.A.A., to appear.

[L-T.4] I. Lasiecka and R. Triggiani, "A lifting theorem for the time regularity of solutions to abstract equations with unbounded operators and applications to hyperbolic equations," Proc. Amer. Math. Soc., to appear.

[L-T.5] I. Lasiecka and R. Triggiani, Sharp regularity theory for second order hyperbolic equations of Neumann type, 1986.

[L-L-T.1] I. Lasiecka, J. L. Lions, and R. Triggiani, "Nonhomogeneous boundary value problems for second order hyperbolic operators," J. de Mathematiques Pures et Appliquees, 65, 149-192 (1986).

[M.1] S. Miyatake, "Mixed problems for hyperbolic equations of second order," J. Math. Kyoto University, 13, 435-487 (1973).

[S.1] R. Sakamoto, "Mixed problems for hyperbolic equatyions," I, II, J. Math.Kyoto University, Vol. 10-2, 343-373 (1970) and Vol. 10-3, 403-417 (1970).

[S.2] W. W. Symes, "A trace theorem for solutions of the wave equation and the remote determination of acoustic sources," Mathematical Methods in the Appled Sciences, 5, 131-152 (1983).

[T.1] R. Triggiani, "A cosine operator approach to modeling $L_2(0,T; L_2(\Gamma))$ - boundary input problems for hyperbolic systems," Proceedings 8th IFIP Conference on Optimization Techniques, University of Würzburg, West Germany 1977, Springer-Verlag, Lecture Notes CIS M6, 380-390 (1978).

CONTROL OF FOLDS

A. Trubuil, M. Seoane and J.P. Kernévez
U.T.C., B.P.233, 60206, Compiègne CEDEX, France

Abstract

We present optimal control problems of folds, Hopf bifurcations and isolas of solutions. Motivated by industrial or biological applications, they constitute a new class of optimization problems, where the goal is no more to optimize the state of a dynamical system, but rather to optimize the behavior of a whole family of states. We use penalty methods, not only for determining starting singular points, but also for minimizing cost functions.

1. Introduction

Let us first give three examples to motivate the study of such problems. The first one is exposed with details in the contribution of G.Joly in these proceedings [1]: the study of heat transfer through a wall from liquid sodium to liquid water and vapor steam shows the existence of folds and multiple steady states as the sodium temperature varies. The aim is to control the relative position of 2 folds. The second example also comes from an engineering situation and concerns the phenomenon of ferroresonance due to transformers in electrical networks, modeled by Duffing equation [2,3]. As one parameter μ varies, two kinds of solutions can coexist: periodic solutions with the same period T than a driving oscillator, and subharmonic solutions of period 3T. Indeed, together with a branch of solutions of period T, there is an isola of solutions of period 3T (Figure 1). Our aim is to "kill" this isola by acting on control parameters, and this will be achieved by minimizing $(\mu_1-\mu_2)^2$ until $\mu_1 = \mu_2$ [4]. Various other examples are biochemical systems [5,7,8], where interacting diffusion and reaction induce interesting behaviors, in particular the coexistence of multiple steady states and the existence of folds, Hopf bifurcations and isolas. For evident biological and medical reasons it is of outmost importance to be able to act upon folds and Hopf points, thus controlling families of solutions. As examples,we will restrict in this paper to such biochemical systems, namely enzyme systems [11].

2. Statement of the problem

2.1. Behavior of an enzyme system

As a specific example we consider the dynamical system

$$\begin{cases} s'(t) = s_0 - s(t) - \rho R(s(t),a(t)) \\ a'(t) = \alpha(a_0 - a(t)) - \rho R(s(t),a(t)) \\ R(s,a) = \dfrac{a}{k_1+a} \quad \dfrac{s}{1+s+k_2s^2} \end{cases} \qquad (2.1)$$

with steady states

$$\begin{cases} s_0 - s - \rho R(s,a) = 0 \\ \alpha(a_0 - a) - \rho R(s,a) = 0 \end{cases}$$

Here s and a are the concentrations of two substates S and A in a cell where an enzyme E catalyzes their reaction with a rate proportional to R(s,a).The cell is separated from an outside reservoir (concentrations s_0 and a_0) by a membrane of diffusion. This system admits both steady state and periodic solutions. Figure 2 represents the norm of these solutions as a_0 varies, the other parameters being fixed. This norm is

$$\begin{cases} (s^2+a^2) & \text{for steady states} \\ (\frac{1}{T}\int_0^T [s^2(t)+a^2(t)]dt)^{\frac{1}{2}} & \text{for a periodic solution of period T} \end{cases}$$

The heavy (resp. light) lines correspond to stable (resp.unstable) steady state solutions, the dashed line to stable periodic solutions.

For another value of the parameter s_0, the diagram of solutions may be like in Figure 3.

Thus we are in presence of a system with multiple behaviors: for given values of the parameters, it can be either oscillating with time or in one of several steady states. Two kinds of control problems arise. The first class is analogous to what is usually considered: one asks how to act on control parameters in order to minimize some cost fonction involving the state-control pair:

2.2. A "classical" optimal control problem

One would like to find periodic solutions with "largest amplitude".
The state equations are

$$\begin{cases} s' = T[s_0 - s - \rho R(s,a)] & t \in (0,1) \\ a' = T[\alpha(a_0 - a) - \rho R(s,a)] & \\ s(0) = s(1) & \\ a(0) = a(1) & \\ s(0) = \int_0^1 s(t)\, dt = s^* & \text{(anchor equation)} \end{cases} \tag{2.2}$$

The objective function to maximize is

$$\omega = \int_0^1 (s(t)-s^*)^2\, dt - T^2 - (a_0 - 500)^2. \tag{2.3}$$

The state u is $u = (s,a,T,s^*)$, the control is $\lambda = (s_0, a_0, \rho)$.

Now here are 3 examples of problems of the 2nd type:

2.3. Coalescence of 2 limit points

In order to avoid the coexistence of "low" and "high" solutions , one would like the two limit points of steady state solutions of (2.1) to coalesce. We adopt the notations:

$$F(\lambda,\mu,y) = \begin{cases} s_0 - s - \rho R(s,a) \\ \alpha(a_0 - a) - \rho R(s,a) \end{cases} \qquad \lambda = (s_0,\rho),\ y = (s,a), \mu = a_0$$

and denote F_y the Jacobian of F. At the limit points we have

$$\begin{cases} F(\lambda,\mu_i,y_i) = 0 \\ F_y(\lambda,\mu_i,y_i)v_i = 0 \\ \|v_i\|^2 = 1 \end{cases} \qquad i = 1 \text{ or } 2. \tag{2.4}$$

Thus the problem to coalesce the two limit points can be posed as:
"minimize $\omega = \frac{1}{2}(\mu_2-\mu_1)^2$, the"state" being $u = (\mu_1,y_1,v_1,\mu_2,y_2,v_2)$,
constrained by (2.4) and the control $\lambda = (s_0,\rho)$". So doing we control a whole family of solutions.

2.4. Coalescence of 2 Hopf bifurcation points

Each Hopf bifurcation point is characterized by the equations

$$\begin{cases} F(\lambda,\mu_i,y_i) = 0 \\ \frac{T_i}{2\pi}F_y\xi_i + \eta_i = 0 \\ -\frac{T_i}{2\pi}F_y\eta_i + \xi_i = 0 \\ \xi_i^*\xi_i + \eta_i^*\eta_i = 1 \end{cases} \qquad (i = 1,2) \tag{2.5}$$

$y,F,\xi_i,\eta_i \in R^n$ $(n = 2)$.
The "state" is $(\mu_1,y_1,T_1,\xi_1,\eta_1,\mu_2,y_2,T_2,\xi_2,\eta_2) = u$, the objective function is $\omega = \frac{1}{2}(\mu_1-\mu_2)^2$. The problem is to minimize ω under the constraints (2.5), thus controlling a whole family of solutions, since the coalescence of the 2 Hopf bifurcation points may give rise to the appearance of an isola of solutions emerging "above" the family of steady state solutions, which all become stable.

2.5. Disappearance of an isola

An example of an isola that one would like to kill arises in the modelling of morphogenesis by the diffusion-reaction model $F(\lambda,\mu,y) = 0$, where

$$F(\lambda,\mu,y) = \begin{cases} -s'' + \mu[R(s,a)-(s_0-s)] \\ -\beta a''+\mu[R(s,a)-\alpha(a_0-s)] \\ s'_0 = \varepsilon \\ s'(0) = s'(1) = a'(0) = a'(1) = 0 \\ s_0(0) = S \end{cases}, \tag{2.6}$$

$\lambda = (S,a_0,\rho,\varepsilon)$, $y = (s,a,s_0)$.
Here s, a, and s_0 are functions of the space variable x, $0 \leq x \leq 1$, and the reaction rate R is defined like in (2.1). We consider s_0 as an unknown function in order to deal with an autonomous system. The limit points which delimit an isola of solutions of (2.6) (like the one described in Figure 1) are defined like in Section (2.3) by equations (2.4), and the cost function is the same: $\omega = \frac{1}{2}(\mu_2-\mu_1)^2$.We wish to

"kill" the isola because we prefer to follow, as μ varies, the "main" branch of solutions and avoid catastrophes (i.e. jumps).

2.6. General framework

Both kinds of optimization problems fall into the same framework: to minimize $g(\lambda,u)$ under the constraint $f(\lambda,u) = 0$, where λ and u are respectively control and state variables, $f: X = \Lambda x U \mapsto Y$, where Λ, U and Y are Banach spaces and $g: X \mapsto R$.

Problem 1: $u = (y,T,s^*), y = (s,a)$, $\lambda = (s_0,a_0,\rho)$.

Problem 2: $u = (\mu_1,y_1,v_1,\mu_2,y_2,v_2)$, $\lambda = (s_0,\rho)$.

Problem 3: $u = (\mu_1,y_1,T_1,\xi_1,\eta_1,\mu_2,y_2,T_2,\xi_2,\eta_2)$, $\lambda = (s_0,\rho)$.

Problem 4: $u = (\mu_1,y_1,v_1,\mu_2,y_2,v_2)$, $\lambda = (S,a_0,\rho,\varepsilon)$, where $y_i = (s_i,a_i,s_{0i})$, $i = 1,2$ and v_i is the variable adjoint to y_i.

3. Existence of an optimal pair (control, state)

We make the following assumptions: X_{ad} is a closed subset of X, $A = \{x \in X_{ad} \mid f(x) = 0\}$ is a non empty, bounded, weakly closed subset of X, X is a reflexive Banach space and $g: A \mapsto \mathbb{R}$ is weakly lower semi continuous. Then there exists $x_0 \in A$ such that $g(x_0) = \min g$ on A.

4. Validity of the optimality system

If f and g are C^1, $x_0 \in A$ (i.e. $f(x_0) = 0$), $g(x_0) = \min g$ on A, $X = X_1 \oplus X_2$ where X_1 and X_2 are 2 closed complementary subspaces in X, $X_1 = N(DF(x_0))$ and $DF(x_0)_{/X_2}$ is a linear homeomorphism on Y, then there exists $p \in Y^*$ and $q \in R$ such that

$$\begin{cases} (Df)^*p+(Dg)^*g = 0 \\ \|p\|^2_{Y^*}+q^2 = 1 \end{cases}.$$

In case Λ (the parameter space) has finite dimension n_λ, the optimality system can be written

$$\begin{cases} f(\lambda,u) = 0 \\ \omega = g(\lambda,u) \\ (D_uf)^*p+(D_ug)^*q = 0 \\ \|p\|^2_{Y^*}+q^2 = \alpha_0 \\ (D_{\lambda i}f)^*p+(D_{\lambda i}g)^*q = \alpha_i \qquad 1 \leq i \leq n_\lambda \end{cases}$$

with $\alpha_0 \neq 0$ and $\alpha_i = 0$ at an optimal point. This is the basis for Doedel's algorithm of successive continuation[5,6,7,8], in which at the beginning $\alpha_0 = 0$, whence p , q and the α_i's are 0, whereas at termination $\alpha_0 = 1$ and the α_i 's are 0. Remark that the coefficients α_i 's are proportional to the gradient.

5. Obtaining starting points for optimization

A limit point with respect to μ is characterized by the equations

$$\begin{cases} F(\lambda,\mu,u) = 0 & \lambda \in R^m \quad \mu \in R \quad u \text{ and } F \in R^n \\ F_u\ v\ = 0 & v \in R^n \\ \|v\| = 1 \end{cases}$$

and can be found by minimizing

$$J(\lambda,\mu,u,v) = \sum_{i=1}^{n} \left\{ F_i^2(\lambda,\mu,u) + \Big(\sum_{j=1}^{n} \frac{\partial F_i}{\partial u_j} v_j \Big)^2 \right\} + (\|v\| - 1)^2$$

A Hopf bifurcation point with respect to μ is characterized by

$$\begin{cases} F(\lambda,\mu,u) = 0 & \lambda \in R^m,\ \mu \in R \text{ and } u,\ F \in R^n \\ \frac{T}{2\pi} F_u \xi + \eta = 0 & \xi \text{ and } \eta \in R^n \\ -\frac{T}{2\pi} F_u \eta + \xi = 0 \\ \xi * \xi + \eta * \eta = 1 \end{cases}$$

and can be found by minimizing

$$J(\lambda,\mu,u,T,\xi,\eta) = \sum_{i=1}^{n} \left\{ F_i^2(\lambda,\mu,u) + \Big(\frac{T}{2\pi} \sum_{j=1}^{n} \frac{\partial F_i}{\partial u_j} \xi_j + \eta_i \Big)^2 + \Big(-\frac{T}{2\pi} \sum_{j=1}^{n} \frac{\partial F_i}{\partial u_j} \eta_j + \xi_i \Big)^2 \right\} + (|\xi| + |\eta| - 1)^2$$

Remark that a limit point is also a global minimum of J.

6. Obtaining optimal control by penalty methods

6.1. Coalescence of 2 Hopf points

Equations (2.5) characterizing 2 Hopf bifurcation points can be rewritten $\boldsymbol{F}(\lambda,\mu_i,y_i,T_i,\xi_i,\eta_i) = 0$, $i = 1,2$ and we wish to minimize the objective function $\omega = \frac{1}{2}(\mu_2 - \mu_1)^2$. For a sequence $\varepsilon_k \to 0$, solve by optimization (2.5) , and $\mu_2 - \mu_1 = \varepsilon_k$ i.e. minimize (by the Levenberg-Marquardt method for example)

$$J = \sum_{i=1}^{2} \| \boldsymbol{F}(\lambda,\mu_i,y_i,T_i,\xi_i,\eta_i) \|^2 + (\mu_2 - \mu_1 - \varepsilon_k)^2$$

6.2. Maximum amplitude periodic solutions (problem 1)

The "maximum amplitude" problem(2.2),(2.3) where $u = (s,a,T,s^*)$ and $\lambda = (s_0,a_0,\rho)$, was solved by 2 methods: Doedel's successive continuation method and a penalty method, i.e.

(i) discretize(2.2), thus obtaining constraints $h_i(x) = 0$, $i = 1,\dots,p$

(ii) to solve min J(x) under these constraints, minimize

$$J_r(x) = J(x) + r\sum_{i=1}^{p} h_i^2(x) \text{ with } r \to \infty.$$

Other methods (based on continuation method and penalty method) are described by Poore [8].

6.3. Ordinary Differential Equations with boundary and integral constraints

Both continuation and penalty methods have been applied to the optimization of systems governed by O.D.E.s with boundary and integral constraints of the form

$$\begin{cases} u'(t) = f(u(t),\mu,\lambda) & t \in (0,1) \quad u \text{ ηand } f \in R^n \ \lambda \in R^l \ \mu \in R^m \\ b(u(0),u(1),\mu,\lambda) = 0 & b \in R^{n_b} \\ \int_0^1 q(u(t),\mu,\lambda)\, dt = 0 & q \in R^{n_q} \end{cases}$$

Examples of the successive continuation method can be found in (7,8) and examples of the penalization method are given in [9] (control of a wave speed in a reaction-diffusion system, control of a period doubling point,...).

6.4. Isolas (problem 4)

Equations (2.4) characterizing 2 limit points can be rewritten $F(\lambda,\mu_i,y_i,v_i) = 0$, $i = 1,2$ and we wish to minimize the objective function $\omega = \frac{1}{2}(\mu_2 - \mu_1)^2$. We tried a projected gradient method: after discretization of the differential equations we are faced to the problem to minimize g: $\mathbb{R}^{m+n} \to \mathbb{R}$ under the constraints $f(x) = 0$ where f: $\mathbb{R}^{m+n} \to \mathbb{R}^n$ (dimension m for the parameter and n for the state). The algorithm is the following: we suppose known a starting point $x^0 = (\lambda^0,u^0)$, $u^0 = (\mu_i^0,y_i^0,v_i^0)$ and in general, knowing x^k, we obtain x^{k+1} by the predictor-corrector scheme

(i) prediction : $x_p^{k+1} = x^k - h_k \nabla g(x^k)$

(ii) correction : $x^{k+1} = P(x_p^{k+1})$, where P is the projection on $f(y) = 0$. This projection $y = P(x)$ is defined by minimizing (by the Levenberg-Marquardt method) the sum of squares $S_\varepsilon(z) = \|x-z\|^2 + \frac{1}{\varepsilon}\|f(z)\|^2$ for values of ε tending to 0: $y = \lim_{\varepsilon \to 0} z_\varepsilon$ where $S_\varepsilon(z_\varepsilon) = \min_z S_\varepsilon(z)$.

This method, admittedly time consuming, has shown the feasibility of controlling isolas of solutions of boundary value problems.

7. Acknowledgements

The Authors would like to thank J.P. Yvon for presenting this paper at the Conference and for valuable suggestions.

9. **Figures**

Figure 1:

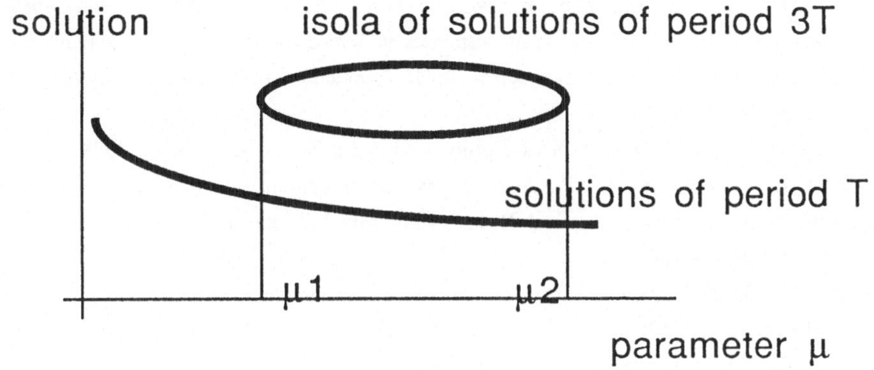

Figure 2:

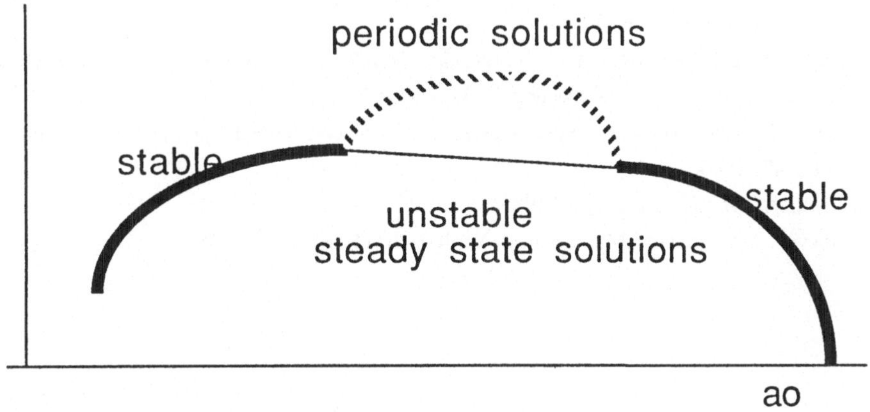

Figure 3:

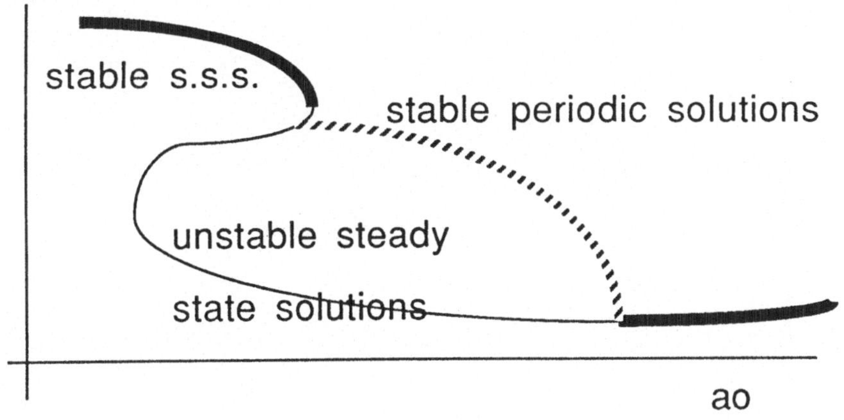

8. References

[1] Joly, G. et al, these proceedings

[2] Hayashi, C., Nonlinear oscillations in physical systems, New York: McGraw-Hill, 1964.

[3] Kawakami, H., Bifurcation of periodic responses in forced dynamic nonlinear circuit's computation of bifurcation values of the system parameters, IEEE Transactions on Circuits and Systems, Vol.CAS-31, N°3, p.248-260, 1984.

[4] Debraux, L., Analysis and control of Duffing's equation, Thesis, Compiegne, 1988.

[5] Doedel, E.J. and Kernevez, J.P., AUTO/software for continuation and bifurcation problems in ordinary differential equations, Technical Report, California Institute of Technology, 1986, 226 pages (includes the AUTO manual)

[6] Kernevez, J.P. and Doedel, E.J., Optimization in bifurcation problems using a continuation method, p.153-160 in: T.Kupper, R.Seydel, H.Troger, edrs, Bifurcation: Analysis, Algorithms, Applications, Birkhauser Verlag, Basel, Boston, 1987.

[7] Kernevez, J.P. and Doedel, E.J., Optimization in bifurcation problems, Part I:Theory and illustration, Proceedings of a meeting in Vorau(1986)

[8] Doedel, E.J. and Kernevez, J.P , Optimization in bifurcation problems, Part II:Numerical method and applications, Proceedings of a meeting in Vorau(1986)

[9] Poore, A.B., The expanded Lagrangian system for constrained optimization problems, ICASE Report n°86-47, 1986.

[10] Trubuil, A., Thesis, Compiegne, 1986.

[11] Kernévez, J.P., Enzyme Mathematics, North-Holland, 1980.

CONTROLLABILITY AND STABILIZATION OF TWO-DIMENSIONAL
ELASTIC VIBRATION WITH DYNAMICAL BOUNDARY CONTROL†

Y. C. You and E. B. Lee
Center for Control Science & Dynamical Systems and Dept. of Electrical Engineering
University of Minnesota

Abstract. Dynamical control from the boundary of a membrane with edge mass and control is mathematically modelled and then the mathematical model is reduced to an abstract evolutionary equation with principal differential operator being Friedrichs extension of the original dissipative and symmetric operator. Using spectral/eigenspace analysis it is established that this evolutionary system is approximately controllable and strongly stabilizable by linear boundary damping feedback control.

1. Introduction and the Mathematical Model

During the last decade there has been extensive research on stabilization and related areas (controllability and optimal control, etc.) for hyperbolic type systems with usual non-dynamical boundary control (e.g. Dirichlet type or Neumann type), cf. [1-5]. Recently hybrid systems with dynamical boundary control have attracted interest, with motivation being practical applications related to stabilization and active control of large elastic structures, cf. [6 -7].

In this paper, we report on our study of stabilization of two-dimensional membrane vibration on a rectangular bounded region with mass/force dynamics along one edge. It is our expectation that the key ideas as reported can be generalized to more general two-dimensional and even higher dimensional "hybrid" systems.

Consider a membrane/edge-mass system as diagrammed in Fig. 1. For simplicity, assume that the elastic membrane is stretched on the unit rectangle $\Omega = (0,1)\times(0,1)$, and has constant mass density $\rho = 1$ and constant tension $\tau = 1$. Denote the boundary Γ by

$$\Gamma = \bigcup_{i=0}^{3} \Gamma_i$$

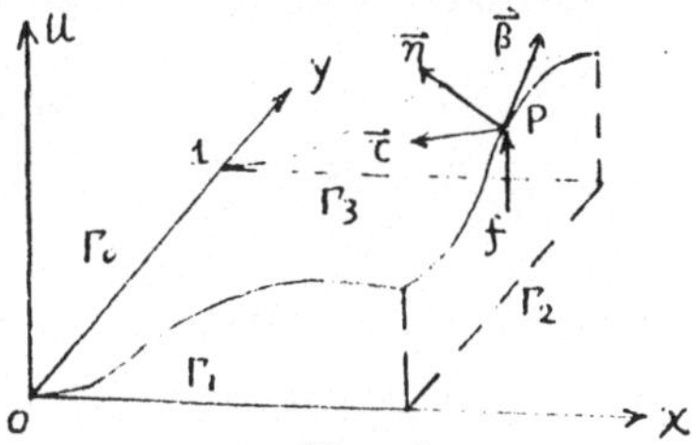

Fig. 1

as shown in Fig. 1. The vertical motion of the membrane can be assumed to satisfy

$$\frac{\partial^2 u}{\partial t^2} = \frac{\partial^2 u}{\partial x^2} + \frac{\partial^2 u}{\partial y^2}, \quad (x,y)\in\Omega,\ t > 0 \text{ (a linear two-dimensional wave equation).} \quad (1.1)$$

Besides, assume initial conditions $u(0,x,y) = g(x,y), u_t(0,x,y) = h(x,y), (x,y)\in\Omega$ (1.2) which will be specified later, and that:

1) on the edge Γ_0, $u(t,0,y) \equiv 0$, i.e. the membrane is clamped;

2) on the edge Γ_1 and Γ_3, $\frac{\partial u}{\partial y}(t,x,0) = \frac{\partial u}{\partial y}(t,x,1) \equiv 0$, i.e. these two membrane sides slide freely along the vertical "wall"; and

3) on the edge Γ_2, the membrane edge is subject to a vertical motion governed by a mass-force balance equation which is derived as follows.

Suppose the linear mass density is constant, $m = 1$, along the edge Γ_2. Let control force $f(t,y)$ be exerted vertically along this edge. Then for any point $P(u,x,y)$ on the edge, $\frac{\partial^2 u}{\partial t^2}(t,1,y) =$ "the tensile force" at $P + f(t,y)$. (1.3)

Denote by $\gamma_2{}^t = \{(u(t,1,y),1,y)\varepsilon R^3 : y\varepsilon[0,1]\}$, the actual configuration of the edge on Γ_2 at any time t. So $P\varepsilon\gamma_2{}^t$ is an arbitrary point. Let arc length parameter along $\gamma_2{}^t$ be denoted by $s>0$. Then parametrically $\gamma_2{}^t$: $u=u(s)$, $x = x(s) \equiv 1$, $y=y(s)$.

Physically it is known that the tensile vector $\vec{\tau}$ at P is perpendicular to the tangent vector $\vec{\beta}$ to $\gamma_2{}^t$ at P and lies in the tangent plane to the membrane at P (See Fig. 1). Since the normal vector $\vec{n}$ and the tangent vector $\vec{\beta}$ at P are given by

$$\vec{n} = \text{grad}_{(u,x,y)}\{u-u(t,x,y)\} = \begin{pmatrix} -\partial u/\partial x \\ -\partial u/\partial y \\ 1 \end{pmatrix}$$

and

$$\vec{\beta} = \begin{pmatrix} dx/ds \\ dy/ds \\ du/ds \end{pmatrix}, \quad \text{with } \|\vec{\beta}(s)\| = 1 \quad (s \text{ is the arc length}),$$

it follows that (noting that the tension coefficient $\tau = 1$)

$$\vec{\tau}(P) = \frac{\vec{n}\times\vec{\beta}}{\|\vec{n}\|} = \frac{1}{\sqrt{1+u_x^2+u_y^2}} \begin{vmatrix} \vec{i} & \vec{j} & \vec{k} \\ -\frac{\partial u}{\partial x} & -\frac{\partial u}{\partial y} & 1 \\ \frac{dx}{ds} & \frac{dy}{ds} & \frac{du}{ds} \end{vmatrix}, \tag{1.4}$$

and the u-directional component of $\vec{\tau}(P)$ is

$$\tau_u = \frac{1}{\sqrt{1+u_x^2+u_y^2}}\left[-\frac{\partial u}{\partial x}\frac{dy}{ds} + \frac{\partial u}{\partial y}\frac{dx}{ds}\right] = \frac{-1}{\sqrt{1+u_x^2+u_y^2}}\frac{\partial u}{\partial x}\frac{dy}{ds}$$
$$= -\frac{1}{\sqrt{1+u_x^2+u_y^2}}\cdot\frac{1}{\sqrt{1+u_y^2}}\frac{\partial u}{\partial x} \approx -\frac{\partial u}{\partial x}, \tag{1.5}$$

where we can write in first approximation since we will only consider deformations for which u_x^2 and u_y^2 are small and neglectible, as is usual.

Now insert (1.5) as tensile force term in (1.3), to obtain boundary equation,

$$\frac{\partial^2 u}{\partial t^2}(t,1,y) = -\frac{\partial u}{\partial x}(t,1,y) + f(t,y), \quad y\varepsilon(0,1),\ t > 0. \tag{1.6}$$

Therefore the mathematical model to be studied is

$$\frac{\partial^2 u}{\partial t^2} = \frac{\partial^2 u}{\partial x^2} + \frac{\partial^2 u}{\partial y^2}, \quad (x,y)\varepsilon\Omega,\ t > 0; \quad \frac{\partial^2 u}{\partial t^2}\Big|_{\Gamma_2} = -\frac{\partial u}{\partial x}\Big|_{\Gamma_2} + f(t,y), \quad y\varepsilon(0,1),\ t > 0;$$

$$u\big|_{\Gamma_0} = 0, \quad \frac{\partial u}{\partial y}\Big|_{\Gamma_1\cup\Gamma_3} = 0,\ t > 0; \quad u(0,x,y) = g(x,y),\ u_t(0,x,y) = h(x,y), (x,y)\varepsilon\Omega. \tag{1.7}$$

2. Abstract Formulation of the Control System

Set $H = L^2(\Omega)xL^2(0,1)$ with the usual inner product. Let

$$v(t) = \begin{pmatrix} u(t,x,y) \\ u_1(t,y) \end{pmatrix} \quad \text{where } u_1(t,y) = u(t,1,y), \tag{2.1}$$

which has two function components for each given $t \geq 0$. Define operators

$$A = \begin{bmatrix} \dfrac{\partial^2}{\partial x^2} + \dfrac{\partial^2}{\partial y^2} & 0 \\ -\dfrac{\partial u}{\partial x}\Big|_{x=1} & 0 \end{bmatrix} \quad \text{and} \quad B = \begin{pmatrix} 0 \\ I \end{pmatrix} \tag{2.2}$$

with the domain of A given by

$$\mathcal{D}(A) = \left\{ \begin{pmatrix} u(x,y) \\ u_1(y) \end{pmatrix} \epsilon H^2(\Omega)xL^2(0,1) \;\middle|\; \begin{array}{l} u(0,y) = 0, \ \dfrac{\partial u}{\partial y}(x,0) = \dfrac{\partial u}{\partial y}(x,1) = 0, \\ u(1,y) = u_1(y) \end{array} \right\} \tag{2.3}$$

so that A: $\mathcal{D}(A)(\subset H) \to H$ is a linear operator, and $I = I_{L^2(0,1)}$ is the identity operator on $L^2(0,1)$, so that $B \epsilon \mathcal{L}(L^2(0,1);\ H)$. Denote by $f(t) = f(t,y)$ for $0 \leq y \leq 1$.

Thus the original control system described by (1.7) can be written as follows,

$$\frac{d^2v(t)}{dt^2} = Av(t) + Bf(t), \quad v(0) = g \ \text{ and } \ v_t(0) = h, \tag{2.4}$$

where $g = \begin{pmatrix} g(x,y) \\ g(1,y) \end{pmatrix}$ and $h = \begin{pmatrix} h(x,y) \\ h(1,y) \end{pmatrix}$.

Before the further reformulation of the evolutionary system (2.4), we need to relate some properties of the unbounded operator A.

<u>Lemma 1</u> A: $\mathcal{D}(A) \to H$ is a coercively dissipative and symmetric operator.

Proof. As a first step, we claim that A is symmetric and $\mathcal{D}(A) \subset \mathcal{D}(A^*)$. In fact, note that $H^2(\Omega) \subset C(\Omega)$ since $2 > k + \frac{n}{2}$ with $k = 0$ and $n = 2$ here, $\mathcal{D}(A)$ is well-defined by (2.3). For any $\begin{pmatrix} u(x,y) \\ u_1(x,y) \end{pmatrix}$ and $\begin{pmatrix} w(x,y) \\ w_1(x,y) \end{pmatrix}$ in $\mathcal{D}(A)$, we have

$$\left\langle A \begin{pmatrix} u(x,y) \\ u_1(y) \end{pmatrix}, \begin{pmatrix} w(x,y) \\ w_1(y) \end{pmatrix} \right\rangle_H = \left\langle \begin{pmatrix} \Delta u \\ -\dfrac{\partial u}{\partial x}\Big|_{x=1} \end{pmatrix}, \begin{pmatrix} w(x,y) \\ w(1,y) \end{pmatrix} \right\rangle$$

$$= \int_\Omega \Delta u(x,y) \cdot W(x,y)dxdy - \int_0^1 \frac{\partial u}{\partial x}(1,y) \cdot W(1,y)dy \quad \text{(by the Green's formula)}$$

$$= -\int_\Omega \nabla u \cdot \nabla W dxdy + \int_\Gamma \frac{\partial}{\partial \nu} u(x,y) \cdot W(x,y)ds - \int_0^1 \frac{\partial u}{\partial x}(1,y) \cdot W(1,y)dy$$

(ds stands for curve element along boundary Γ, and $\frac{\partial}{\partial \nu}$ stands for normal derivative)

$$= -\int_\Omega \nabla u \cdot \nabla W dxdy \ , \tag{2.5}$$

where the last equality follows from $W|_{\Gamma_0} = 0$, $\frac{\partial}{\partial\nu}u|_{\Gamma_1\cup\Gamma_3} = 0$ and the cancellation of two terms of integration along Γ_2. Thus the claim is verified since similarly

$$<\begin{pmatrix} u(x,y) \\ u_1(y) \end{pmatrix}, A\begin{pmatrix} W(x,y) \\ W_1(y) \end{pmatrix}>_H = -\int_\Omega \nabla u\cdot\nabla W dxdy. \tag{2.6}$$

As a result, we have for any $\begin{pmatrix} u(x,y) \\ u_1(y) \end{pmatrix} \varepsilon \mathcal{D}(A)$,

$$<A\begin{pmatrix} u(x,y) \\ u_1(y) \end{pmatrix}, \begin{pmatrix} u(x,y) \\ u_1(y) \end{pmatrix}>_H = -\int_\Omega \|\mathrm{grad}u\|^2 dxdy. \tag{2.7}$$

By the Sobolev imbedding theorem and the trace theorem[8], $H^1(\Omega) \subsetneq L^2(\Omega)$, and $u\to\gamma_0 u$ is continuous from $H^1(\Omega)$ to $H^{1/2}(\Gamma)$, and $H^{1/2}(\Gamma) \subsetneq L^2(\Gamma)$. So there is $c>0$, such that

$$\left\|\begin{pmatrix} u(x,y) \\ u_1(y) \end{pmatrix}\right\|_H^2 = \|u(x,y)\|^2_{L^2(\Omega)} + \|u(1,y)\|^2_{L^2(0,1)} \le c\int_\Omega \|\mathrm{grad}u\|^2 dxdy. \tag{2.8}$$

A is coercively dissipative since (2.7)/(2.8) give a constant $\delta > 0$, such that

$$<A\begin{pmatrix} u(x,y) \\ u_1(y) \end{pmatrix}, \begin{pmatrix} u(x,y) \\ u_1(y) \end{pmatrix}>_H \le -\delta\left\|\begin{pmatrix} u(x,y) \\ u_1(y) \end{pmatrix}\right\|_H^2, \quad \forall\begin{pmatrix} u(x,y) \\ u_1(y) \end{pmatrix} \varepsilon \mathcal{D}(A). \tag{2.9}$$

Q.E.D.

Note that for any $\begin{pmatrix} u(x,y) \\ u_1(y) = u(1,y) \end{pmatrix} \varepsilon \mathcal{D}(A)$, we have $u\varepsilon H^2(\Omega)$ so that $-\frac{\partial u}{\partial\nu} \varepsilon H^{1/2}(\Gamma)$. Thus $-\frac{\partial u}{\partial x}| \varepsilon H^{1/2}(\Gamma_2) \subsetneqq L^2(0,1)$. It is hopeless to show that $\mathrm{Ran}(A) = H = L^2(\Omega)\times L^2(0,1)$. Therefore, A is merely symmetric but not selfadjoint in H.

In order to develop a corresponding theory in this case, we shall now construct the Friedrichs extension[9] of the operator A.

Lemma 2 The operator $A: \mathcal{D}(A)\to H$ admits a self-adjoint extension.

Proof. Since A is densely defined operator in H and by (2.9),

$$<-Au,u>_H \ge \delta\|u\|_H^2, \quad \forall u\varepsilon\mathcal{D}(A) \tag{2.10}$$

we can define a bilinear form

$$a(u,v) = <-Au,v>, \quad \forall u,v\varepsilon\mathcal{D}(A). \tag{2.11}$$

Actually a(u,v) is given by

$$a(u,v) = \int_\Omega <\mathrm{grad}u,\mathrm{grad}v>dxdy. \tag{2.12}$$

Define a set M by

$$M = \{u\varepsilon H: \exists \text{ a sequence } \{u_n\} \subset \mathcal{D}(A) \text{ such that } u_n\to u \text{ in H and } a(u_n-u_m,u_n-u_m)\to 0 \text{ as } m,n\to\infty\} \tag{2.13}$$

By the Schwarz inequality, it is clear that $M \supset \mathcal{D}(A)$ and M is a subspace of H.

For each $u\varepsilon M$, a sequence $\{u_n\} \subset \mathcal{D}(A)$ satisfying the statement in (2.13) will be called an admitting sequence for u. We can show that $\{a(u_n,u_n)\}$ is bounded for an admitting sequence. This follows from

$$|a(u_n,u_n)| \leq |a(u_n-u_{m_0},\ u_n-u_{m_0})| + |a(u_{m_0},u_{m_0})| + 2\|Au_{m_0}\|_H\|u_n\|_H \quad , \tag{2.14}$$

where we can choose m_0 fixed and sufficiently large such that the first two terms on the right side of (2.14) $\leq$ const. for $n \geq m_0$, then (2.14) implies $|a(u_n,u_n)| \leq$ const. Note that for $u \in M$ the admitting sequence is not unique.

If $u,v \in M$, then for any corresponding admitting sequences $\{u_n\}$ and $\{v_n\}$,

$$\alpha(u,v) = \lim_{n\to\infty} a(u_n,v_n) \tag{2.15}$$

exists and is independent of the specific admitting sequences. This is true because

$$|a(u_n,v_n)-a(u_m,v_m)| \leq |a(u_n,v_n-v_m)| + |a(u_n-u_m,v_m)|$$
$$\leq \sqrt{a(u_n,v_n)}\sqrt{a(v_n-v_m,v_n-v_m)} + \sqrt{a(u_n-u_m,u_n-u_m)}\sqrt{a(v_m,v_m)} \to 0 \quad \text{as } n,m\to\infty \tag{2.16}$$

and if $\{\tilde{u}_n\}$ and $\{\tilde{v}_n\}$ are also admitting sequences for u and v, then

$$|a(u_n,v_n)-a(\tilde{u}_n,\tilde{v}_n)| = |a(\Delta u_n,v_n) + a(\tilde{u}_n,\Delta v_n)|$$
$$\leq \sqrt{a(\Delta u_n,\Delta u_n)}\sqrt{a(v_n,v_n)} + \sqrt{a(\tilde{u}_n,\tilde{u}_n)}\sqrt{a(\Delta v_n,\Delta v_n)} \to 0 \quad \text{as } n\to\infty \tag{2.17}$$

where $\Delta u_n = u_n-\tilde{u}_n$ and $\Delta v_n = v_n-\tilde{v}_n$ satisfy

$$a(\Delta u_n,\Delta u_m,\Delta u_n,\Delta u_m) + a(\Delta v_n-\Delta v_m,\Delta v_n-\Delta v_m) \to 0,\ n,m\to\infty$$

and moreover $a(\Delta u_n,\Delta u_n)\to 0$, $a(\Delta v_n,\Delta v_n)\to 0$, $n\to\infty$. (2.18)

Hence $\alpha(u,v)$ is well-defined and is obviously a scalar product on M. Beside $\alpha(u,u) = 0$ if and only if $u = 0$. We can show that M equipped with this inner product $\langle u,v\rangle_M = \alpha(u,v)$ becomes a Hilbert space; the details are omitted.

Next we can define an operator $-\mathcal{A}$ by

$$-\mathcal{A}u = W \quad \text{and} \quad \mathcal{D}(\mathcal{A}) = \{u \in M\colon\ \alpha(u,v) = \langle W,v\rangle_H,\ \forall v \in M, \text{ for some } W \in H.\}. \tag{2.19}$$

It can be shown that $\mathcal{A}$ is an extension of the previous operator A. In fact, if $u \in \mathcal{D}(A)$, then $\alpha(u,v) = a(u,v) = \langle -Au,v\rangle_H$ for all $v \in \mathcal{D}(A)$ and also for all $v \in M$ by taking limits of admitting sequences. Thus $u \in \mathcal{D}(\mathcal{A})$ and $-\mathcal{A}u = -Au$, so $\mathcal{A}u = Au$.

It remains to show that: (1) Ran $(\mathcal{A}) = H$ and (2) $\mathcal{A}$ is symmetric. Then it follows that $\mathcal{A}$ is a self-adjoint operator in H.

The Lax-Milgram Lemma shows Ran $(\mathcal{A}) = H$. That $\mathcal{A}$ is symmetric follows from symmetry of extension form $\alpha(u,v)$ as a limit of $a(u_n,v_n)$, a symmetric form. Q.E.D.

Remark 1. One can show that A: $\mathcal{D}(A)\to H$ is not self-adjoint in H. If A is self-adjoint, then A must be a closed operator. Since $\mathrm{Ran}(A) \supset L^2(\Omega)\times H_0^{1/2}(0,1)$ is dense in H and A is coercively dissipative, the closedness will imply that Ran(A) = H; a contradiction. Therefore, by Lemma 2, A is closable but not a closed operator.

Auxiliary Lemma 1. ([9]) Let A be a semibounded symmetric operator and let D[a] be domain of closure of the form associated with A. If $\tilde{A} \supset A$ satisfies (i) $\tilde{A} = \tilde{A}^*$ and (ii) $\mathcal{D}(\tilde{A}) \subset D[a]$, then $\tilde{A}$ coincides with Friedrichs extension of A.

Lemma 3 The self-adjoint Friedrichs extension $\mathcal{A}$ of the operator A coincides with the smallest closed extension $\overline{A}$ of A, i.e. the closure operator of A.

Proof. Since $\overline{A}$ is a closed operator and A is symmetric, it follows that $\overline{A}$ is also symmetric. On the other hand, $\mathrm{Ran}(\overline{A})$ is dense in H and $-\langle\overline{A}u,u\rangle \geq \delta\|u\|_H$ together with the closedness of $\overline{A}$ imply that $\mathrm{Ran}(\overline{A}) = H$. Hence in turn, the symmetry and surjection of $\overline{A}$ imply that $\overline{A} = A^*$, i.e. $\overline{A}$ is self-adjoint.

Next, by the definition of $\alpha(u,v)$ in (2.15), we see that $D[\alpha] = M$ given by (2.13). By the definition of $\overline{A}$, it is easy to show that $\mathcal{D}(\overline{A}) \subset D[\alpha] = M$. Thus the Auxiliary Lemma can be applied to obtain the conclusion. Q.E.D.

<u>Lemma 4</u> $\mathcal{A}$ is invertible and $\mathcal{A}^{-1}: H \to H$ is compact operator.

Proof. Since $\langle -\mathcal{A}u,u\rangle = \alpha(u,u) \geq \delta\|u\|_H^2$ and $\mathrm{Ran}(\mathcal{A}) = H$, so $\mathcal{A}$ is invertible and $\mathcal{A}^{-1} \in \mathcal{L}(H)$. This also shows that

$$\|\mathcal{A}^{-1}W\|_M = \alpha(\mathcal{A}^{-1}W, \mathcal{A}^{-1}W)^{1/2} \leq \|W\|_H, \tag{2.20}$$

which indicates that $\mathcal{A}^{-1}$ maps any given bounded subset of H into a bounded subset of M. It remains only to show that $M \subset H$ is compact imbedding with respect to corresponding topologies.

By definition, for any $u = \begin{pmatrix} u(x,y) \\ u_1(y) \end{pmatrix} \in M$, there exists an admitting sequence

$u_n = \begin{pmatrix} u_n(x,y) \\ u_n(1,y) \end{pmatrix} \in \mathcal{D}(A) \subset H^2(\Omega)\times H^{3/2}(\Gamma_2) \hookrightarrow H^1(\Omega)\times H^{1/2}(\Gamma_2)$, such that

$$\begin{aligned} &u_n(x,y) \to u(x,y) \quad \text{in } L^2(\Omega), \\ &u_n(1,y) \to u_1(y) \quad \text{in } L^2(0,1), \\ &\|\nabla(u_n(x,y)-u(x,y))\|_{L^2(\Omega)} \to 0. \end{aligned} \tag{2.21}$$

Thus $u_n(x,y) \to u(x,y)$ in $H^1(\Omega)$ and by the trace theorem, $u_n(1,y) \to u_1(y)$ in $H^{1/2}(0,1)$, $u_1(y) = \gamma_0 u(x,y)|_{\Gamma_2}$, where γ_0 is the trace mapping of zero-order. Therefore, $M \subset H^1(\Omega)\times H^{1/2}(0,1)$. Note that $(M,\alpha(\cdot,\cdot))$ is a closed subspace of $H^1(\Omega)\times H^{1/2}(0,1)$. The latter is imbedded compactly into H, so that $M \subset H$ is compact imbedding. Q.E.D.

<u>Corollary 1</u> $\mathcal{A}$ is an infinitesimal generator of a self-adjoint contraction semigroup of bounded linear operators on H.

<u>Corollary 2</u> $\sigma(\mathcal{A}) = \sigma_p(\mathcal{A}) = \{\lambda_n\}_1^\infty \subset R^-$, each λ_n has finite multiplicity $\lambda_1 \geq \lambda_2 \geq \dots \geq \lambda_n \geq \dots$, with $\lim_{n\to\infty} \lambda_n = -\infty$ and without any finite accumulation point. Moreover, the complete system of normalized eigenvectors $\{\phi_n\}_1^\infty$ forms an orthonormal basis for H.

$$V \overset{\Delta}{=} \mathcal{D}(\sqrt{-\mathcal{A}}) \text{ with inner product } \langle v_1,v_2\rangle_V = \langle\sqrt{-\mathcal{A}}\, v_1, \sqrt{-\mathcal{A}}\, v_2\rangle_H \tag{2.22}$$

V is a Hilbert space. Define the state space to be $X = V\times H$ with the usual product topology. Then we can define operators

$$G = \begin{pmatrix} 0 & I \\ \mathcal{A} & 0 \end{pmatrix} : \mathcal{D}(G)(\subset X) \to X, \quad \mathcal{D}(G) = \mathcal{D}(\mathcal{A})\times V, \tag{2.23}$$

and

$$K = \begin{pmatrix} 0 \\ B \end{pmatrix} \in \mathcal{L}(L^2(0,1);X). \tag{2.24}$$

<u>Lemma 5</u> It holds that 1) G is a closed and densely defined operator in X, 2) G is skew-adjoint, i.e. $G = -G^*$, 3) G has compact resolvent $G^{-1}: X \to X$, 4) G is the generator of a Co-unitary group $T(t)$, $t \in R$, on X,

$$T(t) = \begin{pmatrix} \cos\sqrt{-\mathcal{A}}\,t) & \sqrt{-\mathcal{A}}^{-1} \sin(\sqrt{-\mathcal{A}}\,t) \\ -\sqrt{-\mathcal{A}} \sin(\sqrt{-\mathcal{A}}\,t) & \cos(\sqrt{-\mathcal{A}}\,t) \end{pmatrix}, \ t \in R, \text{ and} \tag{2.25}$$

5) $\sigma(G) = \sigma_p(G) = \{\pm j\mu_n,\ n=1,2,\ldots \mid -\mu^2{}_n\varepsilon\sigma_p(\mathcal{A})\}$, where $j = \sqrt{-1}$.

Proof. All these facts are consequences of the results shown by Lemma 1 through Lemma 4. The details are omitted (cf. [6],[7]). Q.E.D.

Now let

$$W(t) = \begin{pmatrix} v(t) \\ \dot{v}(t) \end{pmatrix} \text{ where } v(t) \text{ is defined by (2.1),} \tag{2.26}$$

and

$$W_0 = \begin{pmatrix} g \\ h \end{pmatrix} \text{ where } g = \begin{pmatrix} g(x,y) \\ g_1(y) \end{pmatrix} \text{ and } h = \begin{pmatrix} h(x,y) \\ h_1(y) \end{pmatrix} . \tag{2.27}$$

Then the second-order evolutionary system (2.4) can be reduced to first-order

$$\frac{dW(t)}{dt} = GW(t) + Kf(t), \quad W(0) = W_0\varepsilon X . \tag{2.28}$$

We shall consider the mild solution $W(t)$ of (2.28),

$$W(t) = T(t)W_0 + \int_0^t T(t-s)Kf(s)ds, \quad t > 0, \tag{2.29}$$

for $f(\cdot)\varepsilon L^1_{loc}(R^+;L^2(0,1))$, as the state function of original controlled system (1.7).

3. Stabilization via Boundary Feedback

In order to investigate stabilization of the reduced evolutionary system (2.8) or (2.9), we study relations between spectra of operator A and its extension $\mathcal{A}$.

Lemma 6 $\mathfrak{d}(A) = \sigma_p(A) = \sigma_p(\mathcal{A}) = \sigma(\mathcal{A})$, where $\mathfrak{d}(A)$ represents the core of the spectrum of A, which is the complement of the quasi-regular set $\mathfrak{p}(A)$ of A.

Proof. First we claim that $\mathcal{A} = A^*$, here A^* is the adjoint operator of A. In fact, by Lemma 3, $\bar{A} = \mathcal{A}$, here $\bar{A}$ is the closure operator of A. However, we know that $\bar{A} = A^{**}$. Since $\bar{A} = \mathcal{A}$ is self-adjoint, then $A^{**} = A^{***}$. On the other hand, we know that A^* must be a closed operator. Therefore, it follows that

$$A^* = \bar{A}^* = A^{***} = A^{**} = \bar{A} = \mathcal{A} . \tag{3.1}$$

Next we can use Theorem 2.16.5 in [10] (p. 56) to claim that

$$\mathfrak{p}(A) = \rho(A^*),\ \sigma_p(A) \subset \sigma_p(A^*) \cup \sigma_r(A^*),\ \sigma_c(A) \subset \sigma_r(A^*) \cup \sigma_c(A^*), \tag{3.2}$$

where $\sigma_p(\cdot)$, $\sigma_c(\cdot)$, $\sigma_r(\cdot)$ represent the point, continuous and residual spectrum respectively. Then (3.1) and (3.2) imply that

$$\mathfrak{d}(A) = \mathbb{C} \setminus \mathfrak{p}(A) = \mathbb{C} \setminus \rho(\mathcal{A}) = \sigma(\mathcal{A}), \tag{3.3}$$

and

$$\sigma_p(A) \subset \sigma_p(\mathcal{A}) \cup \sigma_r(\mathcal{A}) = \sigma_p(\mathcal{A}) \quad \text{(since Lemma 5 implies } \sigma_r(\mathcal{A}) = \phi), \tag{3.4}$$

and

$$\sigma_c(A) \subset \sigma_r(\mathcal{A}) \cup \sigma_c(\mathcal{A}) = \phi \qquad \text{(since Lemma 5 implies } \sigma_c(\mathcal{A}) = \phi). \tag{3.5}$$

Thus it remains only to show that $\sigma_r(A) = \phi$. (3.6)

It is easy to see that $\sigma_r(A) \subset \sigma_p(A^*) = \sigma_p(\mathcal{A})$. But on the other hand, we have

$Ran(\lambda I-\overline{A}) = Ran(\lambda I-\mathcal{A}) \subset \overline{Ran(\lambda I-A)}$

and $\overline{Ran(\lambda I-A)} \neq H$, hence it follows that $\sigma_r(A) \subset \sigma_r(\mathcal{A})$. Therefore,

$\sigma_r(A) \subset \sigma_p(\mathcal{A}) \cap \sigma_r(\mathcal{A}) = \phi.$

The conclusion follows from (3.3) through (3.6). Q.E.D.

Lemma 7 $\sigma(\mathcal{A}) = \sigma_p(\mathcal{A}) = \{-\lambda \varepsilon R^-: \lambda = n^2\pi^2 + \mu^2_{mn},\ n=0,1,\ldots,\ m=1,2,\ldots\}$ where $\{\mu_{mn}: m=1,2,\ldots\}$ are the increasing positive roots of the following equation:

$$\frac{\mu}{n^2\pi^2+\mu^2} = \tan\mu; \qquad n=0,1,2,\ldots \tag{3.7}$$

Proof: By Lemma 6, we only need to find the eigenvalues $\sigma_p(A)$ of the operator A. By the definition (2.2) and (2.3), and the property of finite multiplicity, we can use the method of separation of variables as follows.

Let $-\lambda\varepsilon\sigma_p(A)$. Then one has a nonzero vector $\begin{pmatrix} u(x,y) \\ u(1,y) \end{pmatrix} \varepsilon\ \mathcal{D}(A)$, such that

$$\left|\begin{array}{l} \Delta u = -\lambda u \\ \dfrac{\partial u}{\partial x}\Big|_{x=1} = \lambda u\Big|_{x=1}, \\ u(0,y) = 0,\ \dfrac{\partial u}{\partial y}(x,0) = \dfrac{\partial u}{\partial y}(x,1) = 0. \end{array}\right. \tag{3.8}$$

Seek a solution of the form $u(x,y) = \phi(x)\theta(y)$, then (3.8) becomes

$$\frac{\phi''_{xx}}{\phi} + \frac{\theta''_{yy}}{\theta} = -\lambda,$$

so that

$$\left|\begin{array}{l} \phi''_{xx} = \alpha\phi, \\ \phi(0) = 0, \\ \phi'(1) = \lambda\phi(1) \end{array}\right. \quad \text{and} \quad \left|\begin{array}{l} \theta''_{yy} = \beta\theta, \\ \theta'(0) = 0, \\ \theta'(1) = 0; \end{array}\right. \tag{3.9}$$

where α and β are constants satisfying $\alpha+\beta = -\lambda$.

It is easy to compute that $\beta = -\nu^2$ and $\theta(y) = c_1\sin(\nu y)+c_2\cos(\nu y)$ with $|c_1|+|c_2| \neq 0$. The boundary conditions give us $c_1 = 0$, $c_2 \neq 0$, and $\nu = n\pi$, $n = 0,1,\cdots$. Thus

$$\theta(y) = \cos(n\pi y), \quad \beta = -(n\pi)^2,\ n=0,1,2,\ldots \tag{3.10}$$

Then we can calculate that $\alpha = -\mu^2$ and $\phi(x) = c_3\sin(\mu x)+c_4\cos(\mu x)$ with $|c_3|+|c_4| \neq 0$. The boundary conditions give us $c_4 = 0$, $c_3 \neq 0$, and

$$\mu\cos\mu = \lambda\sin\mu = -(\alpha+\beta)\sin\mu = (\mu^2+n^2\pi^2)\sin\mu. \tag{3.11}$$

For any root μ of (3.11) such that $\mu > 0$, then $\cos\mu \neq 0$. Thus for each nonnegative integer n, (3.11) is equivalent to the transcendental equation (3.7). Therefore the eigenvalues of $\mathcal{A}$ and A are

$\{-\lambda = \alpha + \beta = -(n^2\pi^2 + \mu^2_{mn}): ;n=0,1,\ldots,\ m=1,2,\ldots\}$,

and the complete system of eigenvectors of A will be

$$\{u_{n,m} = \begin{pmatrix} u_{n,m}(x,y) \\ u_{n,m}(1,y) \end{pmatrix}: u_{n,m}(x,y) = \sin(\mu_{mn}x)\cos(n\pi y),\ n=0,1,\ldots,\ m=1,2,\ldots\} \tag{3.12}$$

Q.E.D.

Lemma 8 The complete system of eigenvectors of $\mathcal{A}$ is also given by (3.12).

Proof. For any given $\lambda_0 \varepsilon \sigma_p(A) = \sigma_p(\mathcal{A})$, since $\mathcal{A}$ is an extension of A, we have $N_{\lambda_0}(A) \subset N_{\lambda_0}(\bar{A}) = N_{\lambda_0}(\mathcal{A})$, here N_{λ_0} means the eigen-space corresponding to λ_0. By the above results, both dim $N_{\lambda_0}(A)$ and dim $N_{\lambda_0}(\mathcal{A})$ are finite.

Suppose $N_{\lambda_0}(A) \subsetneqq N_{\lambda_0}(\mathcal{A})$ is a proper inclusion. Then, since $N_{\lambda_0}(\mathcal{A})$ is a reducing subspace of $\mathcal{A}$ by the self-adjointness, it follows that

$$\overline{A \mid \mathfrak{D}(A) \cap N_{\lambda_0}(\bar{A})} = \bar{A} \mid N_{\lambda_0}(A) = \mathcal{A} \mid N_{\lambda_0}(\mathcal{A}). \tag{3.13}$$

However $\mathfrak{D}(A) \cap N_{\lambda_0}(\bar{A}) = N_{\lambda_0}(A)$, and $\mathrm{Ran}(\mathcal{A} \mid N_{\lambda_0}(\mathcal{A})) \subset \overline{\mathrm{Ran}(A \mid N_{\lambda_0}(A))} = \mathrm{Ran}(A \mid N_{\lambda_0}(A)) \subset \mathrm{Ran}(\mathcal{A} \mid N_{\lambda_0}(\mathcal{A}))$. Thus

$$N_{\lambda_0}(A) = \mathrm{Ran}(A \mid N_{\lambda_0}(A)) = \mathrm{Ran}(\mathcal{A} \mid N_{\lambda_0}(\mathcal{A})) = N_{\lambda_0}(\mathcal{A}). \tag{3.14}$$

The proper inclusion leads to a contradiction; so the conclusion holds. Q.E.D.

Denote then by $\{e_{n,m} = u_{n,m}/\|u_{n,m}\|_H: n=0,1,\dots; m=1,2,\dots\}$ the complete orthonormal basis for H, where $\{u_{n,m}: n=0,1,\dots; m=1,2,\dots\}$ is given by (3.12).

<u>Lemma 9</u> The evolutionary system (2.28) is strongly stabilizable by bounded linear feedback if and only if it is approximately controllable.

Proof. By the decomposition result[11] of Co contraction semigroups, there is an orthogonal decomposition with respect to T(t) as follows,

$$X = X_{cnu} \oplus W_u \oplus W^{\perp}, \tag{3.15}$$

where the subspaces X_{cnu}, W_u and $W^{\perp}$ reduces T(t) ($t>0$) to be completely nonunitary, unitary and weakly stable, and unitary but weakly unstable respectively. Since T(t) is unitary group and G admits a compact resolvent, one can prove that (cf.[6],[7])

$$X_{cnu} = \{0\} \quad \text{and} \quad W_u = \{0\}, \tag{3.16}$$

so that $X = W^{\perp}$. By Benchimol's Theorem[11], note the compact resolvent property of G, we can conclude that the system (2.28): $\{G,K\}$ is strongly stabilizable if and only if $X = W^{\perp} \subset C\{G,K\} \subset X$, i.e., $X = C\{G,K\}$. Q.E.D.

<u>Lemma 10</u> The evolutionary system (2.28) is approximately controllable if and only if any pair of sequences $\{\sqrt{\lambda_{n,m}}p_{n,m}\}_{\substack{n>0\\m>1}} \varepsilon \ell^2$ and $\{q_{n,m}\}_{\substack{n>0\\m>1}} \varepsilon \ell^2$ (where $\lambda_{n,m} = n^2\pi^2 + \mu^2_{mn}$, $n>0$, $m>1$) satisfying

$$\sum_{\substack{n>0\\m>1}} \{\sqrt{\lambda_{n,m}}p_{n,m}\sin(\sqrt{\lambda_{n,m}}t) + q_{n,m}\cos(\sqrt{\lambda_{n,m}}t)\} \frac{u_{n,m}(1,y)}{\|u_{n,m}\|_H} = 0, \qquad t > 0 \tag{3.17}$$

implies that $\{p_{n,m}\}_{\substack{n>0\\m>1}} = 0$ and $\{q_{n,m}\}_{\substack{n>0\\m>1}} = 0$. Moreover, if the above condition is satisfied, then (2.28) is strongly stabilizable by the linear feedback

$$f(t) = -K^*W(t), \quad t > 0. \tag{3.18}$$

Proof. That the system (2.28) is approximately controllable amounts to

$$C\{G,K\}^{\perp} = \bigcap_{t>0} N[K^*T^*(t)] = \{0\}, \tag{3.19}$$

where $N(\cdot)$ means the nullspace, and $T^*(t) = T(-t)$ since T(t) is a unitary group. If $\binom{p}{q} \varepsilon \bigcap_{t>0} N[K^*T^*(t)]$, then by direct calculation we have

$$0 = [K^*T^*(t)]\binom{p}{q} = (0,B^*)T(-t)\binom{p}{q} = B^*[\sqrt{-\mathcal{A}}\sin(\sqrt{-\mathcal{A}}\,t)p + \cos(\sqrt{-\mathcal{A}}\,t)q]$$

$$= (0,I) \sum_{\substack{n\geq 0\\ m\geq 1}} \{\sqrt{\lambda_{n,m}} \sin(\sqrt{\lambda_{n,m}}t)\langle p,e_{n,m}\rangle_H + \cos(\sqrt{\lambda_{n,m}}t)\langle q,e_{n,m}\rangle_H\} e_{n,m}$$

(denote then by $p_{n,m} = \langle p,e_{n,m}\rangle$ and $q_{n,m} = \langle q,e_{n,m}\rangle$)

$$= \sum_{\substack{n\geq 0\\ m\geq 1}} \{\sqrt{\lambda_{n,m}} \sin(\sqrt{\lambda_{n,m}}t)\, p_{n,m} + \cos(\sqrt{\lambda_{n,m}}t)\, q_{n,m}\} \frac{u_{n,m}(1,y)}{\|u_{n,m}\|_H}, \quad t > 0. \tag{3.20}$$

Then we see that (3.19) is equivalent to that (3.20) implies $\{p_{n,m}\}_{\substack{n\geq 0\\ m\geq 1}} = 0$ and $\{q_{n,m}\}_{\substack{n\geq 0\\ m\geq 1}} = 0$. Thus the first part of the Lemma is proved.

The second part of the Lemma concerning the stabilization by (3.18) follows directly from the first part and the Benchimal theorem in [11]. Q.E.D.

Auxiliary Lemma 2[12]. If a sequence of real numbers $\{\xi_m\}_1^\infty \subset R^+$ satisfies following asymptotic gap condition $\liminf_{m\to\infty} (\xi_{m+1}-\xi_m) > d \text{ (const)} > 0,$ (3.21)

and $T > 2\pi/d$, then for every L^2_{loc} almost periodic function

$$\psi(t) = \sum_{m=1}^{\infty} (a_m \sin(\xi_m t) + b_m \cos(\xi_m t)), \quad t > 0, \tag{3.22}$$

there exist two constants C_1 and C_2, such that $C_1 > 0$, $C_2 > 0$, and

$$C_1 \sum_{m=1}^{\infty} (|a_m|^2 + |b_m|^2) < \frac{1}{T} |\psi(t)|^2 dt < C_2 \sum_{m=1}^{\infty} (|a_m|^2 + |b_m|^2). \tag{3.23}$$

Lemma 11 Let $\{\lambda_{n,m} = n^2\pi^2 + \mu^2_{mn}\}_{m=1}^\infty$ be given in which $\mu_{mn} > 0$ $(m\geq 1)$ are the increasing positive roots of the equation (3.7) for $n\in\{0,1,\ldots\}$. Then,

$$\liminf_{m\to\infty} (\sqrt{\lambda_{n,m+1}} - \sqrt{\lambda_{n,m}}) > \frac{\pi}{16} > 0. \tag{3.24}$$

Proof. For each given n (integer) > 0, we have

$$\sqrt{\lambda_{n,m+1}} - \sqrt{\lambda_{n,m}} = \frac{\mu_{m+1,n} + \mu_{m,n}}{\sqrt{\lambda_{n,m+1}} + \sqrt{\lambda_{n,m}}} (\mu_{m+1,n} - \mu_{mn})$$

$$\text{(for large } m \text{ such that } n^2\pi^2 < \mu^2_{m+1,n}) > \frac{\mu_{m+1,n} + \mu_{mn}}{2\sqrt{2}\,\mu_{m+1,n}} (\mu_{m+1,n}-\mu_{mn}) > \frac{1}{4}(\mu_{m+1,n}-\mu_{mn}) \tag{3.25}$$

For sufficiently large μ, since $\frac{\mu}{n^2\pi^2+\mu^2} \approx \frac{1}{\mu}$, so for very large m, μ_{mn} satisfies

$$\frac{1}{\mu} + \varepsilon = \tan \mu, \tag{3.26}$$

where $|\varepsilon| = |\varepsilon(m)|$ can be arbitrarily small. Inspection of the graph shows that

$$\liminf_{m\to\infty} (\mu_{m+1,n} - \mu_{mn}) > \frac{\pi}{4} \tag{3.27}$$

Thus (3.24) holds. Q.E.D.

4. Main Results

Based on the above Lemmas, we can present the controllability and stabilization results of the evolutionary system (2.28) as follows.

Theorem 1 The evolutionary system (2.28) is approximately controllable.

Proof. Let a pair of sequences $\{\sqrt{\lambda_{n,m}}\, p_{n,m}\}_{\substack{n\geq 0\\ m\geq 1}} \in \ell^2$ and $\{q_{n,m}\}_{\substack{n\geq 0\\ m\geq 1}} \in \ell^2$ satisfy (3.17). It is enough to prove that $\{p_{n,m}\}_{\substack{n\geq 0\\ m\geq 1}} = 0$ and $\{q_{n,m}\}_{\substack{n\geq 0\\ m\geq 1}} = 0$ (cf. Lemma 10).

Now (3.17) can be written as follows,

$$\sum_{n=0}^{\infty}\sum_{m=1}^{\infty}\left\{\frac{\sin(\mu_{mn})}{\|\mu_{mn}\|_H}\left[\sqrt{\lambda_{n,m}}\, p_{n,m}\sin(\sqrt{\lambda_{n,m}}t)+q_{n,m}\cos(\sqrt{\lambda_{n,m}}t)\right]\right\}\cos(n\pi y) = 0,\ t>0. \quad (4.1)$$

Since $\{\cos(n\pi y)\}_{n=0}^{\infty}$ forms an orthogonal basis for the space $L^2(0,1)$, it follows that

$$\sum_{m=1}^{\infty}\left\{\frac{\sin(\mu_{mn})}{\|\mu_{mn}\|_H}\left[\sqrt{\lambda_{n,m}}\, p_{n,m}\sin(\sqrt{\lambda_{n,m}}t) + q_{n,m}\cos(\sqrt{\lambda_{n,m}}t)\right]\right\} = 0,\ t > 0 \quad (4.2)$$

By Lemma 11 and Auxiliary Lemma 2, it follows that for each given n=0,1,...

$$\sum_{m=1}^{\infty}\left|\frac{\sin(\mu_{mn})}{\|\mu_{mn}\|}\right|^2\left[\lambda_{n,m}\, p_{n,m}^2 + q_{n,m}^2\right] = 0,\ n=0,1,\ldots \quad (4.3)$$

From the characteristic equation (3.7), it is easy to see that $\tan(\mu_{m,n}) > 0$ for all $n\geq 0$ and $m\geq 1$. Hence we have $\frac{\sin(\mu_{mn})}{\|\mu_{mn}\|} \neq 0$. Then (4.3) implies that

$$p_{n,m} = q_{n,m} = 0,\ m = 1,2,\ldots;\ n = 0,1,\ldots \quad (4.4)$$

Therefore the conclusion holds. Q.E.D.

<u>Theorem 2</u> The evolutionary system (2.28) is strongly stabilizable by feedback

$$f(t,y) = -\frac{\partial u}{\partial t}(t,1,y),\quad y\in(0,1),\quad t > 0. \quad (4.5)$$

Proof. By Lemma 9 and Lemma 10, the system (2.28) is strongly stabilizable by

$$f(t) = -K^*W(t) = -(0,B^*)\begin{pmatrix} v(t)\\ \dot v(t)\end{pmatrix} = -B^*\dot v(t) = -(0,I)\begin{pmatrix}\dot u_t(t,x,y)\\ \dot u_t(t,1,y)\end{pmatrix} = -\frac{\partial u}{\partial t}(t,1,y),$$
$$y\in(0,1),\quad t > 0 \quad (4.6)$$

Here it can be seen that $f(t,\cdot)\in C([0,\infty);L^2(0,1))$. Q.E.D.

<u>Remark 2.</u> By the theory of abstract linear differential equations, the mild solution (2.29) of the reduced evolutionary system (2.28) with G involving the Friedrichs extension of A is actually the weak solution of (2.28) in following sense

$$\frac{d}{dt}\langle W(t),\xi\rangle \doteq \langle W(t),G^*\xi\rangle+\langle Kf(t),\xi\rangle,\ \lim_{t\to+0}\langle W(t),\xi\rangle = \langle W_0,\xi\rangle,\ \text{(with X-inner product)}, \quad (4.7)$$

provided $f\in L^1_{loc}(R^+;L^2(0,1))$, where $\xi\in \mathcal{D}(G^*) = \mathcal{D}(G)$ (cf. Lemma 5).

Furthermore, note that $w(t) = \begin{pmatrix} v(t)\\ \dot v(t)\end{pmatrix}$ and let $\xi = \begin{pmatrix}\phi\\ \psi\end{pmatrix}\in \mathcal{D}(G) = \mathcal{D}(\mathcal{A})\times V$, then we see that v(t) is actually the weak solution of the second-order evolutionary system (2.4) in following sense,

$$\frac{d^2}{dt^2}\langle v(t),\ \psi\rangle_H \doteq \langle v(t),\ A^*\psi\rangle_H + \langle Bf(t),\ \psi\rangle_H,$$

$$\lim_{t\to+0}\langle v(t),\ \psi\rangle = \langle y,\psi\rangle, \quad (4.8)$$

$$\lim_{t\to+0} \frac{d}{dt} \langle v(t), \psi\rangle = \langle h,\psi\rangle,$$

for all $\psi \varepsilon \mathcal{D}(\mathcal{A}^*) = \mathcal{D}(\mathcal{A}) = \mathcal{D}(A^*)$.

However, since A itself is merely closable but not a closed operator, even for $W_0 = \binom{g}{h} \varepsilon \mathcal{D}(A) \subsetneqq \mathcal{D}(\mathcal{A}) = \mathcal{D}(A^*)$, the mild solution may not be the strong solution of (2.4) with the feedback $f(t) = -B^*v(t)$ shown by (4.6).

Remark 3. (suggested by L. Littman) Let the initial values be

$u(0,x,y) \equiv g(x)$ and $u_t(0,x,y) \equiv h(x)$, $(x,y)\varepsilon\Omega$.

By the uniqueness of solution, (1.7) reduces to the one-dimensional hybrid elastic system as discussed in our previous work [6]. The exponential unstabilizability result shown there can be applied here to assert that the system (2.28) can never be exponentially stabilized by any bounded linear feedback.

† Research Support by NSF Grant Number 8607687.

References

[1] J. P. Quinn and D. L. Russell, Asymptotic stability and energy decay rates for solutions of hyperbolic equations with boundary damping, Proc. Roy. Soc. Edinburgh Sect. A,77, (1977), 97-127.

[2] G. Chen, Energy decay estimates and exact boundary control of the wave equation in a bounded domain, J. Math. Pures Appl., 58(9) (1979), 249-274.

[3] J. M. Ball and M. Slemrod, Feedback stabilization of distributed semilinear controls, Appl. Math. Optim., 5, (1979), 169-179.

[4] J. Lagnese, Decay of solutions of wave equations in a bounded region with boundary dissipation, J. Diff. Eqns, 50(2), (1983), 163-182.

[5] I. Lasiecka and R. Triggiani, Dirichlet boundary stabilization of the wave equation with damping feedback, J. Math. Anal. Appl., 97(1), (1983), 112-130.

[6] E. Bruce Lee and Yuncheng You, Stabilization of a hybrid (string/point-mass) system, Proc. 5th Int. Conf. on Systems Eng., Dayton, Sept. 1987, 109-112.

[7] W. Littman, L. Markus and Y.C. You, A note on stabilization and controllability of a hybrid elastic system with boundary control, Math. #103, 1987, U of MN.

[8] J. L. Lions, Problemes aux Limites dans les Équations aux Dérivées Partielles, Les Presse de L'Université de Montréal, 1965.

[9] M. S. Birman and M. Z. Solomjak, Spectral Theory of Self-Adjoint Operators in Hilbert Space, D. Reidel Publishing Company, 1987.

[10] E. Hille and R. S. Phillips, Functional Analysis and Semi-Groups, AMS Colloquium Publications, Vol. 31, 1948.

[11] C. D. Benchimol, A note on weak stabilizability of contraction semigroup, SIAM J. Control and Optimization, 16(3), (1978), 373-379.

[12] J. M. Ball and M. Slemrod, Non-harmonic Fourier series and the stabilization of distributed semi-linear systems, Comm. Pure and Appl. Math. 32 (1979), 555-587.

SHAPE DERIVATIVES AND SHAPE ACCELERATION

J.P. ZOLESIO
Laboratoire de Physique Mathématique
U.S.T.L. Place Eugène Bataillon
34060 MONTPELLIER Cedex FRANCE

The aim of this paper is to give an introduction to Shape Acceleration study. Ω is a domain in $\mathbb{R}^N$, $V = V(\Omega)$ is a vector field defined on Ω, in general V is the solution of a boundary value problem well posed in Ω. In our example we very briefly consider the Norton Hoff visco-plactic rehology. At any vector field $V \in C^0\left([0,\epsilon], C^k(\mathbb{R}^N)^N\right)$ is associated a Transformation T_t of $\mathbb{R}^N$ (which is the flow of V). Conversely if a one to one transformation T_t of $\mathbb{R}^N$ is given it is built by the field $V = \frac{\partial}{\partial t} T_t \circ T_t^{-1}$; for any x, consider $x(t,x) = x + \int_0^t V(s, x(s,x)\, ds$ then $T_t(x) = x(t,x)$.
The perturbed domain $\Omega_t = T_t(\Omega)$, with boundary $\Gamma_t = T_t(\Gamma)$. We defined [5], [6] shape derivatives of elements $y(\Omega)$ and $z(\Gamma)$ given on Ω and Γ. The shape acceleration is the situation when the function $y(\Omega)$ is the speed field V itself. The domain Ω is supposed bounded, lying on one side of its boundary Γ which a manifold C^k, $k \geqslant 1$.

I. SHAPE DERIVATIVES
II. SHAPE DERIVATIVES FOR NON SMOOTH DOMAIN
III. SHAPE ACCELERATION
IV. CHARACTERIZATION OF THE SHAPE ACCELERATION $V'(\Omega;V)$
V. NON SMOOTH DOMAINS, GENERALIZATION OF THE MEAN CURVATURE H AT A SINGULAR POINT

SHAPE DERIVATIVES

We shall make a distinction between the two following situations :

$$y(\Omega_t) \in W^{r,p}(\Omega_t) \text{ and } z(\Gamma_t) \in W^{s,p}(\Gamma_t)$$

in general r would be taken as $r = s + \frac{1}{p}$ and z as the trace on the boundary of y. In both cases we define the material derivatives as

$$\dot{y}_{r,p}(\Omega;V) = \frac{d}{dt}\,(y(\Omega_t) \circ T_t)_{t=0}\,, \quad \text{derivative in the } W^{r,p}(\Omega) \text{ norm}$$

$$\dot{z}_{s,p}(\Gamma;V) = \frac{d}{dt}\,(z(\Gamma(t) \circ T_t)_{t=0}\,, \quad \text{derivative in the } W^{s,p}(\Gamma) \text{ norm}$$

Obviously we have $(y|_\Gamma)_{s,p}^{\cdot}(\Gamma;V) = \dot{y}_{s+1/p,p}(\Omega;V)|_\Gamma$

Définition 1

$\Omega \mapsto y(\Omega)$ (resp. $\Gamma \mapsto z(\Gamma)$) is shape differentiable at Ω (resp. at Γ) iff $V \mapsto \dot{y}_{r,p}(\Omega;V)$ (resp. $\dot{z}_{s,p}(\Gamma;V)$) is a linear continuous mapping from

$\mathcal{V}^k_{ad}(\mathbb{R}^n;\mathbb{R}^n)$ in $W^{r,p}(\Omega)$ (resp. $W^{s,p}(\Gamma)$)

and the shape derivatives are given by

$$y'_{r-1,p}(\Omega;V) = \dot{y}_{r,p}(\Omega;V) - \nabla y(\Omega).V(0) \text{ element of } W^{r-1,p}(\Omega)$$

$$z'_{s-1,p}(\Gamma;V) = \dot{z}_{s,p}(\Gamma;V) - \nabla_\Gamma z(\Gamma).V(0). \text{ element of } W^{s,1,p}(\Gamma)$$

which are respectively elements of $W^{r-1,p}(\Omega)$ and $W^{s-1,p}(\Gamma)$.
We now recall some basic properties of these derivatives :

Proposition 1. If the field $V(t,x)$ is such that $V(t,x).n(x) = 0$ for $t \geq 0$ and $x \in \Gamma$, then, y and z being shape differentiable, we have

$$y_{r-1,p}(\Omega;V) = 0 \quad \text{and} \quad z'_{s-1,p}(\Omega;V) = 0$$

Proof. for any $t \geq 0$ we have $\Omega_t = \Omega$ then $y(\Omega_t) = y(\Omega)$ and $z(\Gamma_t) = z(\Gamma)$ then $\dot{y}(\Omega;V) = \nabla y(\Omega).V(0)$ then $Y'(\Omega;V) = 0$. Also we get $z'(\Gamma;V) = \nabla Z.V(0) - \nabla_\Gamma z(\Gamma).V(0)$ for any prolongation Z of z, $Z \in W^{s+1/p,p}(\Omega)$ such that $Z|_\Gamma = z$)

Then $z'(\Gamma;V) = \dfrac{\partial z}{\partial n} V(0).n = 0$. ∎

In studying the shape variation of the solution of boundary value problem we are concerned by the following shape derivatives : $n'(\Gamma;V)$ (n being the normal field on Γ), $H'(\Gamma;V)$ (H being the mean curvature of the manifold Γ), $\left(\dfrac{\partial y}{\partial n}\right)'(\Gamma;V)$ where $y = y(\Gamma;V)$ is an element of $W^{s+1\ p,p}(\Omega)$. We are also concerned by the characterization of shape derivatives and namely by expressions such as $\dfrac{\partial}{\partial n}(y'(\Omega;V))$, $\Delta_\Gamma(y'(\Gamma,V))$, ...

For this purpose we recall here, in a compact form, some basic results etablished in [5], [6], [7].

Derivatives of integrals, $y(\Omega)$ and $z(\Gamma)$ being shape differentiable we have

$$\frac{d}{dt}\left(\int_{\Omega_t} y(\Omega_t)\, dx\right)_{t=o} = \int_\Omega y'_{1,1}(\Omega;V)dx + \int_\Gamma y(\Omega_t)\; V(0).n\; d\Gamma$$

$$\frac{d}{dt}\left(\int_{\Gamma_t} z(\Gamma_t)\, d\Gamma_t\right)_{t=o} = \int_\Gamma z'_{1,1}(\Gamma;V)\; d\Gamma + \int_\Gamma (\nabla_\Gamma\; z(\Gamma).V(o) + Hz(\Gamma)V(o).n)d\Gamma$$

The tangential divergence on Γ of a vector field e defined on Γ is $div_\Gamma(e) = (div\ E - \langle DE.n,n\rangle)_\Gamma$ expression which is independant on the choice of E, smooth prolongation of e to a neigbourhood of Γ, e belonging to $W^{1,p}(\Gamma)^n$ and E belonging to $W^{1+1/p,p}(\Omega)^n$, and we get the by part integration formula as

$$\int_\Gamma \nabla\varphi.e\; d\Gamma = -\int_\Gamma \varphi\; div_\Gamma(e)\; d\Gamma + \int_\Gamma H\;\varphi\; e.n\; d\Gamma$$

where H is the mean curvature of the surface Γ, see [6].
(for any field e on Γ, $e \in H^1(\Gamma)^N$, we have $div_\Gamma e = div_\Gamma(e_\Gamma) + H\; e.n$, where $e_\Gamma = e-e.n\; n$)

Proposition 2. Let Y belongs to $W^{r,p}(\mathbb{R}^n)$ and $y(\Omega) = Y|\Omega$ and $z(\Gamma) = Y|\Gamma$. Then $y'_{r-1,p}(\Omega;V) = 0$ and $z'_{r-1-1/p,p}(\Gamma;V) = \frac{\partial Y}{\partial n} V(o).n$

Proof : $\dot{y}(\Omega;V) = (\nabla Y.V(o))|_\Omega$ then $y' = 0$ from the definition ; then

$$z'(\Gamma;V) = \dot{y}(\Omega;V)|_\Gamma - \nabla_\Gamma y.V(o) = \nabla_Y.V(o)|_\Gamma - \nabla_\Gamma y.V(o) = \frac{\partial Y}{\partial n} V(o).n\quad .$$

In calculation of shape derivative we make use of the classical decomposition for the Laplace operator, see [6], if Y belongs to $C^2(\mathbb{R}^n)$ and Ω is a smooth domain of class C^2 then on the boundary Γ we have

$$\Delta Y = \Delta_\Gamma\; Y + H\frac{\partial y}{\partial n} + \frac{\partial^2}{\partial n^2} Y$$

where Δ_Γ is the Laplace Beltrami operator on the manifold Γ which can be defined (see [6]) as $\Delta_\Gamma Y = div_\Gamma\; (\nabla_\Gamma Y)$ where $\nabla_\Gamma Y$ is the tangential componant of the gradient. By the previous by part integration formula on Γ it is characterized by $\int_\Gamma \varphi\,\Delta_\Gamma\; Y\; d\Gamma = -\int_\Gamma \nabla_\Gamma\;\varphi\;\nabla_\Gamma Y\; d\Gamma$ for all φ in $C^1(\Gamma)$.

An important boundary shape dirivative is $\left(\frac{\partial}{\partial n} y(\Omega)\right)'(\Gamma;V)$ for $y(\Omega) \in W^{s,p}(\Omega)$ with $s > 2 + \frac{1}{p}$. Of course this calculation involves

$$\frac{\partial}{\partial n}\,(y'(\Omega;V))$$

Proposition 3.

$$\left(\frac{\partial}{\partial n}\,y(\Omega)\right)'(\Omega;V) = \frac{\partial}{\partial n}\,[y(\Omega)'\,(\Omega;V)] + \frac{\partial y}{\partial n}\,[\mathrm{div}_\Gamma V - H\,V.n] - \nabla_\Gamma(V.n).\nabla_\Gamma$$

$$+ \frac{\partial^2}{\partial n^2}\,y\,V.n$$ where $V=V(o)$. Of course different presentation of this expression could be given.

Proof. Let $\psi \in \mathcal{D}(\mathbb{R}^N)$ with $\frac{\partial\psi}{\partial n} = 0$ on Γ and we consider the Green formula, with $y = y(\Omega_t)$

$$\int_{\Omega_t} \nabla y\,\nabla\psi\,dx = \int_{\Omega_t} -\Delta y\,\psi\,dx + \int_{\Gamma_t} \frac{\partial dy}{\partial n_t}\,\psi\,d\Gamma$$ and we take the derivative with respect to t, at $t = 0$; using the fact that $\psi'(\Omega;V) = \psi'(\Gamma;V) = 0$

$$\int_\Omega \nabla y'(\Omega;V)\,\nabla\psi\,dx + \int_\Gamma \nabla_\Gamma y\,\nabla_\Gamma\psi\,V.n\,d\Gamma = \int_\Omega -\Delta\,y'(\Omega;V)\,\psi\,dx$$

$$+ \int_\Gamma -\Delta y\,\psi\,V.n\,d\Gamma + \int_\Gamma \left(\frac{\partial y}{\partial n}\right)'(\Gamma.V)\,\psi\,d\Gamma + \int_\Gamma \nabla_\Gamma\left(\frac{\partial y}{\partial n}\,\psi\right).V\,d\Gamma$$

$$+ \int_\Gamma H\,\frac{\partial y}{\partial n}\,\psi\,V.n\,d\Gamma\,.$$

Using the by parts integration formula on Ω and on Γ we get

$$\int_\Gamma \frac{\partial}{\partial n}\,(y'(\Omega;V))\,\psi\,d\Gamma - \int_\Gamma \mathrm{div}_\Gamma(V.n\,\nabla_\Gamma y)\,\psi\,d\Gamma = \int_\Gamma -\Delta\,y\,\psi\,V.n\,d\Gamma$$

$$+ \int_\Gamma \left(\frac{\partial y}{\partial n}\right)'(\Gamma;V)\,\psi\,d\Gamma - \int_\Gamma (\mathrm{div}_\Gamma V)\,\frac{\partial y}{\partial n}\,\psi\,d\Gamma + 2\int_\Gamma H\,\frac{\partial y}{\partial n}\,\psi\,V.n\,d\Gamma$$

using the decomposition of Δy with Laplace Beltrani we conclude ∎

Proposition 3 gives a linear relation between $\left(\frac{\partial}{\partial n}\,y\right)'(\Gamma;V)$ and $\frac{\partial}{\partial n}\,(y'(\Omega;V))$. In practice it is used in both sens. In shape optimization we have functional such as $J(\Gamma) = \int_\Gamma \left(\frac{\partial y}{\partial n}\right)^2 d\Gamma$, then the Eulerian derivative is $dJ(\Gamma;V) = 2\int_\Gamma \frac{\partial y}{\partial n}\left(\frac{\partial y}{\partial n}\right)'(\Gamma;V)\,d\Gamma$ and if $\frac{\partial}{\partial n}(y'(\Omega;V))$ is known, by proposition 3, we know $dJ(\Gamma;V)$. Reversely if $\frac{\partial y}{\partial n}$ is a known function on Γ then by proposition 3 we can characterise $\frac{\partial}{\partial n}[y'(\Omega;V)]$. In particular if $\frac{\partial y}{\partial n} g$ where g is a given function, independant on the domain we have from proposition 2 $g'(\Gamma;V) = \frac{\partial y}{\partial n} V(o).n$ and we get the

Corollary 4. Let $y(\Omega) \in W^{s,p}(\Omega)$, $1 \leq p < \infty$, $s >$ and $g \in W(\mathbb{R}^N)$ such that $\frac{\partial y}{\partial n} = g$ on Ω then

$$\frac{\partial}{\partial n}[y'(\Omega;V)] = \nabla_\Gamma y\, \nabla_\Gamma (V.n) - \frac{\partial^2}{\partial n^2} y\, V.n - g\,[\mathrm{div}_\Gamma V - H\, V.n] + \frac{\partial g}{\partial n} V.n \qquad \blacksquare$$

But we also obtain at Proposition 3 a much more general situation for g could depend on Γ, $g = g(\Gamma)$ and if we knows $g'(\Gamma;V)$ we obtain $\frac{\partial}{\partial n}[y'(\Omega';V)]$.

II. SHAPE DERIVATIVES FOR NON SMOOTH DOMAIN

(We recall here results from [5]).We suppose now that Ω is a domain in $\mathbb{R}^N$, an open set in $\mathbb{R}^N$ lying on one side of its boundary Γ which has singularities in a finite number of point $a_1,\ldots,a_m \in \Gamma$. That is that $\tilde{\Gamma} = \Gamma\backslash(a_1,\ldots,a_m)$: a C^k manifold. We just restrict here to such singularities for simplicity. We suppose Γ to be continuous - for any speed field $V \in C^0([0,\epsilon[, C^k(\mathbb{R}^N)^N)$ and y belonging to $W^{2,p}(\mathbb{R}^N)$ we consider $J(\Gamma_t) = \int_{\Gamma_t} y\, d\Gamma_t$, $M(DT_t)$ being the cofactors matrix of DT_t,

$$M(DT_t) = \det(DT_t) * DT_t^{-1}, \text{ with } \gamma(t) = \|M(DT_t).n\|$$

where n is the normal field on $\tilde{\Gamma}$ the change of variable $x = T_t(V)(x)$ leads to $J(\Gamma_t) = \int_\Gamma y \circ T_t\, \gamma(t)\, d\Gamma$ and taking the derivative with respect to t, at $t = 0$, we get the Eulerian Derivative as

$$dJ(\Gamma;V) = \int_\Gamma (\nabla y.V(o) + y[\mathrm{div}\, V(o) - \langle DV(o).n, n\rangle])\, d\Gamma$$

To explicit the last term we introduce a unity decomposition :

r_i, $1 \leq i \leq m$, $r_i \in C^{\infty}(\mathbb{R}^N)$, $\sum_{i=1,\ldots,k} x_i = 1$, $r_i(a_j) = \delta_{ij}$,

support of $r_i \subset\subset U_i$, U_i being a neigborhood of a_i, $a_j \notin U_j$ for $i \neq j$. We introduce $y = \Sigma y_i = r_i y$, $\Gamma_i = \Gamma \cap U_i$, $V = \Sigma V_i$, $V_i = r_i V$. The previous expression of dJ leads to $dJ(\Omega;V) = \sum_i A_i + \sum_{i \neq j} B_{ij}$ with

$$A_i = \int_{\Gamma_i} (\nabla y_i V_i \; div_\Gamma V_i) d\Gamma \;, \quad B_{ij} = \int_{\Gamma_{ij}} (\nabla y_j V_i + y_j div_\Gamma \; V_i) d\Gamma$$

$\Gamma_{ij} = \Gamma_i \cap \Gamma_j$ is of class C^k and functions involves in B_{ij} have compact supports then the classical results holds for $i \neq j$:

$B_{ij} = \int_\Gamma \left(\frac{\partial}{\partial n} y_j + H y_j \right) V_i(o).n \; d\Gamma$. We introduce $W_i = V_i(o) - V_i(o,a_i)$

then $W_i(a_i) = 0$, support of y_i is compact in Γ_i then W_i can be modified in $\tilde{W}_i$ having its support compact in Γ_i and such that

$A_i(\Gamma_i, W_i) = A_i\left(\Gamma_i, \tilde{W}_i\right)$. Also we introduce a function θ_ϵ such that

$\theta_\epsilon = 1$ in $\Gamma_i \cap \{x \mid |a_i - x| > 2\epsilon\}$.

$\theta_\epsilon = 0$ in $\Gamma_i \cap \{x \mid |a_i - x| < \epsilon\}$, $\theta_\epsilon \in C^{\infty}(\gamma_i)$, we introduce $\tilde{W}_i^\epsilon = \tilde{W}_i \theta_\epsilon$

having its compact support in Γ_i and being zero in a neighborhood of a_i then classicaly

$$A_i\left(\Gamma_i, \tilde{W}_i\right) = \text{limit } A_i\left(\Gamma_i, \tilde{W}_i^\epsilon\right) = \int_{\Gamma_i} \left(\frac{\partial}{\partial n} y_i + H y_i \right) W_i.n \; d\Gamma$$

and

$$A_i(\Gamma_i; V_i) = A_i\left(\Gamma_i, \tilde{W}_i\right) + A_i(\Gamma_i, V_i(a_i))$$

But $V_i(a_i)$ being a constant field we easily obtain that the last term is $V_i(a_i).\int_{\Gamma_i} \nabla y_i \; d\Gamma$ and finely we get the

Proposition 4.

$$\frac{d}{dt}\left(\int_{\Gamma_t} y \; d\Gamma_t\right)_{t=o} = \int_\Gamma \left(\frac{\partial y}{\partial n} + Hy \right) V(o).n \; d\Gamma + \sum_{k=1}^{m} V(o,a_i) \int_{\Gamma_i} (\nabla_\Gamma y_i - H y_i n) d\Gamma$$

Corollary 5 : the term $Z_i = \int_{\Gamma_i} (\nabla_\Gamma y - H y_i\, n) d\Gamma$ is independant on the choice of r_i (and of the neighborhood U_i of a_i). It is equal to zero if l is not singular at a_i.
We refer to [4], [5] and [3], where explicit expression are given for Z_i in particular situations. Also numerical experiments give numerical verifications of these results.
For simplicity here we supposed that y is independant of t but as in the previous section we could have considered $z(\Gamma_t)$ and obtain the same regular term.

III. SHAPE ACCELERATION

We are now concerned with the situation of fluid dynamic when the function $y(\Omega_t)$ is itself a vector field over the domain Ω_t. Then we change the notation and we write $V(\Omega_t) \in W^{r,p}(\Omega_t)$ but we suppose that the domain Ω_t is a perturbation of Ω obtained by the field V itself : $\Omega_t = \Omega_t(V)$. It is the situation when one consider some free fluid flow and Ω_t is at time t the domain occuped by the fluid particles having the speed V.

Definition 6 : The shape acceleration of V at Ω is the shape derivative of V in the direction V :

$$V'(\Omega;V) = \dot{V}(\Omega;V) - DV(\Omega).V(\Omega)$$

where $DV(\Omega)$ is the Jacobian matrix of $V(\Omega)$. Of course here we assume that $V(\Omega)$ is smooth enough so that the ordinary differential equation have solution and the transformation $T_t(V)$ is well defined so that the material derivative $\dot{V}(\Omega;V)$ make sens. As an example we consider here the situation of a quasi steady visco-plastic flow. The flow is slow so that the inerty terms (and surface tension) are neglected. Ω_t is the domain occuped by the material at time t and for each $x \in \Omega_t$ the particle lying at x has the speed $V(\Omega_t)(x)$. Now the field $V(\Omega_t) \in W^{1,m+1}(\Omega_t)^3$ (Ω_t lies in $\mathbb{R}^3$) is solution of some variational problem. Considering the Norton Hoff reology one have (the free divergence being approximated by a penalty technique) :

$$\int_{\Omega_t} \|\epsilon(V(\Omega_t))\|^{m-1} \epsilon(V(\Omega_t))..\epsilon(\varphi)\,dx + \rho\int_{\Omega_t} \operatorname{div} V(\Omega_t) \operatorname{div}\varphi\,dx = \int_{\Omega_t} f.\varphi\,dx$$

$\forall \varphi \in W^{1,m+1}(\Omega_t(V))$, $0 < m < 1$
(φ and V being supposed equal to zero on a fixed part Γ_0 of Γ_t) where $2\epsilon(V) = DV + {}^*DV$. Assuming that the material derivative $\dot{V}(\Omega;V)$ is well defined and assuming some smoothness results on $V(\Omega)$ and $\dot{V}(\Omega;V)$ by taking

the derivative with respect to t, at t = o, we obtain the characterization of $V'(\Omega;V)$. Considering $m = 1$ for simplicity here, we have

$$\int_\Omega \epsilon(V'(V))..\epsilon(\varphi)dx + \int_\Gamma \epsilon(V)..\epsilon(\varphi)\ V.nd\Gamma + \rho \int_\Omega div(V'(V))\ div\ \varphi\ dx$$

$$+ \int_\Gamma div\ V\ div\ \varphi\ V.n\ d\Gamma = \int_\Gamma f.\varphi\ V.n\ d\Gamma$$

for any φ in $H^1(\mathbb{R}^N)^N$.

And using the Green formula one get

$$-\Delta V'(V) - \rho\ \nabla(div\ V'(V)) = 0 \quad on \quad \Omega$$

$$\int_\Gamma [\epsilon(V'(V)).\varphi.n + \epsilon(V)..\epsilon(\varphi)\ V.n + \rho\ div(V'(V))\varphi.n + \rho\ div\ V\ div\ \varphi\ V.n]d\Gamma$$

$$= \int_\Gamma f.\varphi\ V.n\ d\Gamma,\ \forall\ \varphi$$

and using the by part integration formula on Γ one can obtain the linear boundary condition for V on Γ.

On numerical wiev point it is not necessary to explicit this relation. We propose to use the expression of $V'(V)$ to obtain a higher accurate explicit method to compute $\Omega_t, y(\Omega_t)$.

The free boundary problem for Norton Hoff flow has been studied in [1], [2]. We introduce the functional $J(\Gamma) = \frac{1}{2}\int_\Gamma (V.n)^2 d\Gamma$ where V is the solution of the Norton Hoff equation and assuming V to be divergence free, div V = 0, (that is $\rho = +\infty$), we obtain the :

Proposition 7 [8] : $dJ(\Gamma,W) = \left(\frac{d}{dt} J(\Gamma_t)\right)_{t=o} = \int_{\partial\Omega} g\ W.n\ d\Gamma$ for any virtual admissible speed W. Where the density gradient g is given by

$$g = 2\ rot\ V\ V.t + H\left(2\ (Vt)^2 - (Vn)^2\right) + \left\|\dot\epsilon(V)\right\|^{m-1}\ \dot\epsilon(V).\dot\epsilon(U)$$

where U is the solution of adjoint problem :

$$div\ U = 0,\quad \int_\Omega \left\|\dot\epsilon(V)\right\|^{m-1}\ \dot\epsilon(U)..\epsilon(\varphi)dx = -\int_{\partial\Omega} V.n\ \varphi.n\ d\Gamma$$

$$\forall\ \varphi \in W^{1,m+1}(\Omega)^2,\ div\ \varphi = 0 \qquad \blacksquare$$

IV. CHARACTERIZATION OF THE SHAPE ACCELERATION $V'(\Omega;V)$

We have to obtain the boundary condition for the linear problem whose $V'(\Omega;V)$ is solution. The previous by part integration formula on Γ leads to

$$\int_\Gamma (\partial_i \varphi\, \psi + \varphi\, \partial_i \psi)\, d\Gamma = \int_\Gamma [\partial_n (\varphi\psi) + H\, \varphi\psi]\, n_i\, d\Gamma$$

In particular if $\partial_n \varphi = 0$ on Γ, $= \int_\Gamma \varphi\, (\partial_n \psi + H\psi)\, n_i\, d\Gamma$ and if B is a symetric 2×2 tensor and $\partial_n \varphi_i = 0$ on Γ,

$$\int_\Gamma B..\epsilon(\varphi)\, d\Gamma = -2 \int_\Gamma \operatorname{div} B.\varphi\, d\Gamma + 2 \int_\Gamma (\partial_n B + HB).n.\varphi\, d\Gamma$$

Then we obtain the

Proposition 8 : In the particular case $m = 1$

$$\epsilon(V').n + \varphi \operatorname{div}(V')\, n = 2 \operatorname{div} \epsilon(V) - 2\, (\partial_n \epsilon(V) + H\, \epsilon(V)).n$$
$$+ \varphi\, \nabla_\Gamma \operatorname{div} V + f\, V.n \quad \text{on } \Gamma$$

For the proof we use $\operatorname{div} \varphi|_\Gamma = \operatorname{div}_\Gamma \varphi_G + 2\, H\, \varphi.n + D\varphi.n.n$ but with $\partial_n \varphi_i = 0$ on Γ then $\operatorname{div} \varphi|_\Gamma = \operatorname{div}_\Gamma \varphi_\Gamma . = 0$

V. NON SMOOTH DOMAINS, GENERALIZATION OF THE MEAN CURVATURE H AT A SINGULAR POINT

In the case of piecewise smooth boundary Γ these results are unchanged on the smooth parts of the boundary but, as in [4], [5], must be augmented by pointwise extra terms arising from Proposition 4. In dimension 2 the term Z_i is the oriented angle of Γ at a_i, in dimension 3, it is the oriented solid angle of Γ at a_i. In fact $Z_i = Z_i(y) = y(a_i)\, Z_i(r_i)$

We propose to define the mean curvature $\tilde{H}$ of Γ : considering the vector $T_i = Z_i(r_i)$ (which is independant on the choice of r_i).

Defintion 9 : The classical mean curvature H and the normal field n are defined almost every where on Γ, namely on $\Gamma-(a_i)$, then we extend the vector Hn to Γ as follows

$$\tilde{H}\, n = H\, n + \sum_{i=1}^{n} T_i\, \delta_{a_i} .$$

Proposition 10 : Let Γ be piecewise C^k, $k \geqslant 1$, but at points a_i, $1 \leqslant i \leqslant m$, and Z_i given at Corollary 5 then we have for any $y \in H^2(\mathbb{R}^N)$,

$$\frac{d}{dt} \int_{\Gamma_t} y\, d\Gamma_t = \int_\Gamma \left(\frac{\partial y}{\partial n} + \tilde{H}n .y \quad V(o) \right) d\Gamma .$$

It turns out that in the result of Proposition 8, if Γ is non smooth we have just to change Hn for its generalization $\tilde{Hn}$. On numerical view point in [4], the smooth boundary Γ has been approach by a piecewise linear boundary and at each point Z_i δ_{ai} turns out to be a good approximation for Hn so that we introduced in [4] a third kind of consideration : the continuous problem associated to the discretized geometry.

REFERENCES

[1] A. Bern, Thesis, Sophia Antipolis, November 1987

[2] A. Bern, J.L. Chenot, Y. Demay, J.P. Zolesio, Communication at INRIA Conference, Antibes, June 1986.

[3] M. Souli, In Proc. of IFIP Workshop "Boundary Control and Boundary Variation", Nice June 1986, J.P. Zolesio Ed. Springer.

[4] M. Souli and J.P. Zolesio, in same proc. as [3].

[5] J.P. Zolesio, CRM repport n° 1111, 1983, Montréal Univ. and proc. of INRIA School on Shape Optimization, Nice 1984.

[6] J.P. Zolesio, The Speed Method in proc. of a Nato Conf. IOWA City, 1980, Ed. Haug and J. Cea, eds., Northof and Sij publishers.

[7] J.P. Zolesio, These de Doctorat d'Etat, Nice 1979.

[8] J.P. Zolesio, Algorithme pour le calcul numérique de la frontière libre dans un écoulement non Newtonien, Cemef, ENSMP, Sophia Antipolis 1984.

Lecture Notes in Control and Information Sciences

Edited by M. Thoma and A. Wyner

Vol. 62: Analysis and Optimization of Systems
Proceedings of the Sixth International Conference on Analysis and Optimization of Systems
Nice, June 19–22, 1984
Edited by A. Bensoussan, J. L. Lions
XIX, 591 pages. 1984.

Vol. 63: Analysis and Optimization of Systems
Proceedings of the Sixth International Conference on Analysis and Optimization of Systems
Nice, June 19–22, 1984
Edited by A. Bensoussan, J. L. Lions
XIX, 700 pages. 1984.

Vol. 64: Arunabha Bagchi
Stackelberg Differential Games in Economic Models
VIII, 203 pages, 1984

Vol. 65: Yaakov Yavin
Numerical Studies in Nonlinear Filtering
VIII, 273 pages, 1985.

Vol. 66: Systems and Optimization
Proceedings of the Twente Workshop
Enschede, The Netherlands, April 16–18, 1984
Edited by A. Bagchi, H. Th. Jongen
X, 206 pages, 1985.

Vol. 67: Real Time Control of Large Scale Systems
Proceedings of the First European Workshop
University of Patras, Greece, Juli 9–12, 1984
Edited by G. Schmidt, M. Singh, A. Titli, S. Tzafestas
XI, 650 pages, 1985.

Vol. 68: T. Kaczorek
Two-Dimensional Linear Systems
IX, 397 pages, 1985.

Vol. 69: Stochastic Differential Systems – Filtering and Control
Proceedings of the IFIP-WG 7/1 Working Conference
Marseille-Luminy, France, March 12-17, 1984
Edited by M. Metivier, E. Pardoux
X, 310 pages, 1985.

Vol. 70: Uncertainty and Control
Proceedings of a DFVLR International Colloquium
Bonn, Germany, March, 1985
Edited by J. Ackermann
IV, 236 pages, 1985.

Vol. 71: N. Baba
New Topics in Learning Automata Theory and Applications
VII, 231 pages, 1985.

Vol. 72: A. Isidori
Nonlinear Control Systems: An Introduction
VI, 297 pages, 1985.

Vol. 73: J. Zarzycki
Nonlinear Prediction Ladder-Filters for Higher-Order Stochastic Sequences
V, 132 pages, 1985.

Vol. 74: K. Ichikawa
Control System Design based on Exact Model Matching Techniques
VII, 129 pages, 1985.

Vol. 75: Distributed Parameter Systems
Proceedings of the 2nd International Conference, Vorau, Austria 1984
Edited by F. Kappel, K. Kunisch, W. Schappacher
VIII, 460 pages, 1985.

Vol. 76: Stochastic Programming
Edited by F. Archetti, G. Di Pillo, M. Lucertini
V, 285 pages, 1986.

Vol. 77: Detection of Abrupt Changes in Signals and Dynamical Systems
Edited by M. Basseville, A. Benveniste
X, 373 pages, 1986.

Vol. 78: Stochastic Differential Systems
Proceedings of the 3rd Bad Honnef Conference, June 3–7, 1985
Edited by N. Christopeit, K. Helmes, M. Kohlmann
V, 372 pages, 1986.

Vol. 79: Signal Processing for Control
Edited by K. Godfrey, P. Jones
XVIII, 413 pages, 1986.

Vol. 80: Artificial Intelligence and Man-Machine Systems
Edited by H. Winter
IV, 211 pages, 1986.

Lecture Notes in Control and Information Sciences

Edited by M. Thoma and A. Wyner

Vol. 81: Stochastic Optimization
Proceedings of the International Conference, Kiew, 1984
Edited by I. Arkin, A. Shiraev, R. Wets
X, 754 pages, 1986.

Vol. 82: Analysis and Algorithms of Optimization Problems
Edited by K. Malanowski, K. Mizukami
VIII, 240 pages, 1986.

Vol. 83: Analysis and Optimization of Systems
Proceedings of the Seventh International Conference of Analysis and Optimization of Systems
Antiba, June 26-27, 1986
Edited by A. Bensoussan, J. L. Lions
XVI, 901 pages, 1986.

Vol. 84: System Modelling and Optimization
Proceedings of the 12th IFIP Conference
Budapest, Hungary, September 2–6, 1985
Edited by A. Prékopa, J. Szelezsán, B. Strazicky
XII, 1046 pages, 1986.

Vol. 85: Stochastic Processes in Underwater Acoustics
Edited by Charles R. Baker
V, 205 pages, 1986.

Vol. 86: Time Series and Linear Systems
Edited by Sergio Bittanti
XVII, 243 pages, 1986.

Vol. 87: Recent Advances in System Modelling and Optimization
Proceedings of the IFIP-WG 7/1 Working Conference
Santiago, Chile, August 27-31, 1984
Edited by L. Contesse, R. Correa, A. Weintraub
IV, 199 pages, 1987.

Vol. 88: Bruce A. Francis
A Course in H_∞ Control Theory
XI, 156 pages, 1987.

Vol. 88: Bruce A. Francis
A Course in H_∞ Control Theory
X, 150 pages, 1987.
Corrected - 1st printing 1987

Vol. 89: G. K. H. Pang/A. G. J. McFarlane
An Expert System Approach to Computer-Aided Design of Multivariable Systems
XII, 223 pages, 1987.

Vol. 90: Singular Perturbations and Asymptotic Analysis in Control Systems
Edited by P. Kokotovic, A. Bensoussan, G. Blankenship
VI, 419 pages, 1987.

Vol. 91 Stochastic Modelling and Filtering
Proceedings of the IFIP-WG 7/1 Working Conference
Rome, Italy, Decembre 10-14, 1984
Edited by A. Germani
IV, 209 pages, 1987.

Vol. 92: L. T. Grujić, A. A. Martynyuk, M. Ribbens-Pavella
Large-Scale Systems Stability Under Structural and Singular Perturbations
XV, 366 pages, 1987.

Vol. 93: K. Malanowski
Stability of Solutions to Convex Problems of Optimization
IX, 137 pages, 1987.

Vol. 94: H. Krishna
Computational Complexity of Bilinear Forms
Algebraic Coding Theory and Applications to Digital Communication Systems
XVIII, 166 pages, 1987.

Vol. 95: Optimal Control
Proceedings of the Conference on Optimal Control and Variational Calculus
Oberwolfach, West-Germany, June 15-21, 1986
Edited by R. Bulirsch, A. Miele, J. Stoer and K. H. Well
XII, 321 pages, 1987.

Vol. 96: H. J. Engelbert/W. Schmidt
Stochastic Differential Systems
Proceedings of the IFIP-WG 7/1 Working Conference
Eisenach, GDR, April 6-13, 1986
XII, 381 pages, 1987.

Lecture Notes in Control and Information Sciences

Edited by M. Thoma and A. Wyner